Electromagnetic Processes

T0252111

PRINCETON SERIES IN ASTROPHYSICS

EDITED BY DAVID N. SPERGEL

Theory of Rotating Stars, *by Jean-Louis Tassoul*

Theory of Stellar Pulsation, *by John P. Cox*

Galactic Dynamics, *by James Binney and Scott Tremaine*

Dynamical Evolution of Globular Clusters, *by Lyman Spitzer, Jr.*

Supernovae and Nucleosynthesis: An Investigation of the History of Matter, from the Big Bang to the Present, *by David Arnett*

Unsolved Problems in Astrophysics, *edited by John Bahcall and Jeremiah P. Ostriker*

Galactic Astronomy, *by James Binney and Michael Merrifield*

Active Galactic Nuclei: From the Central Black Hole to the Galactic Environment, *by Julian H. Krolik*

Plasma Physics for Astrophysics, *by Russell M. Kulsrud*

Electromagnetic Processes, *by Robert J. Gould*

Electromagnetic Processes

ROBERT J. GOULD

PRINCETON UNIVERSITY PRESS

PRINCETON AND OXFORD

Copyright © 2006 by Princeton University Press

Published by Princeton University Press, 41 William Street,
Princeton, New Jersey 08540
In the United Kingdom: Princeton University Press, 3 Market Place,
Woodstock, Oxfordshire OX20 1SY

All Rights Reserved

Library of Congress Cataloging-in-Publication Data
Gould, Robert J. (Robert Joseph), 1935-
Electromagnetic processes / Robert J. Gould
p. cm.—(Princeton series in astrophysics)
Includes bibliographical references and index.
ISBN-13: 978-0-691-12443-8 (acid-free paper)
ISBN-10: 0-691-12443-4 (acid-free paper)
ISBN-13: 978-0-691-12444-5 (pbk. : acid-free paper)
ISBN-10: 0-691-12444-2 (pbk. : acid-free paper)
1. Electromagnetic theory. 2. Quantum electrodynamics. 3. Nonrelativistic quantum mechanics. 4. Scattering (Physics) I. Title. II. Series.

QC670.G616 2006
530.14′33—dc22 2005040562

British Library Cataloging-in-Publication Data is available

Printed on acid-free paper. ∞

pup.princeton.edu

Printed in the United States of America

10 9 8 7 6 5 4 3 2 1

Contents

Preface ix

Chapter 1. Some Fundamental Principles 1

1.1 Units and Characteristic Lengths, Times, Energies, Etc. 1
1.2 Relativistic Covariance and Relativistic Invariants 5
1.2.1 Spacetime Transformation 5
1.2.2 Other Four-Vectors and Tensors—Covariance 8
1.2.3 Some Useful and Important Invariants 10
1.2.4 Covariant Mechanics and Electrodynamics 13
1.3 Kinematic Effects 15
1.3.1 Threshold Energies in Non-Relativistic and Relativistic Processes 15
1.3.2 Transformations of Angular Distributions 17
1.4 Binary Collision Rates 18
1.5 Phase-Space Factors 21
1.5.1 Introduction 21
1.5.2 Simple Examples 23
1.5.3 General Theorems—Formulation 26
1.5.4 General Formulas—Evaluation of Multiple Integrals 28
1.5.5 One-Particle Distributions 32
1.5.6 Invariant Phase Space 34

Chapter 2. Classical Electrodynamics 37

2.1 Retarded Potentials 37
2.1.1 Fields, Potentials, and Gauges 37
2.1.2 Retarded Potentials in the Lorentz Gauge 39
2.2 Multipole Expansion of the Radiation Field 41
2.2.1 Vector Potential and Retardation Expansion 41
2.2.2 Multipole Radiated Power 43
2.3 Fourier Spectra 46
2.4 Fields of a Charge in Relativistic Motion 49
2.4.1 Liénard-Wiechert Potentials 49
2.4.2 Charge in Uniform Motion 51
2.4.3 Fields of an Accelerated Charge 53
2.5 Radiation from a Relativistic Charge 54
2.6 Radiation Reaction 57
2.6.1 Non-Relativistic Limit 57
2.6.2 Relativistic Theory: Lorentz-Dirac Equation 60

2.7 Soft-Photon Emission 61
2.7.1 Multipole Formulation 61
2.7.2 Dipole Formula 62
2.7.3 Emission from Relativistic Particles 63
2.8 Weizsäcker-Williams Method 65
2.8.1 Fields of a Moving Charge 66
2.8.2 Equivalent Photon Fluxes 68
2.9 Absorption and Stimulated Emission 70
2.9.1 Relation to Spontaneous Emission 71
2.9.2 General Multiphoton Formula 72
2.9.3 Stimulated Scattering 73

Chapter 3. Quantum Electrodynamics 75
3.1 Brief Historical Sketch 76
3.2 Relationship with Classical Electrodynamics 78
3.3 Non-Relativistic Formulation 80
3.3.1 Introductory Remarks 80
3.3.2 Classical Interaction Hamiltonian 80
3.3.3 Quantum-Mechanical Interaction Hamiltonian 83
3.3.4 Perturbation Theory 84
3.3.5 Processes, Vertices, and Diagrams 88
3.4 Relativistic Theory 94
3.4.1 Modifications of the Non-Covariant Formulation 94
3.4.2 Photon Interactions with Charges without Spin 97
3.4.3 Spin-$\frac{1}{2}$ Interactions 103
3.4.4 Invariant Transition Rate 107
3.5 Soft-Photon Emission 109
3.5.1 Non-Relativistic Limit 109
3.5.2 Emission from Spin Transitions 113
3.5.3 Relativistic Particles without Spin 116
3.5.4 Relativistic Spin-$\frac{1}{2}$ Particles 119
3.6 Special Features of Electromagnetic Processes 123
3.6.1 "Order" of a Process 123
3.6.2 Radiative Corrections and Renormalization 127
3.6.3 Kinematic Invariants 130
3.6.4 Crossing Symmetry 132

Chapter 4. Elastic Scattering of Charged Particles 135
4.1 Classical Coulomb Scattering 135
4.1.1 Small-Angle Scattering 135
4.1.2 General Case 138
4.1.3 Two-Body Problem—Relative Motion 139
4.1.4 Validity of the Classical Limit 141
4.2 Non-Relativistic Born Approximation and Exact Treatment 142
4.2.1 Perturbation-Theory Formulation 142
4.2.2 Sketch of Exact Theory 145
4.2.3 Two-Body Problem 148
4.2.4 Scattering of Identical Particles 150
4.2.5 Validity of the Born Approximation 154

4.3 Scattering of Relativistic Particles of Zero Spin 156
4.3.1 Coulomb Scattering 156
4.3.2 Scattering of Two Distinguishable Charges 158
4.3.3 Two Identical Charges 162
4.3.4 Scattering of Charged Antiparticles 163
4.4 Scattering of Relativistic Spin-$\frac{1}{2}$ Particles 166
4.4.1 Spin Sums, Projection Operators, and Trace Theorems 166
4.4.2 Coulomb Scattering 170
4.4.3 Møller and Bhabha Scattering 171

Chapter 5. Compton Scattering 177
5.1 Classical Limit 177
5.1.1 Kinematics of the Scattering 177
5.1.2 Derivation of the Thomson Cross Section 178
5.1.3 Validity of the Classical Limit 181
5.2 Quantum-Mechanical Derivation: Non-Relativistic Limit 182
5.2.1 Interactions and Diagrams 182
5.2.2 Calculation of the Cross Section 184
5.3 Scattering by a Magnetic Moment 186
5.4 Relativistic Spin-0 Case 188
5.5 Relativistic Spin-$\frac{1}{2}$ Problem: Klein-Nishina Formula 191
5.5.1 Formulation 191
5.5.2 Evaluation of the Cross Section 193
5.5.3 Invariant Forms 194
5.5.4 Limiting Forms and Comparisons 195
5.6 Relationship to Pair Annihilation and Production 197
5.7 Double Compton Scattering 199
5.7.1 Non-Relativistic Case. Soft-Photon Limit 199
5.7.2 Non-Relativistic Case. Arbitrary Energy 202
5.7.3 Extreme Relativistic Limit 207

Chapter 6. Bremsstrahlung 211
6.1 Classical Limit 211
6.1.1 Soft-Photon Limit 211
6.1.2 General Case: Definition of the Gaunt Factor 214
6.2 Non-Relativistic Born Limit 217
6.2.1 General Formulation for Single-Particle Bremsstrahlung 217
6.2.2 Coulomb (and Screened-Coulomb) Bremsstrahlung 222
6.2.3 Born Correction: Sommerfeld-Elwert Factor 223
6.2.4 Electron-Positron Bremsstrahlung 226
6.3 Electron-Electron Bremsstrahlung. Non-Relativistic 228
6.3.1 Direct Born Amplitude 228
6.3.2 Photon-Emission Probability (without Exchange) 232
6.3.3 Cross Section (with Exchange) 234
6.4 Intermediate Energies 236
6.4.1 General Result. Gaunt Factor 236
6.4.2 Soft-Photon Limit 239

6.5 Relativistic Coulomb Bremsstrahlung 240
6.5.1 Spin-0 Problem 241
6.5.2 Spin-$\frac{1}{2}$: Bethe-Heitler Formula 244
6.5.3 Relativistic Electron-Electron Bremsstrahlung 248
6.5.4 Weizsäcker-Williams Method 251
6.6 Electron-Atom Bremsstrahlung 254
6.6.1 Low Energies 254
6.6.2 Born Limit—Non-Relativistic 256
6.6.3 Intermediate Energies—Non-Relativistic 257
6.6.4 Relativistic Energies—Formulation 259
6.6.5 Relativistic Energies—Results and Discussion 264

Index 269

Preface

The aim of this book is to provide an understanding of processes that are electromagnetic in nature. That is, they take place as a result of the interaction of a particle's charge or magnetic moment with the electromagnetic field. It is the subject of Quantum Electrodynamics (QED), but the more general designation "Electromagnetic Processes" is adopted for a title. The reason for doing this is that for some processes there is a limit where a classical treatment is valid in the sense that both the charge motion and the electromagnetic field can be treated classically. There is also the limit where the charge motion must be described quantum mechanically but a non-relativistic Born-approximation treatment is adequate. In fact, usually calculations in this domain are simpler than those in a purely classical treatment. For a full understanding of the subject the more general relativistic covariant treatment is necessary, but a detailed systematic formulation of this topic requires a lengthy formal development and this is done in a number of fine textbooks. The covariant theory is what we generally refer to as QED, and it would include the non-relativistic (NR) limit. For problems involving scattering in a Coulomb field the classical domain corresponds to low energy and there is a gap between the classical and non-relativistic Born domains. That is, the two domains do not overlap, but for some processes there are ways of bridging the gap in approximate treatments.

However, non-relativistic, non-covariant QED is a valid limit as is, even, the classical domain. We shall try to make an intuitive transition from the NR theory to the covariant treatment, instead of giving a systematic formulation of the latter. Actually, for some processes the covariant calculations are very difficult and a general formula for, say, a cross section cannot be computed from a relativistic expression from which the NR limit would be taken. In some such cases the correct analytic expressions for cross sections can be obtained through a non-relativistic treatment. The NR theory can introduce perturbation "diagrams" just like Feynman diagrams are used in the covariant theory. While these diagrams are not necessary in the calculations, they are useful as guides to outline the computation of the effects of perturbations that cause processes to take place. Although the diagrams in the NR theory do not have the same exact meaning as in the covariant theory, they are helpful in making the transition from the NR theory to the covariant formulation. The covariant theory is much simpler for spin-0 charged particles than it is for the more important case of spin-$\frac{1}{2}$, and for the processes considered it is treated first in the relativistic calculations.

In the interest of brevity and simplicity, the scope of this book is limited to only a few processes, although some phenomena that are closely related to the ones considered are discussed. For example, in the chapter on Compton scattering, pair production in photon-photon collisions and pair annihilation to two photons are brought in. For the processes considered the treatment follows the same path. We go from the classical theory to NR QED to the relativistic covariant QED. Moreover, in the latter we first consider the case of spin-0 before going on to the spin-$\frac{1}{2}$ calculation.

The first chapter of this book gives a summary of some important principles, including a review of special relativity. In particular, some emphasis is given to the method or device of inferring a correct relativistic covariant formula from its non-relativistic limiting expression. Also, certain kinematic effects are discussed, such as the use of invariant momentum-space volume and the formula for the binary collision rate for distributions of relativistic particles. The chapter also gives a fairly complete account of "phase-space effects," showing how in some cases the phase-space integral should involve only energy conservation and not momentum conservation. Some simple examples are given for cases where a combination of zero-mass and NR particles are outgoing. Although we do not make extensive use of the results on phase space integrals, the treatment of the topic could be useful to people working in other areas, for example, in problems of multiparticle production in high-energy and cosmic-ray physics. Chapter 2 gives a treatment of some features of classical electrodynamics (CED), especially for its application in radiation problems. The third chapter is on QED, treating both the non-relativistic and covariant theories.

In an attempt to make the book as self-contained as possible, certain basic developments are formulated from first principles. For example, time-dependent perturbation theory is treated, including the derivation of the Fermi Golden Rule. The Dirac equation is introduced, as well, with its modern covariant notation. However, parts of some calculations for spin-$\frac{1}{2}$ particles are not given in detail, such as the calculation of the corresponding "trace" for a particular process. Reference to textbooks is given, instead, since the task is sometimes lengthy (but not difficult). Processes involving bound states are not treated extensively. The various types of radiative transitions in atoms, molecules, and nuclei constitute an enormous subject. Nevertheless, at least the foundations are developed in the simple treatment of non-relativistic QED, which yields the necessary forms for the interaction Hamiltonian for couplings to charges and magnetic moments and for the two-photon coupling.

I am very grateful to the staff at Princeton University Press for their help and encouragement in seeing the project of this book come to completeness. In particular, my sincere thanks go to Ingrid Gnerlich and Carmel Lyons, without whom it would not have been accomplished. The work by Ginny Dunn and Mark Bellis was also indespensible. I am grateful, as well, to two referees, who read the original manuscript carefully and made useful suggestions.

La Jolla,
December, 2004

Electromagnetic Processes

Chapter One

Some Fundamental Principles

1.1 UNITS AND CHARACTERISTIC LENGTHS, TIMES, ENERGIES, ETC.

In the measurement of quantities by laboratory instruments, both the c.g.s. and m.k.s. units are convenient. However, for the description of particle and atomic processes, the c.g.s. system is preferable in that equations and formulas are sometimes simpler in form; for this reason, the c.g.s. system will be employed throughout this book. At the same time, it is often useful to express quantities in dimensionless units in terms of certain "fundamental" values defined in terms of the fundamental physical constants. Different fundamental quantities—for example, a characteristic length—can be formed from different combinations of physical constants, and the particular choice appropriate for the description of some process is dictated by the nature of the process.

Concerning the physical constants themselves, the most fundamental one is perhaps the "velocity of light" (c). The constant is of more general significance than the name given to it, since it is the characteristic parameter of spacetime, and its value is relevant to all dynamical processes in physics. Our fundamental theory of spacetime is special relativity, and we shall review certain basic features of the theory in the following section. Considerations of some general consequences of special relativity are extremely powerful, in particular, as a guide in formulating the fundamental equations of physics.

After c, the most fundamental physical constant is probably Planck's constant ($\hbar$). Loosely put, this constant might be designated as a "quantization parameter," but this is probably not a good description. Another try at description might be to call it the "fundamental indeterminacy parameter," but it is questionable whether the "uncertainty relations" deserve the title of *principle*, since they follow from the superposition principle (which really *is* a principle). Given that discrete particle motion is to be described in terms of an associated wave or propagation vector $\boldsymbol{k}$ and frequency ω, Planck's constant is then the proportionality factor between $\boldsymbol{k}$ and the particle momentum:

$$\boldsymbol{p} = \hbar \boldsymbol{k}. \tag{1.1}$$

The uncertainty relations for an individual particle follow from this relation and the superposition principle. If momentum is to be regarded as a particle dynamical property and the wave propagation vector a kinematical variable, we might designate $\hbar$ more descriptively as a parameter of particle dynamics. However, we shall, as usual, refer to $\hbar$ simply as Planck's constant like everyone else.

The third most fundamental physical constant may be the "electronic charge" (e), since it seems to be a fundamental unit common to the various charged elementary particles. That is, although there is a spectrum of masses for the particles, except for the fractionally charged "quarks," the particle charges are multiples of e.

From the three physical constants c, $\hbar$, and e, it is not possible to construct a fundamental length by various combinations of products. From e and $\hbar$ it is possible to form a characteristic velocity

$$v_0 = e^2/\hbar, \tag{1.2}$$

and this velocity is of significance for particle processes. Combining the fundamental physical constants, a dimensionless number

$$\alpha = e^2/\hbar c \approx 1/137 \tag{1.3}$$

can be formed that is of great importance, especially for electromagnetic processes. This number is called the "fine structure constant" because of its role in determining the magnitude of the small relativistic level shifts in atomic hydrogen; it can also be regarded as a dimensionless coupling constant for electromagnetic processes. Because of its small value, these processes can be calculated well by perturbation theory.

The masses of the various elementary particles play a major role in particle processes. The electron (and positron) mass (m), being the smallest of all, is of great importance because the particle is easily perturbed by an electromagnetic field. In particular, a variety of radiative (photon-producing) processes are associated with the electron and its interactions. A description of these processes is the principal task of this book. Almost all of our knowledge about the world outside our solar system comes from the analysis of the spectral distribution of radiation from distant sources. Our understanding of the details of the microscopic photon-producing processes allows us to interpret these source spectra and learn something of the nature of the sources. Fortunately, the electromagnetic processes are very well understood, and they can be calculated to high accuracy by perturbation theory.

The nucleon mass (M)—say, the mass of the proton, which is stable—is significant in that it is much larger (about $1836m$) than that of the electron. Along with their corresponding antiparticles, the electron and proton are the only stable "particles." In fact, there is now good evidence, from inelastic scattering of very high energy electrons off protons, that the latter are not "elementary" or "fundamental" particles. Instead, protons are thought to be composites, built from quarks, and they have, as a consequence, *structure*. For example, protons have a characteristic size and charge distribution that can be measured. Pions (and also kaons) are also quark composites, and the pion is especially important as the least massive of the strongly interacting species. In the older theory of strong interactions, the pion was treated as a fundamental particle and its mass (m_π) determined the characteristic range of the interaction. These ideas are still useful in understanding certain particle and nuclear processes.

The masses of the elementary particles determine the various fundamental or characteristic lengths, all of which are inversely proportional to the mass value. There are different kinds of lengths, each having a different physical meaning and playing a different role in determining the characteristic magnitude of importance of various processes. Along with $\hbar$ and e, the electron mass determines the characteristic atomic size

$$a_0 = \hbar^2/me^2. \tag{1.4}$$

This is the Bohr radius, and it is one of the triumphs of quantum mechanics that the atomic radius ($\sim a_0 \sim 10^{-8}$ cm) is explained by physical principles. Classical physics had no explanation for the characteristic size of atoms as determined in the last century. The basic physical meaning of the characteristic length a_0 can be indicated through considerations of atomic binding. The classical total energy of an electron of momentum p in the neighborhood of a proton is

$$E_{\rm cl} = p^2/2m - e^2/r. \tag{1.5}$$

In a quantum-mechanical description, the spectra of position and momentum values are such that there is a spread in each, determined by the uncertainty relation. Setting $pr \sim \hbar$ as a constraint condition added to Equation (1.5), we see that $E_{\rm cl}$ is minimized at a value

$$(E_{\rm cl})_{\rm min} \sim -e^2/2a_0 \equiv -{\rm Ry} \tag{1.6}$$

for the r-value

$$r_{\rm min} \sim a_0. \tag{1.7}$$

This little analysis shows, very simply, why atoms have a ground state or state of minimum energy. In a classical model with $\hbar \to 0$ the electron "orbit" size could be infinitesimally small and the energy would be infinitely negative.

A characteristic length that does not involve $\hbar$ is the "classical electron radius" r_0. If the electron mass is attributed to its electrostatic self-energy ($\sim e^2/r_0 \sim mc^2$), the result is

$$r_0 = e^2/mc^2. \tag{1.8}$$

This is a very small distance ($\sim 3 \times 10^{-13}$ cm), and the quantity really has no physical meaning, because the classical self-energy considerations are not valid. However, the combination e^2/mc^2 appears often to various powers in expressions for parameters for electromagnetic quantities. Thus, it is still designated r_0 and called by its original name.

The erroneous nature of the classical model for electromagnetic self-energy is clear through considerations that introduce another characteristic length. If we attempt to localize an electron to a very small distance, of necessity we introduce a spectrum of momentum states extending to high values. For $p \sim mc$, the energy values become large enough to produce $e^{\pm}$ pairs, which affect and limit the localization. The uncertainty relation then suggests a minimum localization distance

$$r_{\rm loc} \sim \hbar/mc \equiv \Lambda. \tag{1.9}$$

Again for historical reasons, the quantity Λ is called the *electron Compton wavelength.* It appears often as a factor in formulas for cross sections for electromagnetic processes and, in general, in many equations describing phenomena involving electrons.

The three lengths a_0, r_0, and Λ are related through a linear equation with the fine-structure constant as a proportionality factor:

$$r_0 = \alpha\Lambda = \alpha^2 a_0. \tag{1.10}$$

Although the three lengths are connected by means of the factor α, only a_0 and Λ have a useful physical meaning, and most formulas given throughout this work will not be expressed in terms of r_0.

It might be noted that each of the lengths a_0, Λ, and r_0 is inversely proportional to the electron mass. For some problems it is convenient to consider corresponding lengths involving masses of other particles. While a_0 determines the characteristic (electron) atomic unit of length, and $E_0 = e^2/a_0 (= 2\text{Ry})$ the atomic unit of energy, the electron mass can be replaced by the nucleon (proton) mass M to introduce a "nucleon atomic unit" of distance

$$a_M = (m/M)a_0 \tag{1.11}$$

and a characteristic "nucleon Rydberg energy"

$$\text{Ry}_M = (M/m)\text{Ry}. \tag{1.12}$$

These units are convenient, for example, in the treatment of proton-proton scattering; in that problem, in which the nuclear and Coulomb forces contribute, the Coulomb force plays the major role (except at very high energy).

Another important characteristic distance is the particle Compton wavelength associated with the least massive of the strongly interacting particles (i.e., the pion). The quantity

$$\Lambda_\pi = \hbar/m_\pi c \tag{1.13}$$

determines the range of the strong interaction and the magnitude of characteristic cross sections associated purely with this interaction. The cross section is

$$\sigma_s \sim \Lambda_\pi^2 \sim 20 \text{ mb}, \tag{1.14}$$

where 1 mb = 10^{-3} b, the *barn* (b), defined as 10^{-24}cm^2, being a cross section unit common in nuclear and strong-interaction physics (1 barn is a large cross section for nuclear processes: "as big as a barn").

The choice of units for the description of some particular phenomenon is dictated not just by considerations of characteristic numerical values of relevant quantities. Depending on the type of units chosen, the equations describing a process take on slightly different form. When formulated in terms of the "most natural" units, the equations are more transparent in exhibiting the nature of the physics involved. In problems of atomic structure or in the description of the scattering of non-relativistic electrons by atomic systems or by a pure Coulomb field, the so-called atomic or "hartree" units are natural. In these units e, $\hbar$, and m are each set equal to unity, and lengths are in units of the Bohr radius a_0, cross sections are in units of a_0^2, and energies are in units of $e^2/a_0 = 2\text{Ry}$. The atomic units are, however, not as convenient in problems involving relativistic particles; then the more useful choice is $\hbar = c = 1$ for which $e^2 = \alpha$ is fixed by the dimensionless fine structure constant [Equation (1.3)]. These units are particularly useful in describing electromagnetic phenomena. Further, if the process involves electrons or positrons, the rest energy $m(c^2)$ is a natural characteristic energy. Throughout this book, certain important results will often be expressed in forms that exhibit dimensions clearly by collecting products of factors that are dimensionless ratios. For example, if a cross section for some electromagnetic process at energy E is expressed in terms of a factor Λ^2, a function of E/mc^2, and a factor α^n, we immediately identify n as the "order" of the process. Higher order electromagnetic processes have cross sections down by powers of α. Equations expressed in this manner are preferable to those in which numerical values of physical constants are substituted in.

1.2 RELATIVISTIC COVARIANCE AND RELATIVISTIC INVARIANTS

Ideas of covariance are extremely powerful as a guide in formulating basic physical laws and in the derivation of results in mathematical descriptions of certain physical processes. Considerations of covariance can even provide a path to the discovery of new fundamental laws and then to the development of these new areas of physics. In the description of physical processes, it is often possible to simplify derivations by imposing conditions of relativistic covariance as a trick to arrive at formulas of general validity. We shall often make use of this kind of device.

1.2.1 Spacetime Transformation

The basic laws of physics are generally expressed as differential equations with space and time coordinates as independent variables. The spacetime coordinates refer, in some cases, to "events" such as the position (or possible position) of a particle or of a particle process. Further, the properties of spacetime are described in terms of its "structure" or its transformation properties, and this is the theory of special relativity. For spacetime reference frames K and K' whose spatial coordinate axes are moving with constant relative velocity, the relationship between the coordinates of events in the two frames is the Lorentz transformation

$$x'_\mu = \sum_\nu a_{\mu\nu} x_\nu. \tag{1.15}$$

Here, x_ν, with $\nu = 0, 1, 2, 3$, represents the time ($\nu = 0$) and space ($\nu = 1, 2, 3$) coordinates. Because of the fundamental isotropy of space, it is convenient to choose Cartesian coordinates ($x_1, x_2, x_3 = x, y, z$) for the spatial coordinate description. These are thus "natural" or "preferred" coordinates for formulating the basic equations of physics. It is, in one sense, convenient to choose an imaginary component $x_0 = ict$ for the time variable. This is because the fundamental property of spacetime can then be described by, in addition to the property (1.15), the equation

$$ds^2 \equiv \sum_\mu dx'_\mu\, dx'_\mu = \sum_\nu dx_\nu\, dx_\nu = \text{invariant (inv)}, \tag{1.16}$$

in which dx_μ are the differential coordinate separations between two spacetime events.

Because of the choice of an imaginary time component, it has not been necessary to introduce a "metric" or metric tensor. The spacetime is essentially four-dimensional cartesian, and the metric tensor ($g_{\mu\nu}$) is identical to the Kronecker δ-function

$$\delta_{\mu\nu} = \begin{cases} 1 & \text{for } \mu = \nu \\ 0 & \text{otherwise.} \end{cases} \tag{1.17}$$

It is somewhat a matter of taste whether this notation procedure is adopted. Actually, the trend in physics seems to be away from the use of the imaginary zero component of spacetime and other four-vectors, at the expense of the introduction of a metric tensor. There is then also the necessity of the introduction of "covariant" and "contravariant" vectors and tensors, etc. with component indices appearing as superscripts and subscripts. Perhaps the resurgence of interest in general relativity during the past forty years has led to this fashion, and it is a necessity in that subject. However, for the treatment of physical processes occurring in a localized region, even if a strong gravitational field is present, coordinates can be chosen corresponding to a "flat" or Minkowskian spacetime. We can then do without a metric and reduce the notation complexity by employing $x_0 = ict$ and indices that only appear as subscripts.

The summation convention notation henceforth is adopted, in which summation over an index is always implied if it appears twice either on a single symbol or in a product of factors with subscripts. For example, in Equations (1.15) and (1.16), the summation sign could simply be left off. This procedure saves space and is employed extensively with no confusion or difficulty.

It is a fundamental assumption that all inertial systems are equally good for formulating a mathematical description of physical phenomena, and in terms of coordinates and other quantities (momenta, fields, etc.), the equations must have the same form whether expressed in terms of "unprimed" or "primed" quantities. Ideas like this—a *Principle of Relativity*—really go back to Newton's time. The principle would imply that, if we transform from the primed to the unprimed coordinates, the transformation in the inverse direction with an inverse matrix should have the same form as Equation (1.15):

$$x_\nu = a^{-1}_{\nu\mu} x'_\mu. \tag{1.18}$$

The invariance of the quantity ds^2 implies the existence of sets of orthogonality relations for the transformation coefficients:

$$a_{\mu\rho} a_{\mu\lambda} = a^{-1}_{\nu\rho} a^{-1}_{\nu\lambda} = \delta_{\rho\lambda}. \tag{1.19}$$

A double transformation from the primed to unprimed to primed coordinates (or vice versa) can also be considered, leading to the set of relations

$$a^{-1}_{\rho\mu} a_{\mu\lambda} = \delta_{\rho\lambda}, \tag{1.20}$$

and, through comparison with (1.19), to the identity

$$a^{-1}_{\rho\mu} = a_{\mu\rho}; \tag{1.21}$$

that is, the inverse matrix is equal to the transposed matrix.

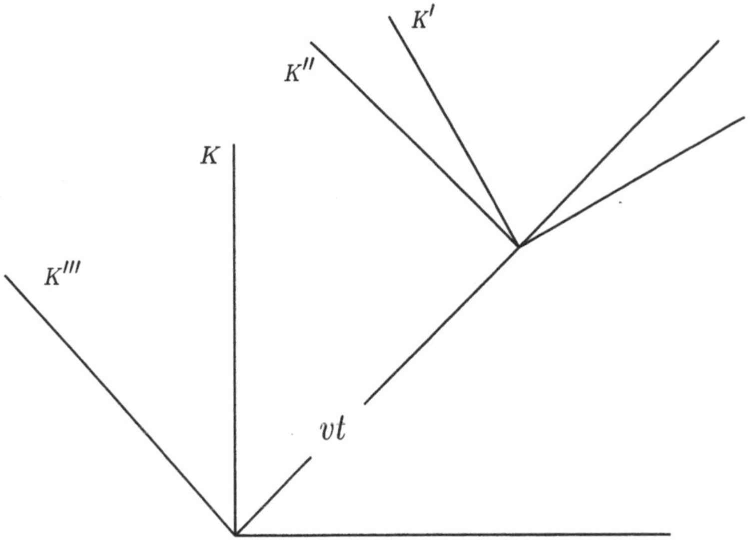

Figure 1.1 The reference frames K, K', K'', and K'''.

The spatial axes of K and K' can have any orientation with respect to the direction of relative motion. However, the transformation coefficients are particularly simple for the case where[1] the x- and x'-axes are aligned and the relative motion of the two frames is along this axis. Since there is no relative motion in the y and z directions, these coordinated must transform as $y' = y$ and $z' = z$, and there can be no dependence of x' and t' on y and z. The Lorentz transformation matrix must then be of the form

$$a_{\mu\nu} = \begin{pmatrix} a_{00} & a_{01} & 0 & 0 \\ a_{10} & a_{11} & 0 & 0 \\ 0 & 0 & 1 & 0 \\ 0 & 0 & 0 & 1 \end{pmatrix}. \tag{1.22}$$

If $x_1' = x' = 0$ represents the coordinate of the origin of K', which moves with a velocity $v = \Delta x_1/\Delta t = ic\Delta x_1/\Delta x_0 = \beta c$ with respect to K', the $\mu = 1$ component of Equation (1.15) then yields the result $a_{10}/a_{11} = i\beta$, and the identities (1.19) and (1.21) can be employed to obtain the solutions

$$\begin{aligned} a_{00} &= a_{11} = \gamma, \\ a_{10} &= -a_{01} = i\beta\gamma, \end{aligned} \tag{1.23}$$

where

$$\gamma = (1 - \beta^2)^{-1/2} = (1 - v^2/c^2)^{-1/2}. \tag{1.24}$$

[1]If K and K' are not oriented this way we can consider two other frames K'' and K''' that are and such that, say, K'' is not in motion with respect to K' and has the same origin as K''. Then the $K - K''$ and $K' - K'''$ transformations are simple spatial rotation operations and the $K'' - K'''$ transformation would be an ordinary Lorentz transformation with relative velocity along, say, x'' and x'''. See Figure 1.1.

1.2.2 Other Four-Vectors and Tensors—Covariance

Because the basic equations of physics correspond to a description of phenomena in space and time, other quantities appearing as symbols in the equations must have transformation properties that are determined by those of the spacetime four-vector x_μ. In fact, the equations are such that other quantities should have the *same* transformation properties as that of x_μ. That is, a general four-vector transforms as

$$V'_\mu = a_{\mu\nu} V_\nu, \tag{1.25}$$

with the $a_{\mu\nu}$ as the same transformation coefficients as in Equation (1.16). A product of two four-vectors would transform as

$$V'_\mu U'_\nu = a_{\mu\lambda} a_{\nu\rho} V_\lambda U_\rho, \tag{1.26}$$

where, again, the transformation coefficients are the same as that for x_μ. A tensor[2] is a double index quantity that transforms in the same manner as the product of two four-vectors:

$$T'_{\mu\nu} = a_{\mu\lambda} a_{\nu\rho} T_{\lambda\rho}. \tag{1.27}$$

The Kronecker δ-function $\delta_{\mu\nu}$, although it does not represent a physical quantity, can be considered, mathematically, a tensor, because its components, which are the same in all Lorentz frames, satisfy the transformation law (1.27).

Entities with more than two indices can be introduced, defined in terms of a transformation law that is an obvious generalization of Equations (1.25) and (1.27). However, almost all physical quantities are either scalars, vectors, or tensors, depending on whether they possess 0, 1, or 2 indices, respectively. Scalars are numbers, equal to the same value when evaluated in any Lorentz frame. The clever use of certain scalars can often simplify the derivation of results, and we shall see many examples of this throughout our treatment of physical processes. The invariant differential quantity (1.16) is a scalar, and can be employed to introduce an "invariant proper time" defined by[3]

$$d\tau^2 = -ds^2/c^2 = dt^2 - dx_j dx_j/c^2, \tag{1.28}$$

where the index j on the spatial Cartesian coordinates runs from 1 to 3. If the spacetime four-vector x_μ refers to events designated by the coordinates of a particle moving with velocity $v_j = dx_j/dt$, then $d\tau = dt/\gamma$, with

$$\gamma^{-2} = 1 - v_j v_j/c^2 = 1 - v^2/c^2. \tag{1.29}$$

The particle four-vector velocity is then defined as

$$v_\mu = dx_\mu/d\tau = \gamma(ic, \mathbf{v}), \tag{1.30}$$

---

[2]Here the designation "tensor" is simply that for a two-index quantity satisfying the transformation law (1.27) and is not to be confused with the more general entity introduced in differential geometry (and general relativity) and called by the same name.

[3]This quantity is often defined with a sign difference. The definition (1.28) is convenient in that, as $c \to \infty$, $d\tau \to dt$.

and we can also introduce a momentum four-vector by multiplying by the particle mass[4]:

$$p_\mu = mv_\mu = \gamma m(ic, \boldsymbol{v}). \tag{1.31}$$

In Section 1.2.3 we shall consider a number of important scalars or invariants. One very useful scalar is formed from the four-dimensional "dot product" of two four-vectors. From the orthogonality relations (1.19) for the transformation coefficients we find that

$$A'_\mu B'_\mu = a_{\mu\nu} a_{\mu\lambda} A_\nu B_\lambda = \delta_{\nu\lambda} A_\nu B_\lambda = A_\nu B_\nu = \text{inv}. \tag{1.32}$$

Two correlaries of this result are that (i) the sum of the squares of the components of any four-vector is an invariant ($A_\mu A_\mu = \text{inv}$) and (ii) the trace of a tensor is an invariant:

$$\text{Tr}\, T_{\mu\nu} = T_{\mu\mu} = \text{inv}. \tag{1.33}$$

By the general *Principle of Covariance*, all equations of physics must be covariant in that they must have the same form in any Lorentz frame. This principle is extremely powerful as a guide to formulating a description of basic physical laws as well as in the calculation of particular processes. For example, if a physical law cannot be expressed mathematically in a form where it is manifestly covariant, the law cannot be correct. The equations must always be expressed in terms of four-vectors, four-tensors, scalars, etc., such that their covariant nature is evident. That is, every term in an equation must have the same transformation property. Often it is possible to guess the correct relativistic law or expression for some quantity by constructing a covariant combination of factors such that the Newtonian formula is obtained in the non-relativistic limit. This is a convenient procedure, since the Newtonian limit is a valid asymptotic domain of physics for which certain physical laws were first established. It is significant that the more general covariant equations, of more general validity, always have a more simple and compact form. The mathematical simplicity of the covariant equations can be considered as strong evidence for their fundamentally correct nature.

Considerations of covariance can even be useful as a guide to formulating basic physical laws where even the non-relativistic limit had not been established. Quantum mechanics is a good example. The description of a propagating plane wave is always in terms of a Fourier component proportional to an amplitude of the form

$$\text{amplitude} \propto e^{i(\boldsymbol{k}\cdot\boldsymbol{r}-\omega t)}. \tag{1.34}$$

Through the introduction of the propagation four-vector

$$k_\mu = (i\omega/c, \boldsymbol{k}), \tag{1.35}$$

the amplitude (1.34) can be written in covariant form in terms of an invariant phase:

$$\text{amplitude} \propto e^{ik_\mu x_\mu}. \tag{1.36}$$

[4]By mass we mean "rest mass," which is an invariant parameter of a particle. To then define γm as a "relativistic mass," which varies with velocity—as is done in many textbooks—is inappropriate and misleading in that it amounts to introducing an unnecessary concept that can only lead to confusion and misunderstanding. There is no reason why quantities and equations in relativistic mechanics have to have a form identical to those in Newtonian mechanics.

The most critical step in quantum mechanics was made by de Broglie in 1924, when he wrote down the relationship between the particle propagation vector and momentum:

$$\boldsymbol{p} = \hbar \boldsymbol{k}, \tag{1.37}$$

that is, a relationship between a mechanical property ($\boldsymbol{p}$) and a wave kinematic property ($\boldsymbol{k}$). But if we write the de Broglie relation in the covariant form

$$p_\mu = \hbar k_\mu, \tag{1.38}$$

since $p_0 = iE/c$, we realize that the zero component of the relation ($E = \hbar\omega$) was actually written down for photons by Einstein in 1905 in his description of the photoelectric effect. One of the properties of covariant laws is that if they are valid for one component (of μ), they are valid for the other components. Thus, the de Broglie relation is a natural one, and the proportionality constant ($\hbar$) is fixed, essentially by the Einstein relation. The generalization of the de Broglie relation to applicability for any kind of particle would follow from the introduction of the concept of the photon, since it had already been clear that electrons, photons, etc. were fundamental particles. The basic germs of both special relativity and modern quantum mechanics were introduced in 1905—and by the same man.[5]

Special relativity can also be given *too much* credit for explaining something. For example, it is stated in many textbooks that "the existence of the spin-$\frac{1}{2}$ value is a consequence of a relativistic quantum-mechanical description of the electron" (or words to that effect). This is nonsense. Spin, including spin-$\frac{1}{2}$, has nothing to do with relativity; it is associated with the fundamental isotropy of space—and *not* relativity. Another misconception is that the value of the electron gyromagnetic ratio (or intrinsic magnetic moment) needs relativistic quantum mechanics for an understanding. It does not, and the literature is filled with wrong statements on this question.[6]

1.2.3 Some Useful and Important Invariants

We have already noted the important invariant (1.32), of which a special example is that associated with the momentum four-vector:

$$p_\mu p_\mu = \text{inv.} \tag{1.39}$$

However, there are a number of important invariants associated with particle distribution functions and volume elements. A very important invariant is the four-dimensional spacetime volume element

$$dX = \prod_{\mu=0}^{3} dx_\mu. \tag{1.40}$$

It is easy to prove that dX is an invariant. Consider two reference frames K and K' with the origin of K' moving with a velocity $\boldsymbol{v}$ with respect to that of K (see Figure 1.1). The axes of K and K' are not necessarily aligned, but they do not

[5]Although "quantization" was first introduced in 1900 by Planck, the particle or "corpuscular" character of light was suggested first by Einstein. Quantization effects are imposed as a result of applications of the Schrödinger equation, which is a natural extension of the Einstein–de Broglie work.

[6]*cf.* R. J. Gould, *Am. J. Phys.* **64**, 597 (1996); **64(E)**, 1202 (1996).

rotate since K and K' are inertial frames. We introduce two other frames K'' and K''', whose x-axes are aligned along $\boldsymbol{v}$ and which are just orientations of K' and K, respectively. Since there is no motion between K and K''' and between K' and K'',

$$dX = dX'''; \; dX' = dX''. \tag{1.41}$$

The K''–K''' transformation is the special type with coefficients given by Equations (1.22)–(1.24), and the volume elements are related by the Jacobian of the transformation:

$$\begin{aligned} dX'' &= J\left(\frac{x_0'', x_1''}{x_0''', x_1'''}\right) dX''' \\ &= \begin{vmatrix} \gamma & -i\beta\gamma \\ i\beta\gamma & \gamma \end{vmatrix} dX''' = dX''', \end{aligned} \tag{1.42}$$

since the Jacobian is unity. Then, by Equation (1.41),

$$dX = dX' = \text{inv}. \tag{1.43}$$

In an exactly similar way, it can be shown that the four-dimensional momentum space volume element is an invariant:

$$dP = \prod_{\mu=0}^{3} dp_\mu = \text{inv}. \tag{1.44}$$

A four-vector three-dimensional "surface element" can be introduced, defined by

$$d\sigma_\mu = dX/dx_\mu = (dx_1 dx_2 dx_3, \; dx_0 dx_2 dx_3, \ldots) \tag{1.45}$$

That this quantity is a four-vector can be seen from the identity

$$d\sigma_\mu dx_\mu = 4\,dX = \text{inv}. \tag{1.46}$$

From what is sometimes called the "quotient rule" in general tensor analysis,[7] since the result (1.46) is true for arbitrary dx_μ, it follows that $d\sigma_\mu$ is a four-vector with, in fact, components proportional to those of the "missing" differential dx_μ.

An important and useful invariant is the "invariant phase-space[8] element"

$$d^3\boldsymbol{p}/E = \text{inv}, \tag{1.47}$$

which is the ratio of the ordinary momentum-space volume element to the particle energy corresponding to the momentum value defined by the element $d^3\boldsymbol{p}$. That is, if the differential dp defines the magnitude of $p = |\boldsymbol{p}|$ and the solid angle element $d\Omega$ defines the direction of $\boldsymbol{p}$, then $d^3\boldsymbol{p} = p^2 dp\, d\Omega$ and $E = (p^2c^2 + m^2c^4)^{1/2} = E(p)$.

---

[7]If A_μ is a four-vector and it is established that

$$A_\nu B_\nu = \text{inv} = A'_\mu B'_\mu = a_{\mu\nu} A_\nu B_\mu,$$

then

$$(a_{\mu\nu} B'_\mu - B_\nu) A_\nu = 0.$$

For this relation to be true for A_ν with arbitrary components, the parentheses must be identically zero; then B_ν has the correct four-vector transformation properties.

[8]The element $d^3\boldsymbol{p}$ is only the momentum part of what is meant, strictly speaking, by *phase space*, which is a product of pairs of differential Hamiltonian coordinates ($\prod dp\, dq$). In considerations of particle processes "phase space" generally refers to momentum space volume, which is a measure of the number of single-particle quantum states available.

We shall prove a more general theorem for which the result (1.47) would be a special case for the space of the momentum four-vector.

As with the four-dimensional momentum space element (1.44) we can consider the space element

$$dU = \prod_{\mu=0}^{3} dU_\mu = dU_0\, d^3\boldsymbol{U} \tag{1.48}$$

associated with an arbitrary four-vector U_μ. Since it is a four-vector, its squared "length" is an invariant: $U_\nu U_\nu = C = \text{inv}$. We can then form the manifestly invariant quantity

$$\int dU\, \delta(U_\nu U_\nu - C) = \int dU_0\, d^3\boldsymbol{U}\, \delta(U_0^2 + U_j U_j - C) = \text{inv}, \tag{1.49}$$

where $U_j U_j$ is the sum over the three "space" components ($j = 1, 2, 3$). Integration over dU_0 with the δ-function yields the result

$$\frac{d^3\boldsymbol{U}}{2(C - U_j U_j)^{1/2}} = \frac{d^3\boldsymbol{U}}{2U_0} = \text{inv}, \tag{1.50}$$

for which the invariant (1.47) is a special case for $U_\mu = p_\mu$. Further, by the relation (1.39), $p\,dp = E\,dE/c^2$ and $p/E = v/c^2$, where v is the particle velocity. The invariant (1.47) can then also be expressed in the following terms:

$$(p^2/E)\,dp\,d\Omega = \text{inv}, \quad pv\,dp\,d\Omega = \text{inv}, \quad p\,dE\,d\Omega = \text{inv}. \tag{1.51}$$

Another useful invariant associated with distributions of particles involves the differential number density of the particles having a particular energy E and moving in a particular direction. For example, suppose that dn is the differential number density of particles having energies within dE and moving in a direction defined by the solid angle element $d\Omega$. Since this density refers to a given (invariant) number dN of particles, $dn = dN/d^3\boldsymbol{r} = (dN/dX)dx_0$ is the zero component of a four-vector.[9] Now consider the transformation properties of dn/E, that is, of dn/p_0. This quantity transforms as

$$\frac{dx_0'}{p_0'} = \frac{a_{0\mu}dx_\mu}{a_{0\mu}p_\mu} = \frac{dx_0}{p_0}\,\frac{a_{00} + a_{0j}dx_j/dx_0}{a_{00} + a_{0j}p_j/p_0}. \tag{1.52}$$

But when the spacetime coordinates refer to the particle of given p_μ, both p_j/p_0 and dx_j/dx_0 are equal to $-i\beta_j$ and it follows that $dx_0/p_0 = \text{inv}$. Since dn can be written as proportional to dx_0 (see above), we then have

$$dn/E = \text{inv}. \tag{1.53}$$

A related invariant is the *occupation number* describing a distribution of particles. If dn_s is the differential density per polarization (spin) state, the occupation number is (defined as)

$$\overline{n} = (2\pi\hbar)^3 dn_s/d^3\boldsymbol{p}. \tag{1.54}$$

By virtue of the invariance of the differentials (1.47) and (1.53), it follows that

$$\overline{n} = \text{inv}. \tag{1.55}$$

---

[9]The four-vector is the current density $j_\mu = (icn, \boldsymbol{j})$, where $\boldsymbol{j}$ is the (three-dimensional) current density. The continuity equation in covariant form is then $\partial j_\mu/\partial x_\mu = \text{Div}\, j_\mu = 0$.

1.2.4 Covariant Mechanics and Electrodynamics

Because Newtonian mechanics is covariant under Galilean spacetime transformations, it is inconsistent with Lorentz-Einstein spacetime. Electromagnetism is, however, Lorentz covariant when written in terms of a four-dimensional spacetime formulation, and it can be helpful as a guide in reformulating mechanics. Because of the validity of the Newtonian limit, to every basic equation in non-relativistic mechanics there must be a corresponding relativistic covariant one having, in fact, a similar form. Actually, the relativistic equations are more compact and elegant, and it is not difficult to guess their form on the basis of their non-relativistic limits. For example, we expect the covariant form of Hamilton's Principle to be

$$\delta \int L(x_\mu, v_\mu, \tau)\, d\tau = 0, \tag{1.56}$$

where $v_\mu = dx_\mu/d\tau$. The Lagrangian must be an invariant function of the four-vectors x_μ and v_μ and of other field parameters, and the resulting Euler-Lagrange equation will be

$$\frac{d}{d\tau}\frac{\partial L}{\partial v_\mu} - \frac{\partial L}{\partial x_\mu} = 0. \tag{1.57}$$

The correct relativistic Lagrangian for a charged[10] particle in an electromagnetic field may be obtained easily. For fields derived from a vector potential $\boldsymbol{A}$ and scalar potential Φ, the non-relativistic Lagrangian for the particle motion is

$$L_{\mathrm{NR}} = \tfrac{1}{2}mv_jv_j + (q/c)A_jv_j - q\Phi. \tag{1.58}$$

If we define the quantity $A_\mu = (i\Phi, \boldsymbol{A})$, we can easily establish its four-vector character. The Lorentz gauge condition is then simply Div $A_\mu = 0$, which is manifestly covariant. Also, the transformation relation to other gauges is covariant: $A_\mu \to A_\mu + \partial_\mu \Lambda$, where Λ is an arbitrary scalar function.[11] Thus, we are immediately led to suggest the covariant relativistic Lagrangian

$$L = \tfrac{1}{2}mv_\mu v_\mu + (q/c)A_\mu v_\mu. \tag{1.59}$$

The Lagrangian itself has, basically, no special physical meaning; rather it is a *function*, which, when employed in Equation (1.57), yields the correct equation of motion. For example, from the definition of v_μ, the factor $v_\mu v_\mu$ in the Lagrangian (1.59) actually equals $-c^2$, but we do not write it that way, since the important character of the Lagrangian is its functional form.

With the Lagrangian (1.59), the equation of motion (1.57) is

$$\frac{d}{d\tau}(mv_\mu) = \frac{q}{c}\left[\frac{\partial}{\partial x_\mu}(A_\nu v_\nu) - \frac{dA_\mu}{d\tau}\right], \tag{1.60}$$

and the right-hand side of this equation is clearly some kind of four-vector force. We expect a relativistic equation of motion to be of the form

$$dp_\mu/d\tau = K_\mu, \tag{1.61}$$

[10]Effects of the interaction of the particle's magnetic moment (if it has one) with the field are neglected.
[11]The convenient compact notation $\partial_\mu = \partial/\partial x_\mu$ is introduced here.

where $K_\mu = (K_0, \boldsymbol{K})$, and the space part ($\boldsymbol{K}$) of this four-vector must be related to the Newtonian notion of the force ($\boldsymbol{F}$). In fact, this relationship can be discovered very easily through a comparison with the result (1.60) for the special case where $A_\mu = (i\Phi, 0)$, corresponding to $\boldsymbol{A} = 0$ and there is only the presence of an electrical component of the Lorentz force $\boldsymbol{F} \to q\boldsymbol{E} = -q\nabla\Phi$, for which $A_\nu v_\nu \to -\gamma c\Phi$. Since $dt = \gamma\, d\tau$, the space part of the covariant equation (1.61) can then be written (in non-covariant form)

$$d(\gamma m\boldsymbol{v})/dt = \boldsymbol{F} = \boldsymbol{K}/\gamma, \tag{1.62}$$

which also allows the identification of $\boldsymbol{K}$ in terms of $\boldsymbol{F}$.

The meaning of the zero or "time" component of K_μ becomes clear when we form the invariant

$$K_\mu v_\mu = m v_\mu dv_\mu/d\tau = \tfrac{1}{2} m\, d(v_\mu v_\mu)/d\tau. \tag{1.63}$$

But $v_\mu v_\mu = -c^2$, so that

$$K_\mu v_\mu = 0, \tag{1.64}$$

which must hold for all types of four-vector forces. Written explicitly in terms of the components, the identity (1.64) is

$$i\gamma c K_0 + \gamma^2 \boldsymbol{F} \cdot \boldsymbol{v} = 0. \tag{1.65}$$

Maintaining the same concept of energy in relativistic mechanics as in Newtonian mechanics,

$$\boldsymbol{F} \cdot \boldsymbol{v} = dT/dt, \tag{1.66}$$

where T is the particle kinetic energy. Then

$$K_0 = (i/c)\gamma\, dT/dt = (i/c)\, dT/dt, \tag{1.67}$$

and the zero component of the equation of motion (1.61) is a relation for the particle energy. The relation also allows the identification of the zero component of the momentum four-vector p_μ; that is,

$$p_0 = m v_0 = i m\gamma c = iE/c, \tag{1.68}$$

where E is the particle energy. The identity $p_\mu p_\mu = m^2 v_\mu v_\mu = -m^2c^2$ then gives a relativistic relation between energy and momentum:

$$E^2 = p^2c^2 + m^2c^4, \tag{1.69}$$

where p is the magnitude of the (relativistic) momentum.

The equations of classical electrodynamics are elegant when expressed in covariant form. The relationships between the vector and scalar potentials and the sources of the fields (charges and currents) are very simple:

$$\Box^2 A_\mu = -(4\pi/c)\, j_\mu, \tag{1.70}$$

where $\Box^2 = \partial_\mu \partial_\mu$ is the invariant D'Alembertian operator. In terms of the antisymmetric electromagnetic field tensor,

$$F_{\mu\nu} = \partial A_\nu/\partial x_\mu - \partial A_\mu/\partial x_\nu = \partial_\mu A_\nu - \partial_\nu A_\mu, \tag{1.71}$$

the components of the electric and magnetic fields can be written

$$\begin{aligned} E_j &= iF_{j0}, \\ B_j &= \varepsilon_{jkl}\partial_k A_l = \tfrac{1}{2}\varepsilon_{jkl}(\partial_k A_l - \partial_l A_k), \\ &= \tfrac{1}{2}\varepsilon_{jkl}F_{lk}. \end{aligned} \tag{1.72}$$

Here, ε_{jkl} is the Levi-Civita or Ricci symbol

$$\varepsilon_{jkl} = \begin{cases} +1 & jkl = \text{even permutation of 123} \\ -1 & jkl = \text{odd permutation of 123} \\ 0 & \text{otherwise.} \end{cases} \tag{1.73}$$

The four Maxwell equations in non-covariant form are condensed into two covariant equations:

$$\partial_\nu F_{\mu\nu} = (4\pi/c)j_\mu, \tag{1.74}$$

$$\partial_\lambda F_{\mu\nu} + \partial_\mu F_{\nu\lambda} + \partial_\nu F_{\lambda\mu} = 0 \quad (\lambda \neq \mu \neq \nu). \tag{1.75}$$

There are four relations associated with each of the above equations; four are from the four values of μ in the inhomogeneous equation (1.74), and there are four possible values for the index not appearing in the cyclic permutation in the homogeneous equation (1.75). Without the radiation reaction force term, the covariant equation of motion for a particle in an electromagnetic field is

$$dp_\mu/d\tau = (q/mc)F_{\mu\nu}p_\nu. \tag{1.76}$$

In treating the various electromagnetic processes, we shall make extensive use of covariance considerations.

1.3 KINEMATIC EFFECTS

Certain important aspects of particle processes have nothing to do with the nature of the interactions involved but are, rather, a consequence of the kinematics. That is, the effects are associated with energy and momentum conservation or with the availability of continuum states for the process ("phase-space effects"). The kinematic effects often dominate the character of a process, and a few examples will be mentioned briefly in this section. The examples will be of a general nature and have applications to a variety of processes.

1.3.1 Threshold Energies in Non-Relativistic and Relativistic Processes

In descriptions of many inelastic processes, it is convenient to consider the processes in a center-of-mass (c.m.) frame for which the total momentum is zero. For example, in a collision of two particles, it may be possible to produce another particle or to excite a higher bound energy level in one of the colliding particles or systems. The minimum energy for such a phenomenon would correspond to the case where the outgoing particles have zero kinetic energy in the c.m. frame.

Let us consider first the case of the collision of two non-relativistic particles or systems (such as an atom) of mass m_1 and m_2 and initial velocity $\mathbf{v}_1$ and $\mathbf{v}_2$ in the lab frame (K). Then the c.m. frame (K') moves with velocity

$$\mathbf{V} = (m_1\mathbf{v}_1 + m_2\mathbf{v}_2)/M, \tag{1.77}$$

with

$$M = m_1 + m_2. \tag{1.78}$$

In K' the particle velocities are $\mathbf{v}' = \mathbf{v} - \mathbf{V}$, each particle having a magnitude of momentum

$$p' = |\mathbf{p}_1| = |\mathbf{p}_2| = \mu v_r, \tag{1.79}$$

where

$$\mu = m_1 m_2/M \tag{1.80}$$

is the reduced mass and

$$v_r = |\mathbf{v}_1 - \mathbf{v}_2| \tag{1.81}$$

is the magnitude of the relative velocity in the *lab frame* (K). The total kinetic energy in the c.m. frame (K') is

$$E' = p'^2/2m_1 + p'^2/2m_2 = p'^2/2\mu = \tfrac{1}{2}\mu v_r^2. \tag{1.82}$$

Suppose that we are interested in some inelastic process whereby the energy E' is to be used in the excitation of some level (which may be in the continuum) or in producing some new particle. If the c.m. energy required for this excitation is χ, setting $E' \geq \chi$ yields, by Equation (1.82), a condition on the relative velocity v_r:

$$\tfrac{1}{2}\mu v_r^2 \geq \chi. \tag{1.83}$$

For the case, say, $\mathbf{v}_2 = 0$ (one particle initially at rest in the lab frame), we have a condition on the required kinetic energy of the incident particle:

$$E_1 = \tfrac{1}{2}m_1 v_1^2 = (m_1/\mu)\chi. \tag{1.84}$$

But $\mu < m_1$, and we see that the kinetic energy required is always *greater than* the excitation energy χ needed in K'. The physical reason for this result is very simple; to satisfy momentum conservation, the products of the collision process must, at threshold, all be moving in the direction of the incident particle at a velocity $\mathbf{V} = (m_1/M)\mathbf{v}_1$. The kinetic energy (in K) of the outgoing M is just $E_M = (m_1/M)E_1$ at threshold, and the difference $(E_1 - E_M)$ equals χ.

The kinematics of relativistic particle collisions can be handled with about the same degree of simplicity as the non-relativistic case. Instead of transforming the particle energy and momentum between the lab frame (K) and the c.m. frame (K') by means of the four-vector Lorentz transformation, it is more convenient to employ the invariant (1.39). Also, in manipulations involving relativistic kinematics, to simplify the algebra somewhat, we can set $c = 1$; if necessary, in final formulas the factors of c to various powers can be reinserted through considerations of dimensionality.

The invariant

$$E^2 - \mathbf{p}^2 = \text{inv} \tag{1.85}$$

can refer to either the components of the total energy and momentum or those of an individual particle, for which $E_1^2 - p_1^2 = m_1^2$. When E and $\boldsymbol{p}$ refer to the total energy and momentum of a two-particle system, since the individual energy and momentum are related by $\boldsymbol{p}_1 = E_1 \boldsymbol{v}_1$, application of the invariant (1.85) yields

$$m_1^2 + m_2^2 + 2E_1E_2(1 - \boldsymbol{v}_1 \cdot \boldsymbol{v}_2) = E'^2 \tag{1.86}$$

where E' is the total energy in the c.m. frame. This simple and general formula has many important applications. In the calculation of binary-collision processes, generally the cross section is most conveniently represented in terms of the energy E', and the expression (1.86) provides the explicit relationship to the lab-frame quantities.[12] Equation (1.86) can also be applied to determine (lab-frame) threshold energies as in the non-relativistic limit. If $m_1 = m_2 = m$ and $\boldsymbol{v}_2 = 0$, for a process corresponding to a c.m. threshold energy $E' = \chi$, the lab-frame kinetic energy required is

$$T_1 = E_1 - m = \chi^2/2m - 2m. \tag{1.87}$$

For example, in producing particle-antiparticle pairs of the same type, the c.m. threshold energy would be $\chi = 4m$ and the lab-frame value would be $T_1 = 6m$. As in our non-relativistic problem, because of the necessity of satisfying momentum conservation, much more lab-frame energy is required than in the c.m. frame.

In the case of photon-photon collisions or for collisions of highly relativistic particles, the relation (1.86) is further simplified:

$$2E_1E_2(1 - \cos\theta_{12}) = E'^2, \tag{1.88}$$

where θ_{12} is the angle between the (massless) particle directions of motion. For example, in the process of pair production in photon-photon collisions, E' could be set equal to $2E'_e$, where E'_e is the c.m. energy of the electron or positron. For a head-on photon-photon collision ($\theta_{12} = \pi$), the threshold condition would be, very simply, $E_1E_2 = m^2$.

1.3.2 Transformations of Angular Distributions

In the evaluation of particle processes, it is often necessary to consider the relationship between the direction of particle motion in different reference frames. For relativistic particle processes, there is an especially important and interesting phenomenon associated with the directions of motion of other particles (with which it interacts) in the lab frame and in the rest frame of the relativisitic particle.

Consider the motion of a particle (P_v) of velocity $\boldsymbol{v}$ in the lab frame (K). Let the axes of K be such that $\boldsymbol{v} = v(1, 0, 0)$; that is, $\boldsymbol{v}$ is along the x-axis. Now consider the motion of another particle (P_u) of velocity $\boldsymbol{u}$; let this velocity lie in K's x–y-plane so that $\boldsymbol{u} = u(\cos\theta, \sin\theta, 0)$. In P_v's rest frame (K'), P_u will be moving at an angle θ' with respect to the x- and x'-axes. From the velocity transformation,

$$\tan\theta' = u'_y/u'_x = u_y/\gamma(u_x - v) = \sin\theta/\gamma(\cos\theta - v/u), \tag{1.89}$$

[12] Often it is necessary to integrate over a spectrum of lab-frame energies and directions of motion.

where $\gamma = (1 - v^2/c^2)^{-1/2}$. The transformation from K' to K is, of course, of the same form except that the sign of v_x (= v) is reversed. For the case where P_u is non-relativistic and $v/u \gg \cos\theta$, $\tan\theta' \to -(u/\gamma v)\sin\theta$, for which $\pi/2 < \theta' < \pi$. In fact, if $v \to c$ ($\gamma \gg 1$), θ' will be very close to π. This means that in P_v's frame (K') every other particle is moving toward it head on. Even if the P_u's are photons ($u \to c$), this will be true as long as $\theta \gtrsim \gamma^{-1}$. A highly relativistic particle moving through a gas "sees" particles incident head on (like a beam).

We could also consider some process involving a highly relativistic P_v in which, in K', there is some outgoing particle P_u (for example, a scattered photon) moving at an angle θ'. Then in the lab frame (K) the corresponding angle would be θ with

$$\tan\theta = \sin\theta'/\gamma(\cos\theta' + v/u'); \tag{1.90}$$

unless θ' is very close to π, θ' will be very small: $\theta' \to (u/c\gamma)\sin\theta' \ll 1$. In other words, the process involves a beam of associated particles (scattered or produced, for example) in the forward direction.

1.4 BINARY COLLISION RATES

An important relativistic kinematic effect is associated with the formula for collision rates for two particles. The relativistic features come in when we consider rates associated with a gas of particles and when we do covariant calculations of cross sections for various processes. The collision rate formula can be derived[13] through the use of kinematic invariants and relativistic transformations for quantities in convenient Lorentz frames. First, suppose we consider the non-relativistic expression for the collision rate for a particle of type a passing through a gas of particles of type b in which both types of particle have finite mass. The collision rate, per particle a, is

$$(dN/dt)_a = \iint |\boldsymbol{v}_a - \boldsymbol{v}_b| d\sigma dn_b \quad \text{(non-relativistic)}, \tag{1.91}$$

where the double integration is over the differential cross section $d\sigma$ and the distribution of particles of number density dn_b. The relative velocity $|\boldsymbol{v}_a - \boldsymbol{v}_b|$ multiplied by the differential density dn_b is the flux of particles of type b incident on particle a. If we have a gas of particles of type a the number of collisions per unit volume per unit time is

$$(dN/dVdt)_{ab} = (1 + \delta_{ab})^{-1} \iiint |\boldsymbol{v}_a - \boldsymbol{v}_b| d\sigma dn_a dn_b, \tag{1.92}$$

with the factor involving δ_{ab} correcting for counting the same type of collision twice if the particles a and b are identical.

We want the relativistic generalizations of the above formulas, including the class of collision in which either or both of the particles have, like photons, zero mass. In particular, the relative velocity factor must be replaced by a more general relativistic expression, and it is not hard to derive the correct formula. First, we note that dN and $dVdt$ are relativistic invariants. The simplest case to consider is that where particle

[13] See, for example, R. J. Gould, *Am. J. Phys.* **39**, 911 (1971).

a has a finite mass and particles b have zero mass, with the real life application being the case of electron-photon collisions. For the rate dN_a/dt we go from the "lab" frame (K) to the rest frame (K') of particle a:

$$dN_a/dt = (dN_a/dt)'(dt'/dt). \tag{1.93}$$

Since a is at rest in K', $dt = \gamma dt'$, in terms of the particle's Lorentz γ in K. In K',

$$(dN/dt)' = \iint c\, d\sigma dn_b', \tag{1.94}$$

where $c\, dn_b'$ is just the incident photon flux (in K'). The result can be written in terms of the lab-frame (K) distribution dn_b by making use of the invariance of dn/E and by writing

$$E' = \gamma E(1 - \beta \cos\theta), \tag{1.95}$$

where θ is the lab-frame angle between the directions of motion of a and b. We then have

$$dN_a/dt = c \iint (1 - \beta\cos\theta) d\sigma dn_b \quad (b \text{ massless}). \tag{1.96}$$

The factor $c(1 - \beta\cos\theta)$ is *not* the relative velocity; instead it is the projection of the relative velocity along the direction of motion of the zero-mass particle b.

When both particles a and b are massless, we cannot go to one or the other's rest frame and the lab-frame rate must be derived by going, instead, to the c. m. frame. Since the cross section $d\sigma$ is an area perpendicular to the direction of relative motion between the rest frame (for a finite-mass particle) and the c. m. frame, it is an invariant for transformations between these two frames; that is, $d\sigma' = d\sigma$. We can take the limit $\beta \to 1$ in the above formula to get the result for the collision rate for a massless particle a traversing a gas of massless particles b:

$$dN_a/dt = c \iint (1 - \cos\theta) d\sigma dn_b \quad (a \text{ and } b \text{ massless}). \tag{1.97}$$

Similarly, the collision rate for the two gasses, if one is massless, would be

$$(dN/dVdt)_{ab} = (1 + \delta_{ab})^{-1} c \times \iiint (1 - \beta\cos\theta) d\sigma dn_a dn_b \quad (a \text{ or } b \text{ massless}), \tag{1.98}$$

and we can obtain the result when both particles are massless by taking the limit $\beta \to 1$.

Now we look for the relativistic generalization of the non-relativistic results (1.91) and (1.92), that is, the case where both particles have finite mass. The derivation of the results starts from the collision rate for particle a in the lab frame (K), relating it to the rate in a's rest frame (K') as is done in (1.93). In K', the rate from the incident flux of particles of type b is

$$dN_a/dt' = dN_a'/dt' = \iint d\sigma |\boldsymbol{v}_b'| dn_b'. \tag{1.99}$$

The factors $|\boldsymbol{v}_b|$ and dn_b' are then written in terms of lab-frame quantities. To do this it is convenient to take an orientation of the frame K such that $\boldsymbol{v}_a$ is along the x-axis and $\boldsymbol{v}_b$ is in the x–y-plane. When the final result is obtained we can recognize rotational invariants equal to these expressions for the special case of this orientation of K's axes, yielding a formula for general direction of motion for $\boldsymbol{v}_a$ and $\boldsymbol{v}_b$. The number densities dn_a and dn_b are differential in the velocity or energy interval and differential in, say, a solid angle element designating direction. That is, the magnitude and direction of the velocities are specified. We then write

$$dn'_b = dn_b(E'_b/E_b) \tag{1.100}$$

The relative velocity for the frames K and K' is just v_{ax}, and

$$E'_b/E_b = \gamma(1 - v_{ax}v_{bx}/c^2), \tag{1.101}$$

with

$$1/\gamma^2 = 1 - v_{ax}^2/c^2. \tag{1.102}$$

Since the position coordinates of particle a are constant in K', the time intervals are related by $dt = \gamma dt'$. The flux velocity $|\boldsymbol{v}'_b|$ in (1.99) is given by

$$|\boldsymbol{v}'_b| = (v'^2_{bx} + v'^2_{by})^{1/2}, \tag{1.103}$$

and the two velocities here can be related to their lab-frame variables by simple Lorentz transformations. Specifically, we have

$$v'_{bx} = \frac{v_{bx} - v_{ax}}{1 - v_{ax}v_{bx}/c^2}, \quad v'_{by} = \frac{v_{by}}{\gamma(1 - v_{ax}v_{bx}/c^2)}. \tag{1.104}$$

The result for the velocity $|\boldsymbol{v}'_b|$ can be put in general form by recognizing that terms from components for the particular choice of orientation of axes can be written as special cases of a formula that is a rotational (in velocity space) scalar. We have

$$v'^2_{bx} + v'^2_{by} = \frac{(\boldsymbol{v}_a - \boldsymbol{v}_b)^2 - (\boldsymbol{v}_a \times \boldsymbol{v}_b)^2/c^2}{(1 - \boldsymbol{v}_a \cdot \boldsymbol{v}_b/c^2)^2}, \tag{1.105}$$

where the expression on the right is symmetric in a and b and can be regarded as the squared magnitude of the velocity of either a or b in the rest frame (K') of the other. The differential densities in K' and in the general lab frame K are related by

$$dn'_b = \gamma(1 - \boldsymbol{v}_a \cdot \boldsymbol{v}_b/c^2)dn_b, \tag{1.106}$$

$$dn'_a dn'_b = (1 - \boldsymbol{v}_a \cdot \boldsymbol{v}_b/c^2)dn_a dn_b. \tag{1.107}$$

Further, we can introduce a fundamental velocity

$$v_0 = [(\boldsymbol{v}_a - \boldsymbol{v}_b)^2 - (\boldsymbol{v}_a \times \boldsymbol{v}_b)^2/c^2]^{1/2}, \tag{1.108}$$

which is a relativistic generalization of the non-relativistic expression $|\boldsymbol{v}_a - \boldsymbol{v}_b|$. We then have

$$dN_a/dt = \iint v_0\, d\sigma\, dn_b, \tag{1.109}$$

$$(dN/dVdt)_{ab} = (1 + \delta_{ab})^{-1} \iiint v_0\, d\sigma\, dn_a dn_b. \tag{1.110}$$

These formulas apply, in fact, to the cases where one or both particles are massless. In these cases we would take the limits $m_a \to 0$ and/or $m_b \to 0$ for which $v_a \to c$ and/or $v_b \to c$. Note how the factors γ and $(1 - \boldsymbol{v}_a \cdot \boldsymbol{v}_b/c^2)$ cancel to give formulas for the rates only in terms of the kinematic factor v_0.

1.5 PHASE-SPACE FACTORS

1.5.1 Introduction

There are in physics many phenomena or processes that are essentially "phase-space effects." By this is meant that the magnitude of the effect is proportional to the available number of states. Actually, the factor that usually comes in is the number of states per unit energy available, and this can be seen from the Fermi Golden Rule formula. From time-dependent perturbation theory (see Chapter 3), the transition rate from an initial state (0) to a final continuum state (f) is given by

$$\frac{\Delta W_{f0}}{\Delta t} = \frac{2\pi}{\hbar}\Big(\sum_f\Big)|H'_{f0}|^2\left(\frac{dN}{dE}\right)_f = \frac{2\pi}{\hbar}\Big(\sum_f\Big)|H'_{f0}|^2\delta(E_f - E_0); \qquad (1.111)$$

here H'_{f0} is the matrix element of the perturbation Hamiltonian that causes the transition, and there is usually a sum over final states. The δ-function expresses energy conservation; that is, the final states are restricted to the "energy shell" determined by the total energy ($E_0 = E$) available to the various final states. For processes between initial and final free-particle (continuum) states there is also conservation of momentum:

$$\sum_f \boldsymbol{p}_f = \sum_0 \boldsymbol{p}_0 \equiv \boldsymbol{P}, \qquad (1.112)$$

and $\boldsymbol{P} = 0$ in the c.m. frame. The condition (1.112) also follows from the evaluation of H'_{f0} in the position representation, employing plane-wave states for the particle wave functions in the initial and final states (see Chapter 3).

Aside from spin states, the number of momentum ($\boldsymbol{p}$) or wave-vector ($\boldsymbol{k}$) states available in a spatial volume L^3 and momentum-space volume $d^3\boldsymbol{p}$ is[14]

$$dN = L^3 d^3\boldsymbol{p}/(2\pi\hbar)^3 = L^3 d^3\boldsymbol{k}/(2\pi)^3. \qquad (1.113)$$

In the description of particle processes the factors involving L^3 always cancel, since L^3 appears in inverse powers in the matrix element H'_{f0} as normalization factors in the individual one-particle wave functions. As remarked earlier in this chapter, although $L^3 d^3\boldsymbol{p}$ is, strictly speaking, what is known as phase space, in the description of particle processes we often refer to the momentum-space part as the phase-space factor. If N particles are being described, the associated phase-space volume, with both energy and momentum-space conservation, is then, in the c.m. frame,

$$\Phi^{(N)} = \int \cdots \int \prod_{j=1}^{N} d^3\boldsymbol{p}_j\, \delta^{(3)}\Big(\sum_j \boldsymbol{p}_j\Big)\,\delta\Big(\sum_j E_j - E\Big). \qquad (1.114)$$

For many processes this factor determines the magnitude of the transition rate or, for example, the cross section for the process. Of special importance is the dependence of $\Phi^{(N)}$ on the available energy E, which would determine the behavior of, say, a cross section in the neighborhood of the threshold energy for the process. This

[14] See Equation (1.54) and any book on quantum mechanics treating the elementary "particle-in-a-box" problem.

dependence can be derived very readily, the result being that $\Phi^{(N)}$ has a power-law form

$$\Phi^{(N)} \propto E^q, \tag{1.115}$$

for certain energy domains [outgoing particles non-relativistic (NR) or extreme relativistic (ER)], with q increasing with N. We shall derive these results, obtaining general formulas for $\Phi^{(N)}$ in the NR and ER limits. In the derivation of the general results, the mathematical techniques and tricks are interesting in themselves, and several such methods will be introduced or, at least, mentioned. Although very general formulas can be obtained, usually the most important applications are for $N \sim 2$ or 3, say, and these specific expressions will be given first, since the mathematics is simple.

There are many important applications or uses for the expressions for the $\Phi^{(N)}$. They are fundamental for general processes in which particles are produced, and often the energies involved are relativistic. Unfortunately, the problem for (general) outgoing particle energies $E_j \sim m_j c^2$ is complex except for small N and, in fact, in this case the invariant expression

$$\phi^{(N)} = \int \cdots \int \prod_{j=1}^{N} (d^3 p/2E)_j \, \delta^{(4)}\Big(\sum_j (p_u)_j - P_\mu\Big) \tag{1.116}$$

is sometimes more useful. Considerations of particle processes in terms of invariant phase-space factors are employed extensively in high-energy physics. However, a covariant description[15] is not always most appropriate, especially if the outgoing particles are non-relativistic. This is usually the case in applications to problems involving nuclear reactions ($E_0 \sim$ MeV energies), and always in the area of chemical kinetics where the reaction products are atoms, molecules, ions, and electrons, at $\sim$ eV energies. In some processes there might be outgoing particles both NR and ER, the latter type being photons, for example, as in some nuclear reactions; thus, it is useful to have expresssions for the phase-space factor in these "mixed" cases. To describe the cases we shall employ the notation N and N' for the total number of NR and ER particles, respectively, and the indices j and k to designate the particles among the N and N'. We shall see that it is possible to obtain exact formulas for Φ for arbitrary N and N' in this limit, where

$$\begin{cases} E_j \to p_j^2/2m_j & (j = 1 \text{ to } N), \\ E_k = p_k c & (k = 1 \text{ to } N'). \end{cases} \tag{1.117}$$

There is an important theorem involving the phase-space volume (1.114) for the case where one of the outgoing particles (say, m_N) has a large mass. In this limit its energy would have to be non-relativistic and the arguments of the δ-functions in $\Phi^{(N)}$ can be written

$$\sum_{j=1}^{N} \boldsymbol{p}_j = \boldsymbol{p}_N + \sum_{j=1}^{N-1} \boldsymbol{p}_j, \tag{1.118}$$

$$\sum_{j-1}^{N} E_j - E = \boldsymbol{p}_N^2/2m_N + \sum_{j=1}^{N-1} E_j - E. \tag{1.119}$$

[15]In Section 1.5.6 we shall return to the use of the invariant expression (1.116). We shall also briefly discuss the relative merits of the covariant and non-covariant formulations and their applications.

For very large m_N the first term on the right of Equation (1.119) is negligible. The δ-function $\delta^{(3)}$ in $\Phi^{(N)}$ can then be eliminated by integrating over $d^3\boldsymbol{p}_N$ using the argument (1.118), and we have the result

$$\Phi^{(N)} \xrightarrow[m_N \text{large}]{} \Psi^{(N-1)}, \tag{1.120}$$

where Ψ is a phase-space volume like $\Phi^{(N)}$ but *without* the momentum-conservation δ-function:

$$\Psi^{(N)} = \int \cdots \int \prod_{j=1}^{N} d^3\boldsymbol{p}_j \delta\Big(\sum_j E_j - E\Big). \tag{1.121}$$

That is, when one of the outgoing particles is very massive, it performs the function of carrying away momentum but not energy. The energy-conservation restriction is still imposed for the other particles and, for the validity of the theorem (1.120), no assumption has been made about the characteristics or energies of the other $N-1$ particles. Incidentally, the form $\Psi^{(N)}$ is the one generally introduced in classical statistical mechanics in which the basic assumption is that equal phase-space volumes within the energy shell have equal weighting or probability. In that subject it is inherently assumed that there is momentum exchange to some (heavy) containment vessel or thermalizing agent.

Similar remarks can be made concerning angular momentum. No restriction on total angular momentum is made in the integrals (1.114) and (1.121). It is assumed that there is some mechanism, perhaps associated with the interaction H', in which angular momentum is conserved without affecting the momentum-space integrations $\Phi^{(N)}$ or $\Psi^{(N)}$. Actually, the angular momentum constraints are quite complex, since the spatial integrations over the $d^3\boldsymbol{r}_j$ are involved as well as particle spin states. Other complicating factors will also be omitted here, such as the spin degeneracy factor

$$g_{\text{spin}} = \prod_j g_j. \tag{1.122}$$

When some of the outgoing particles are identical there is also a "symmetry factor" $1/\sigma$ that must multiply $\Phi^{(N)}$ or $\Psi^{(N)}$ to avoid counting equivalent states more than once.[16] There is also a restriction associated with baryon and lepton conservation that must be applied in considerations of possible combinations of outgoing particles. All of these effects will be omitted here and only formulas for the classical kinematic factors $\Phi^{(N)}$ and $\Psi^{(N)}$ will be given.

1.5.2 Simple Examples

Let us take the very simplest example, which will immediately yield an interesting and instructive result. We consider some process in which the product is a very massive particle (like a nucleus) plus a single additional particle (perhaps a pion or a photon). The relevant phase-space factor would then be

$$\Psi^{(1)} = \int d^3\boldsymbol{p}_1\, \delta(E_1 - E) = 4\pi \int p_1^2\, dp_1\, \delta(E_1 - E). \tag{1.123}$$

[16] The factor is, in fact, given by $\sigma = N!$ for N identical particles or, for the case where there are several kinds (α) of identical outgoing particles, $\sigma = \prod_\alpha (N_\alpha)!$.

Employing the relation (1.69), we obtain

$$\Psi^{(1)} = (4\pi/c^3)E(E^2 - m^2c^4)^{1/2} \tag{1.124}$$

for a particle of arbitrary mass. If the particle is massless or is highly relativistic,

$$\Psi^{(1)} \to (4\pi/c^3)E^2 \quad \text{(ER)}. \tag{1.125}$$

On the other hand, if $E = mc^2 + T$, with $T \ll mc^2$, corresponding to an outgoing low-energy particle with mass,

$$\Psi^{(1)} \to 4\sqrt{2}\pi m^{3/2}T^{1/2} \quad \text{(NR)}. \tag{1.126}$$

In the limit $T \ll mc^2$, the *ratio* of the two formulas above is

$$\frac{\text{(NR)}}{\text{(ER)}} \to \left(\frac{2T}{mc^2}\right)^{1/2} \ll 1. \tag{1.127}$$

The result tells us, for example, that phase-space favors photon production over particle (with mass) production in this application.

Now consider the case of two outgoing particles accompanying a heavy one, so that the relevant factor is

$$\Psi^{(2)} = \iint d^3\boldsymbol{p}_1\, d^3\boldsymbol{p}_2 \delta(E_1 + E_2 - E). \tag{1.128}$$

When both particles are NR or both are ER, the integration is elementary and yields

$$\Psi^{(2)} \to \begin{cases} 4\pi^3(m_1m_2)^{3/2}E^2 & \text{(NR)}, \\ (8\pi^2/15c^6)E^5 & \text{(ER)}. \end{cases} \tag{1.129}$$

The case where one particle is NR and the other is ER is also easy to evaluate:

$$\Psi^{(2)} \to \frac{256\sqrt{2}\pi^2}{105}\frac{m_1^{3/2}}{c^3}E^{7/2} \quad \text{(1 NR, 1 ER)}. \tag{1.130}$$

In the result (1.129) the energy E is the (maximum) *kinetic* energy available to the NR particles; that is, it is the available energy over and above the (rest) energies necessary to produce the particles. The different dependence on E in the above three results should, in particular, be noted.

For the case of arbitrary energy and finite m_1 and m_2, the integral $\Psi^{(2)}$ is fairly complicated. Setting $d^3\boldsymbol{p}_2 = (4\pi/c^2)p_2E_2dE_2$, integration over dE_2 with the δ-function is performed, giving

$$\begin{aligned} \Psi^{(2)} &= (4\pi/c^2)\int d^3\boldsymbol{p}_1\, p_2\,(E - E_1) \\ &= (4\pi/c^2)^2\int p_1p_2E_1(E - E_1)dE_1, \end{aligned} \tag{1.131}$$

with

$$p_1c = (E_1^2 - m_1^2c^4)^{1/2}, \quad p_2c = [(E - E_1)^2 - m_2^2c^4]^{1/2}. \tag{1.132}$$

The lower and upper limits on the integral (1.131) are, respectively, m_1c^2 and $E - m_2c^2$, in which E ($> m_1c^2 + m_2c^2$) is the total energy. The integral is difficult to evaluate,[17] except for $m_2 = 0$; in that case, the result is

$$\Psi^{(2)}(m_2 = 0) = (4\pi c^2)^2 m_1^5 I, \tag{1.133}$$

where I is the integral

$$\begin{aligned} I &= \int_0^{\eta_0} \eta^2[(1+\eta_0^2)^{1/2} - (1+\eta^2)^{1/2}]^2 d\eta \\ &= -\tfrac{1}{4}\eta_0 - \tfrac{1}{12}\eta_0^3 + \tfrac{1}{30}\eta_0^5 + \tfrac{1}{4}(1+\eta_0^2)^{1/2}\ln[(1+\eta_0^2)^{1/2} + \eta_0], \end{aligned} \tag{1.134}$$

η_0 being related to E by

$$\eta_0 = [(E/m_1c^2)^2 - 1]^{1/2}. \tag{1.135}$$

The result has an important application in nuclear β-decay, essentially determining the radioactive lifetime of a nucleus. In that application, particle 2 would be the (massless) neutrino and particle 1 an electron or positron; the heavy residual nucleus carries away momentum, yielding $\Psi^{(2)}$ rather than $\Phi^{(3)}$ as the relevant phase-space volume.

The evaluation of $\Phi^{(N)}$ in simple cases is readily accomplished. With $N = 2$ the expression (1.114) yields, in the NR and ER limits,

$$\Phi^{(2)} \to \begin{cases} 4\sqrt{2}\pi(m_1m_2/M)^{3/2}E^{1/2} & \text{(NR)}, \\ (\pi/2c^3)E^2 & \text{(ER)}, \end{cases} \tag{1.136}$$

where $M = m_1 + m_2$. In the mixed case,

$$\Psi^{(2)} \to (4\pi/c^3)E^2 \quad \text{(1 NR, 1 ER)}, \tag{1.137}$$

which is identical to the result (1.125); in this case, the massless particle carries off negligible momentum, an important result to keep in mind.

For general energy and general m_1, m_2, the relativistically correct expression for $\Phi^{(2)}$ is obtained:

$$\begin{aligned} \Phi^{(2)} &= \frac{\pi}{2c^3E^2}\left[(E^2 - m_1^2c^4 - m_2^2c^4)^2 - 4m_1^2m_2^2c^8\right]^{1/2} \\ &\quad \times \frac{E^4 - (m_2^2 - m_1^2)^2c^8}{E^2}, \end{aligned} \tag{1.138}$$

in which $E > (m_1 + m_2)c^2$ is the total energy including rest energy. The result (1.138), which contains the expressions (1.136) and (1.137) in the corresponding limits, has been written in the above form for comparison with the covariant expression ($\phi^{(2)}$), which will be given later.

It is obvious from, for example, the complexity of the formula (1.138), that the relativistic problem gets complicated for $N > 2$. However, it is possible to derive expressions for $\Psi^{(N)}$ and $\Phi^{(N)}$ for arbitrary N when the outgoing particles are NR or ER or a mixture of each. The result ($M = m_1 + m_2 + m_3$)

$$\Phi^{(3)} \to 4\pi^3(m_1m_2m_3/M)^{3/2}E^2 \quad \text{(NR)} \tag{1.139}$$

[17] The integrand itself provides the distribution ($d\Psi^{(2)}/dE_1$) in outgoing energies for particle 1 (see Section 4.5).

can be obtained directly through some cumbersome algebra. About as much time is spent deriving the expression (NR limit) for general N, and we now turn to this task. Incidentally, for $m_3 \gg m_1$ and m_2, we see that the formula (1.139) reduces to the expression (1.129) for $\Psi^{(2)}$—a specific application of the theorem (1.120). In fact, we shall see that in the NR limit the result for $\Phi^{(N)}$ is very closely related to $\Psi^{(N-1)}$ for *general* $m_1, m_2, \ldots$.

1.5.3 General Theorems—Formulation

Some features of the formulas for $\Phi^{(N)}$ and $\Psi^{(N)}$, such as the very important dependence of E, can be derived without an evaluation of the multiple integrals that are involved in the formulas. This can be achieved through convenient and natural changes of variables to dimensionless form, such that the resulting integrals are functions only of N. The formulation given here is for the special cases where the particles are either NR or ER or a mixture of the two. That is, we consider these three cases with the third obviously containing the first two in a general treatment. To designate the individual and total number of NR and ER particles the notation (1.117) will be employed for their index and total:

$$\text{NR: } j = 1, 2, \ldots, N,$$
$$\text{ER: } k = 1, 2, \ldots, N'.$$

The phase-space integrals will be given first in the c.m. system ($\boldsymbol{P} = 0$); later some generalization for finite $\boldsymbol{P}$ will be given.

In the NR limit, $E_j = \boldsymbol{p}_j^2/2m_j$, and the integral (1.121) for $\Psi^{(N)}$ can be transformed through the variable change

$$\boldsymbol{p}_j = (2Em_j)^{1/2}\boldsymbol{x}_j. \tag{1.140}$$

The integral is then given by

$$\Psi^{(N)} = 2^{3N/2}\Big(\prod_j m_j\Big)^{3/2} E^{3N/2-1} J_N, \tag{1.141}$$

where

$$\begin{aligned} J_N &= \int \cdots \int \prod_{j=1}^{N} d^3\boldsymbol{x}_j\, \delta\Big(\sum_{j=1}^{N} \boldsymbol{x}_j^2 - 1\Big) \\ &= \int \cdots \int \prod_{j=1}^{3N} dx_j\, \delta\Big(\sum_{j=1}^{3N} x_j^2 - 1\Big) \end{aligned} \tag{1.142}$$

is a function of $3N$. This integral occurs in classical statistical mechanics and there are various ways to evaluate it; these methods will be outlined later in this section. However, even without evaluating the integral, we see [Equation (1.141)] that the energy dependence of $\Psi^{(N)}$ has been obtained.

In the ER limit the particle energy is $E_k = p_k c$, and it is convenient to make the variable change

$$p_k c = E\boldsymbol{y}_k. \tag{1.143}$$

The Ψ-integral becomes

$$\Psi^{(N')} = c^{-3N'} E^{3N'-1} J_{N'}, \tag{1.144}$$

where

$$J_{N'} = \int \cdots \int \prod_{k=1}^{N'} d^3\mathbf{y}_k \, \delta\Big(\sum_{k=1}^{N'} y_k - 1\Big), \tag{1.145}$$

in which $y_k = |\mathbf{y}_k|$. This integral (1.145) will be evaluated shortly, but we again see that the energy dependence of $\Psi^{(N')}$ has already been obtained in Equation (1.144).

If we have N outgoing NR particles and N' ER particles, the above variable changes yield the general result

$$\Psi^{(NN')} = 2^{3N/2} c^{-3N'} \Big(\prod_j m_j\Big)^{3/2} E^{3N/2+3N'-1} J_{NN'}, \tag{1.146}$$

where

$$J_{NN'} = \int \cdots \int \prod_{j=1}^{N} d^3\mathbf{x}_j \prod_{k=1}^{N'} d^3\mathbf{y}_k \, \delta\Big(\sum_{j=1}^{N} \mathbf{x}_j^2 + \sum_{k=1}^{N'} y_k - 1\Big) \tag{1.147}$$

Evaluation of this general expression contains the results for J_N and $J_{N'}$ for, respectively, N' and N equal to 0.

Now we turn to the phase-space integrals Φ that involve both energy and momentum conservation. For NR particles it is again convenient to make the variable change (1.140) and, in addition, to express the individual masses m_j as fractions of the total mass M:

$$m_j = \nu_j^2 M, \tag{1.148}$$

$$\sum_{j=1}^{N} \nu_j^2 = 1. \tag{1.149}$$

We then have

$$\Phi^{(N)} = (2M)^{3(N-1)/2} \Big(\prod_{j=1}^{N} \nu_j\Big)^3 E^{3(N-1)/2-1} I_N, \tag{1.150}$$

where the dimensionless integral is now

$$I_N = \int \cdots \int \prod_{j=1}^{N} d^3\mathbf{x}_j \, \delta^{(3)}\Big(\sum_{j=1}^{N} \nu_j \mathbf{x}_j\Big) \, \delta\Big(\sum_{j=1}^{N} \mathbf{x}_j^2 - 1\Big). \tag{1.151}$$

Although the integral appears to be a function of the parameters ν_j, we shall show that it is not, and that, in fact, $I_N = J_{N-1}$. Inspection of the formulas (1.150) and (1.151) and comparison with the results (1.141) and (1.142) shows that the limit (1.120) is satisfied.

For ER particles the transformation (1.143) casts the corresponding $\Phi^{(N')}$ formula into the form

$$\Phi^{(N')} = c^{-(3N'-3)} E^{3N'-4} I_{N'}, \tag{1.152}$$

with

$$I_{N'} = \int \cdots \int \prod_{k=1}^{N'} d^3\mathbf{y}_k \, \delta^{(3)}\Big(\sum_{k=1}^{N'} \mathbf{y}_k\Big) \, \delta\Big(\sum_{k=1}^{N'} y_k - 1\Big). \tag{1.153}$$

The integral (1.153) is a little difficult to evaluate, but the technique is outlined in the following subsection.

In the mixed case the corresponding expression for $\Phi^{(NN')}$ can be simplified[18] because most of the momentum is carried by the NR particles. This is easily seen, since, if $E_j = p_j^2/2m_j \sim E_k = p_k c$, then $p_k/p_j \sim p_j/m_j c \ll 1$. Then in the momentum-conservation δ-function only the contribution from the NR particles must be included. The result is then

$$\Phi^{(NN')} = (2M)^{3(N-1)/2} c^{-3N'} \Big(\prod_{j=1}^{N} \nu_j\Big)^3 E^{3(N-1)/2+3N'-1} I_{NN'}, \tag{1.154}$$

with

$$\begin{aligned} I_{NN'} = \int \cdots \int \prod_{j=1}^{N} d^3\boldsymbol{x}_j \prod_{k=1}^{N} d^3\boldsymbol{y}_k \\ \times \delta^{(3)}\Big(\sum_{j=1}^{N} \nu_j \boldsymbol{x}_j\Big) \delta\Big(\sum_{j=1}^{N} \boldsymbol{x}_j^2 + \sum_{k=1}^{N'} y_k - 1\Big). \end{aligned} \tag{1.155}$$

We shall show that $I_{NN'} = J_{N-1,N'}$.

1.5.4 General Formulas—Evaluation of Multiple Integrals

Here the various J- and I-integrals will be evaluated. There are different ways of calculating these quantities and, to illustrate the methods, we employ the simplest ones for the individual cases. When possible, short cuts will be taken to arrive at the desired result, and the significance of these simplifications will be discussed. In some of the cases, the mathematical methods are very similar, so that most of the details of certain derivations need not be given.

The most elementary general method[19] of evaluating the integral (1.142) obtains a recursion relation. If we write $3N = \nu$, the integral to be evaluated is ($J_N = \mathcal{J}_\nu$)

$$\mathcal{J}_\nu = \int \cdots \int dx_1 \cdots dx_\nu \, \delta(x_1^2 + \cdots + x_\nu^2 - 1). \tag{1.156}$$

The recursion relation is obtained by expressing the integral in terms of the integral over the "last" variable x_1 from -1 to 1 and by making the variable changes

$$x_2 = (1 - x_1^2)^{1/2} \xi_2, \quad x_3 = (1 - x_1^2)^{1/2} \xi_3, \quad \text{etc.} \tag{1.157}$$

Further, if write $u = x_1^2$, the integral can be written

$$\begin{aligned} \mathcal{J}_\nu &= \int_0^1 u^{-1/2} (1-u)^{(\nu-3)/2} du \int \cdots \int d\xi_2 \cdots d\xi_\nu \\ &\quad \times \delta(\xi_2^2 + \cdots + \xi_\nu^2 - 1) \\ &= B\Big(\frac{1}{2}, \frac{\nu-1}{2}\Big) \mathcal{J}_{\nu-1}, \end{aligned} \tag{1.158}$$

[18] See E. Fermi, *Prog. Theor. Phys.* **5**, 570 (1950).

[19] Actually, there is a simpler method described in a number of books on statistical mechanics: cf. C. Kittel, *Elementary Statistical Physics*, p. 37, New York: John Wiley and Sons, Inc., 1958.

in terms of the beta function[20]

$$B(m,n) = \int_0^1 u^{m-1}(1-u)^{n-1}du = \frac{\Gamma(m)\Gamma(n)}{\Gamma(m+1)}. \tag{1.159}$$

Since $\Gamma(\frac{1}{2}) = \pi^{1/2}$, we have, by the results (1.158) and (1.159),

$$\Gamma\left(\frac{\nu}{2}\right)\mathcal{J}_\nu = \pi^{1/2}\Gamma\left(\frac{\nu-1}{2}\right)\mathcal{J}_{\nu-1}, \tag{1.160}$$

or

$$\mathcal{J}_\nu \propto \pi^{\nu/2}/\Gamma(\nu/2) \tag{1.161}$$

Direct calculation of the case $\nu = 1$ yields unity for the proportionality constant, so that

$$\mathcal{J}_\nu = \pi^{\nu/2}/\Gamma(\nu/2) = J_N = \pi^{3N/2}/\Gamma(3N/2). \tag{1.162}$$

The calculation of the ER integral (1.145) for $J_{N'}$ can be carried out in a very similar way. Writing $d^3\boldsymbol{y}_k = 4\pi y_k^2\, dy_k$, with the last integral over dy_1 from 0 to 1, we are then led to make the transformation

$$y_2 = (1-y_1)\xi_2, \quad y_3 = (1-y_1)\xi_3, \quad \text{etc.} \tag{1.163}$$

A recursion relation is again obtained, involving another B-function integral of the type (1.159), and we find

$$J_{N'} = 2^{3N'}\pi^{N'}/\Gamma(3N'). \tag{1.164}$$

The evaluation of the more general integral (1.147) for $J_{NN'}$ can be accomplished in a similar manner, by making the transformations (1.157) and (1.163), yielding a two-index recursion relation and then the result

$$J_{NN'} = \frac{2^{3N'}\pi^{3N/2+N'}}{\Gamma(3N/2+3N')}. \tag{1.165}$$

This expression obviously contains the special cases (1.162) and (1.164). We can, moreover, easily check that formulas (1.125), (1.126), (1.129), and (1.130) are recovered in this general result. However, let us now introduce another method for evaluating the phase-space integrals; this procedure is more powerful and direct, and makes use of the integral representation of the δ-function:

$$\delta(\omega) = (2\pi)^{-1}\int d\alpha\, e^{i\alpha\omega}. \tag{1.166}$$

Here the integration is over all values of α ($-\infty$ to ∞) and, for the purposes of evaluating subsequent integrals, it is convenient to give α a small imaginary part $i\epsilon$. This makes the integral well defined and dictates the way of evaluation of the subsequent integrals in terms of contour integration. With the form (1.166) for the δ-function, the energy-shell restriction is imposed in a convenient way; the integrations over the $d^3\boldsymbol{x}_j$ and $d^3\boldsymbol{y}_k$ are performed first, followed by the final integration

---

[20]For an extensive discussion of the beta and gamma functions, including the identity (1.159), see E. T. Whittaker and G. N. Watson, *Modern Analysis*, 4th ed., Cambridge, UK: Cambridge University Press, 1927.

over α. The method works well for the former integrations, essentially because in the exponential term in the integrand (1.166) we can write

$$\exp\Big[i\alpha\Big(\sum_j \boldsymbol{x}_j^2 + \sum_k y_k - 1\Big)\Big] = e^{-i\alpha} \prod_j \exp(i\alpha \boldsymbol{x}_j^2) \prod_k \exp(i\alpha y_k). \qquad (1.167)$$

There results a product of individual $d^3\boldsymbol{x}_j$ and $d^3\boldsymbol{y}_k$ integrations, each equal to, respectively,

$$I_j = 4\pi \int_0^\infty x^2 e^{i\alpha x^2} dx = e^{-i\pi/4}(\pi/\alpha)^{3/2}, \qquad (1.168)$$

$$I_k = 4\pi \int_0^\infty y^2 e^{i\alpha y} dy = 8\pi e^{-i\pi/2}\alpha^{-3}. \qquad (1.169)$$

These are then raised to the powers N and N', respectively, because of the product, and there remains a final integration over α of the form

$$I_\alpha = \int d\alpha \, \frac{e^{-i\alpha}}{\alpha^{P+1}}. \qquad (1.170)$$

This can be evaluated by contour integration, making use of the integral representation of the Γ-function,[21] giving

$$I_\alpha = 2\pi i^{3P-1}/\Gamma(P+1). \qquad (1.171)$$

The general result (1.165) is then obtained for $J_{NN'}$.

Now let us turn to the problem of evaluating the general phase-space integrals (1.151), (1.153), and (1.155) containing both the energy- and momentum-conservation δ-functions. First we consider the NR integral (1.151) and the relationship (identity) to J_{N-1}, alluded to earlier. The identity can be established easily and directly for the case $N = 2$ and 3, but it is of interest to prove it for general N. Basically, the identity follows as a result of the relation (1.149) involving the dimensionless mass parameters. We consider the indices $j = 1$ to N, labeling the particles, as designating an N-dimensional space. If $\hat{e}_j$ is a unit vector along the jth axis in this space, the axes are taken as orthogonal such that $\hat{e}_j \cdot \hat{e}_k = \delta_{jk}$. Because of the relation (1.149), the collection of parameters

$$\hat{\nu} = (\nu_1, \nu_2, \ldots, \nu_N) \qquad (1.172)$$

can be described as a vector of unit length in the space:

$$\hat{\nu} \cdot \hat{\nu} = \sum_{j=1}^{N} \nu_j^2 = 1. \qquad (1.173)$$

The integral (1.151) can then be written

$$I_N = \int d^3\hat{\boldsymbol{x}} \, \delta(\hat{\boldsymbol{x}}^2 - 1) \, \delta^{(3)}(\hat{\nu} \cdot \hat{\boldsymbol{x}}), \qquad (1.174)$$

where

$$\hat{\boldsymbol{x}} = (\boldsymbol{x}_1, \boldsymbol{x}_2, \ldots, \boldsymbol{x}_N), \qquad (1.175)$$

[21] See Whittaker and Watson (loc. cit.), p. 245.

and

$$d^3\hat{x} = dx_1 \cdots dx_N dy_1 \cdots dy_N dz_1 \cdots dz_N. \tag{1.176}$$

The terms

$$\hat{x}^2 = x_1^2 + \cdots + x_N^2 + y_1^2 + \cdots + y_N^2 + z_1^2 + \cdots + z_N^2 \tag{1.177}$$

and

$$\hat{\nu} \cdot \hat{x} = \sum_{j=1}^{N} \nu_j \boldsymbol{x}_j \tag{1.178}$$

are invariant under spatial rotation[22] of the axes of the space, as would be the volume (1.176). If the result is invariant to an orientation of these axes, it is convenient to take an orientation such that, say, $\hat{\nu} = (0, 0, \ldots, 1)$, corresponding, mathematically, to the case where all masses except m_N are zero. The δ-function $\delta^{(3)}$ is then simply $\delta^{(3)}(\boldsymbol{x}_N)$, which can be eliminated by integration over $d^3\boldsymbol{x}_N$. The resulting integral is then just J_{N-1}, and we have

$$I_N = J_{N-1}. \tag{1.179}$$

A more conventional method of proving the identity (1.179) and of evaluating other general phase-space integrals employs the integral form (1.166) for both the energy- and momentum conservation δ-functions. That is, for the latter in Equation (1.151),

$$\delta^{(3)}\Big(\sum_{j=1}^{N} \nu_j \boldsymbol{x}_j\Big) = (2\pi)^{-3} \int d^3\boldsymbol{\xi}\ \exp\Big(i \sum_{j=1}^{N} \nu_j \boldsymbol{x}_j \cdot \boldsymbol{\xi}\Big). \tag{1.180}$$

The individual $d^3\boldsymbol{x}_j$ integrations of the exponentials from the δ and $\delta^{(3)}$ functions are now [see Equation (1.168)]

$$I_j = 2\pi \int_0^\infty x^2 dx \int_0^\pi \sin\theta\, d\theta\ \exp[i(\alpha x^2 + \xi \nu_j x \cos\theta)]. \tag{1.181}$$

After convenient changes of variables, the integral (1.181) is evaluated; it is found, for example, that the ν_j dependence is

$$I_j \propto \exp[-i(\xi^2/4\alpha)\nu_j^2]. \tag{1.182}$$

The independence of the integral I_N on the distribution of ν_j is then seen when the product of the I_j is taken:

$$\prod_j I_j \propto \exp\Big[-i(\xi^2/4\alpha) \sum_{j=1}^{N} \nu_j^2\Big] = \exp[-i(\xi^2/4\alpha)]. \tag{1.183}$$

This proves directly the independence of the orientation of the vector (1.172), allowing the simplified derivation of the theorem (1.179).

The evaluation of the integral (1.153) for massless particles proceeds in the above manner by taking integral representations for the two δ-functions. Integration over $d^3\boldsymbol{\xi}$ and $d\alpha$ is carried out by contour integration. The result is[23]

$$I_{N'} = \left(\frac{\pi}{2}\right)^{N'-1} \frac{(4N'-4)!}{(2N'-1)!(2N'-2)!(3N'-4)!}. \tag{1.184}$$

---

[22]The $\hat{e}_j$ would transform as $\hat{e}'_k = \gamma_{kj}\hat{e}_j$, a unitary transformation that can be considered as a way of relabeling the particles.

[23]Note the correction to the formula given by J. V. Lepore and R. N. Stuart, *Phys. Rev.* **94**, 1724 (1954), also noted by R. H. Milburn, *Rev. Mod. Phys.* **27**, 1 (1955).

For the derivation of the result for the integral (1.155), we can employ the tedius method involving integral representations of the δ-functions, or we can employ the same trick that led to the establishment of the theorem (1.179). The latter method is much simpler and leads directly to the result

$$I_{NN'} = J_{N-1,N'}. \tag{1.185}$$

That is, the $I_{NN'}$ formula can be evaluated directly from the expression (1.165) with N replaced by $N-1$.

Finally, concerning general formulas, we might ask about the case where the motion is in a space of arbitrary dimensions $n \neq 3$. For example, reactions and phase space-effects on a solid surface ($n = 2$) could be of interest. If, for example,

$$\Psi_n^{(N)} = \int \cdots \int \prod_j (dp_1 \cdots dp_n)_j \, \delta\Big(\sum_{j=1}^{N} E_j - E\Big), \tag{1.186}$$

with

$$E_j = (p_1^2 + \cdots + p_n^2)_j / 2m_j, \tag{1.187}$$

we have

$$\Psi_n^{(N)} = 2^{nN/2}\Big(\prod_{j=1}^{N} m_j\Big)^{n/2} E^{nN/2-1} J_{N(n)}, \tag{1.188}$$

where

$$J_{N(n)} = \int \cdots \int \prod_{j=1}^{N} (dx_1 \cdots dx_n)_j \, \delta\Big(\sum_{j=1}^{N} (x_1^2 + \cdots + x_n^2)_j - 1\Big). \tag{1.189}$$

We obtain

$$J_{N(n)} = \frac{\pi^{nN/2}}{\Gamma(nN/2)}. \tag{1.190}$$

1.5.5 One-Particle Distributions

Sometimes the angular or energy distribution of one particle (or one type of particle) is of interest for a process involving a number of outgoing particles. Although the interaction H'_{f0} can play some role in this, it is principally the phase-space factor that determines the distribution. For some simple cases the distributions can be obtained from the corresponding explicit formulas already introduced. For example, if we have two outgoing particles with no momentum-conservation restriction (third heavy particle involved), the formula (1.128) can be employed, integrating over $d^3\boldsymbol{p}_2 = (4\pi/c^2) p_2 E_2 dE_2$:

$$d\Psi^{(2)}/d^3\boldsymbol{p}_1 = (4\pi/c^2) p_2 E_2 \tag{1.191}$$

or

$$d\Psi^{(2)}/dp_1 = (4\pi/c)^2 p_1^2 p_2 E_2, \tag{1.192}$$

with

$$E_2 = E - E_1, \; p_2 c = (E_2^2 - m_2^2 c^4)^{1/2}. \tag{1.193}$$

Depending on the nature of the particles involved, the distribution has different forms:

$$d\Psi^{(2)}/dp_1 = 16\pi^2(m_2^{3/2}/m_1^{1/2})p_1^2(2m_1E - p_1^2)^{1/2} \quad \text{(NR)}, \tag{1.194}$$

$$d\Psi^{(2)}/dE_1 = (4\pi/c^3)^2E_1^2(E - E_1)^2 \quad \text{(ER)}. \tag{1.195}$$

Some general formulas can be obtained easily. Without momentum conservation, from

$$\Psi^{(N+1)} = \int d^3\boldsymbol{p}_1 \int \cdots \int \prod_{j=2}^{N+1} d^3\boldsymbol{p}_j\, \delta\Big(\sum_{j=2}^{N+1} E_j + E_1 - E\Big), \tag{1.196}$$

we have

$$d\Psi^{(N+1)}/d^3\boldsymbol{p}_1 = \Psi^{(N)}(E \to E - E_1). \tag{1.197}$$

In other words, the one-particle distribution is determined by the phase-space volume available to the *other* particles but with the total energy available being reduced to $E - E_1$. The formula (1.197) is very general[24] and not restricted to NR or ER particles. Another general result is

$$d\Phi^{(N,N'+1)}/d^3\boldsymbol{p}_1 = \Phi^{(NN')}(E \to E - E_1), \tag{1.198}$$

when $\boldsymbol{p}_1$ refers to a massless particle; as we have seen, momentum is carried by the NR particles in this case.

When $\boldsymbol{p}_1$ is an NR particle and momentum conservation is applied, we must know the expression for Φ for arbitrary total momentum $\boldsymbol{P}$, that is, not just for the c.m. frame. The general formula for $\Phi^{(N)}$ for all NR particles is

$$\Phi^{(N)}(E, \boldsymbol{P}) = \int \cdots \int \prod_{j=1}^{N} d^3\boldsymbol{p}_j\, \delta^{(3)}\Big(\sum_{j=1}^{N} \boldsymbol{p}_j - \boldsymbol{P}\Big)\, \delta\Big(\sum_{j=1}^{N} E_j - E\Big). \tag{1.199}$$

Now we can transform the integration to c.m. frame variables ($\boldsymbol{p}'_j$):

$$\boldsymbol{p}_j = \boldsymbol{p}'_j + (m_j/M)\boldsymbol{P}, \tag{1.200}$$

$$\sum_j E_j = \sum_j \boldsymbol{p}_j^2/2m_j = \sum_j E'_j + \boldsymbol{P}^2/2M. \tag{1.201}$$

Thus, we see that

$$\Phi^{(N)}(E, \boldsymbol{P}) = \Phi^{(N)}(E - \boldsymbol{P}^2/2M, 0). \tag{1.202}$$

Similarly,

$$\Phi^{(NN')}(E, \boldsymbol{P}) = \Phi^{(NN')}(E - \boldsymbol{P}^2/2M, 0). \tag{1.203}$$

---

[24]The formula yields, in statistical mechanics, for example, the Maxwellian distribution for $N \gg 1$ and more general such distributions for arbitrary N. The Maxwellian result follows immediately from Equations (1.141) and (1.197) if we set $E = (3N/2)kT$ and make use of the limit $(1 - x)^{1/x} \to 1/e$ for $x \to 0$.

We then find, in the c.m. frame,

$$d\Phi^{(N+1)}/d^3\boldsymbol{p}_1 = \int \cdots \int \prod_{j=2}^{N+1} d^3\boldsymbol{p}_j \times \delta^{(3)}\Big(\sum_{j=2}^{N+1} \boldsymbol{p}_j + \boldsymbol{p}_1\Big)\,\delta\Big(\sum_{j=2}^{N+1} E_j + E_1 - E\Big) = \Phi^{(N)}(E - E_1, -\boldsymbol{p}) = \Phi^{(N)}(E - E_r, 0), \tag{1.204}$$

where the reduced energy is

$$E_r = E_1 - \boldsymbol{p}_1^2/2M = \boldsymbol{p}_1^2/2\mu, \tag{1.205}$$

with

$$\mu = m_1 M/(m_1 + M), \; M = \sum_{j=2}^{N+1} m_j. \tag{1.206}$$

1.5.6 Invariant Phase Space

We have introduced the invariant phase-space factor $\phi^{(N)}$ in Equation (1.116). In modern high-energy physics, it is this quantity that is employed much more extensively than the non-covariant expression $\Phi^{(N)}$; we have examined the latter in more detail in this section because of the relative lack of emphasis it has received in the past few decades. There is no doubt that, in relativistic formulations valid for arbitrary energy, the invariant $\phi^{(N)}$ is more appropriate and convenient. However, it is not possible to derive explicit expressions for arbitrary N as we have done in the NR and ER limits for $\Phi^{(N)}$ and $\Psi^{(N)}$. Whether $\phi^{(N)}$ or $\Phi^{(N)}$ is inherently more fundamental and useful would depend on, for example, the formulation and results for the matrix element H'_{f0}. Sometimes there are reasons for expressing this quantity in terms of certain kinematic invariants,[25] since we expect that these are "natural" quantities classifying the kinematics of a process; the invariant $\phi^{(N)}$ is then clearly more appropriate in measuring the phase space for outgoing particles. On the other hand, the non-covariant $\Phi^{(N)}$ is directly proportional to the number of final states (per unit energy) and, in this sense, it is more fundamental than $\phi^{(N)}$. We shall be very brief here,[26] working out the simplest case ($N = 2$); we now also set $c = 1$, although in final formulas it is clear where factors of c must be inserted.

The two-particle invariant phase-space factor is, when computed in the c.m. frame,

$$\phi^{(2)} = \iint (d^3\boldsymbol{p}_1/2E_1)\, d^3\boldsymbol{p}_2/2E_2)\, \delta^{(3)}(\boldsymbol{p}_1 + \boldsymbol{p}_2)\, \delta(E_1 + E_2 - E) = \int d^3\boldsymbol{p}_1\, [4E_1E_2(p_1)]^{-1}\, \delta[E_1 + E_2(p_1) - E], \tag{1.207}$$

where

$$E_1 = (p_1^2 + m_1^2)^{1/2}, \; E_2(p_1) = (p_1^2 + m_2^2)^{1/2}. \tag{1.208}$$

---

[25] See Section 3.6.3.

[26] For a more complete discussion of invariant phase space, including applications and use of "Dalitz plots," see the article by P. Nyborg and O. Skjeggestad in *Kinematics and Multiparticle Systems*, M. Nikolic, Ed., New York: Gordon and Breach, 1968.

The evaluation of the integral (1.207) yields the extremely simple result

$$\phi^{(2)} = \pi p_1(E)/E \tag{1.209}$$

where $p_1(E)$ is the solution of the equation $E_1 + E_2(p_1) = E$, or

$$p_1(E) = \left[\left(E^2 - m_1^2 - m_2^2\right)^2 - 4m_1^2 m_2^2\right]^{1/2}/2E. \tag{1.210}$$

The phase-space factor is then determined by the total c.m. energy E and the masses:

$$\phi^{(2)} = \phi^{(2)}(E; m_1, m_2). \tag{1.211}$$

Since $\phi^{(2)}$ is an invariant, it is then also equal to the same value in another frame where the total momentum is $\boldsymbol{P}' \neq 0$ and the energy is E', with these quantities related to E by

$$E^2 = E'^2 - \boldsymbol{P}'^2. \tag{1.212}$$

The result can also be expressed explicitly in terms of Lorentz-invariant quantities.[27] The only kinematic invariant involved is $s = -E^2$ so that $\phi^{(2)} = \phi^{(2)}(s)$.

Finally, let us compare the invariant phase-space factor with the corresponding non-invariant expression (1.138); the two are related by

$$\Phi^{(2)} = 4E_1 E_2(p_1)\phi^{(2)}, \tag{1.213}$$

where

$$4E_1 E_2(p_1) = \left[E^4 - \left(m_2^2 - m_1^2\right)^2\right]/E^2. \tag{1.214}$$

The non-invariant expression is more complicated, precisely by the factor (1.214). Further, it should be noted that the formula (1.213) relates only the $\phi^{(2)}$ and $\Phi^{(2)}$ in the c.m. frame. On the other hand, the invariant $\phi^{(2)}$ is easily evaluated for any frame with arbitrary E' and $\boldsymbol{P}'$ because of the simple relation (1.212).

[27]See Section 3.6.3.

Bibliographical Notes

A discussion of fundamental lengths, energies, etc. is given, at least for electromagnetic phenomena, in the first chapter of

1. Thirring, W. E., *Principles of Quantum Electrodynamics*, New York: Academic Press Inc., 1958.

Special relativity is often introduced in books on classical mechanics and electrodynamics:

2. Corben, H. C. and Stehle, P., *Classical Mechanics*, 2nd ed., New York: John Wiley and Sons, Inc., 1960.

3. Goldstein, H., *Classical Mechanics*, 2nd ed., Reading, MA: Addison-Wesley Publ. Co., 1980.

4. Jackson, J. D., *Classical Electrodynamics*, 2nd ed., New York: John Wiley and Sons, Inc., 1975.

5. Panofsky, W. K. H. and Phillips, M., *Classical Electricity and Magnetism*, 2nd ed., Reading, MA: Addison-Wesley Publ. Co., 1962.

6. Landau, L. D. and Lifshitz, E. M., *The Classical Theory of Fields*, 3rd Ed., Reading, MA: Addison-Wesley Publ. Co., 1971.

A more complete treatment of special relativity and of the characteristics of Lorentz transformations and the four-dimensional formulation may be found in

7. Møller, C., *The Theory of Relativity*, Oxford, UK: Oxford University Press, 1957.

Kinematics of relativistic particle collisions are also discussed in References 2–6. Phase space effects in particle collisions are given an elementary discussion in the beautiful little book by Fermi:

8. Fermi, E., *Elementary Particles*, New Haven, CT: Yale University Press, 1951.

See also:

9. Marshak, R. E., *Meson Physics*, New York: McGraw-Hill Book Co., 1952.

A more modern discussion of phase space effects in particle physics, with further references, may be found in

10. Perl, M. L., *High Energy Hadron Physics*, New York: John Wiley and Sons, Inc., 1974.

Chapter Two

Classical Electrodynamics

Classical electrodynamics is contained within quantum electrodynamics as a limiting case. It has a domain of validity and applicability to certain problems for which the classical treatment is clearly preferable to the more general quantum-mechanical approach. There is also a close relationship between a number of results in classical radiation theory and corresponding expressions derived in quantum electrodynamics. In fact, sometimes classical formulas have a range of validity greater than that expected on the basis of elementary considerations. The relationship between classical and quantum electrodynamics will be discussed briefly in Chapter 3. The quantum-mechanical formulation relies heavily on the classical theory as a guide, starting with the classical field Hamiltonian.

This chapter will give a purely classical treatment of radiation. It will not attempt to be a complete description of classical electrodynamics, since that general subject is treated well in several textbooks. However, starting from basic principles, a number of useful and general results will be derived. Applications to specific radiative processes will be given in later chapters.

2.1 RETARDED POTENTIALS

2.1.1 Fields, Potentials, and Gauges

In non-covariant form, the four Maxwell equations are

$$\begin{aligned}
\nabla \times \boldsymbol{B} &= \frac{4\pi}{c}\boldsymbol{j} + \frac{1}{c}\frac{\partial \boldsymbol{E}}{\partial t}, \\
\nabla \cdot \boldsymbol{E} &= 4\pi\rho, \\
\nabla \cdot \boldsymbol{B} &= 0, \\
\nabla \times \boldsymbol{E} + \frac{1}{c}\frac{\partial \boldsymbol{B}}{\partial t} &= 0.
\end{aligned} \tag{2.1}$$

These are the "microscopic" Maxwell equations in terms of the electric and magnetic fields $\boldsymbol{E}$ and $\boldsymbol{B}$. The sources of the fields are the charge and current densities and would include contributions from the polarization and magnetization of the medium. In the "macroscopic" form of Maxwell's equations, only the conduction charge densities and currents appear in the equations, which are now equations for $\boldsymbol{D}$ and $\boldsymbol{H}$ and also involve the dialectric constant and magnetic permeability. The forms (2.1) are more convenient for our purposes.

The fields $\boldsymbol{E}$ and $\boldsymbol{B}$ are physical variables in the sense that they are observable, being directly connected to physical quantities like forces. For convenience in the mathematical description of electromagnetic processes, it is useful to introduce *potentials* from which the fields are derived. Since div curl $\boldsymbol{A} = \nabla\cdot(\nabla\times\boldsymbol{A}) = 0$, the third of the equations (2.1) is satisfied automatically if we write

$$\boldsymbol{B} = \nabla\times\boldsymbol{A}, \tag{2.2}$$

where $\boldsymbol{A}$ is some vector function of $\boldsymbol{r}$ and t. The fourth equation (2.1) is satisfied if we introduce a scalar function $\Phi(\boldsymbol{r}, t)$, such that

$$\boldsymbol{E} = -\nabla\Phi - \frac{1}{c}\frac{\partial\boldsymbol{A}}{\partial t}, \tag{2.3}$$

since curl grad $\Phi = \nabla\times(\nabla\Phi) = 0$. It is not *necessary* to introduce these functions $\boldsymbol{A}(\boldsymbol{r}, t)$ and $\Phi(\boldsymbol{r}, t)$; rather, it is simply convenient to do so. The vector and scalar potentials are not physical quantities in the sense that they can be measured. In fact, they are not unique, since the same $\boldsymbol{E}$ and $\boldsymbol{B}$ fields are obtained from the potentials if they are replaced by

$$\begin{aligned} \boldsymbol{A} \to \hat{\boldsymbol{A}} &= \boldsymbol{A} + \nabla\Lambda, \\ \Phi \to \hat{\Phi} &= \Phi - \frac{1}{c}\frac{\partial\Lambda}{\partial t}, \end{aligned} \tag{2.4}$$

where Λ is any arbitrary function of $\boldsymbol{r}$ and t.

The equation (2.4) is called a gauge transformation, and the invariance of the fields $\boldsymbol{E}$ and $\boldsymbol{B}$ under this transformation is called gauge invariance. Although it introduces subtleties and complications in the general formulation of both classical and quantum electrodynamics, at the same time the gauge invariance can be used to facilitate calculations of electromagnetic phenomena. Because of the freedom of choice in the potentials allowed by the invariance under the transformation (2.4), a subsidiary condition can be imposed on $\boldsymbol{A}$ and Φ. The form of the subsidiary condition establishes the "choice of gauge." For example, if the *Lorentz condition*

$$\nabla\cdot\boldsymbol{A} + \frac{1}{c}\frac{\partial\Phi}{\partial t} = 0 \tag{2.5}$$

is imposed, the inhomogeneous Maxwell equations [first two equations (2.1)] reduce to two equations that are separable in $\boldsymbol{A}$ and Φ:

$$\begin{aligned} \nabla^2\boldsymbol{A} - \frac{1}{c^2}\frac{\partial^2\boldsymbol{A}}{\partial t^2} &= -\frac{4\pi}{c}\boldsymbol{j}, \\ \nabla^2\Phi - \frac{1}{c^2}\frac{\partial^2\Phi}{\partial t^2} &= -4\pi\rho. \end{aligned} \tag{2.6}$$

In terms of $A_\mu = (i\Phi, \boldsymbol{A})$ and $j_\mu = (ic\rho, \boldsymbol{j})$, these equations are manifestly covariant:

$$\Box^2 A_\mu = -(4\pi/c)j_\mu. \tag{2.7}$$

The Lorentz gauge condition is also covariant; $\partial_\mu A_\mu = 0$.

The class of gauges satisfying the condition (2.5) is called the Lorentz gauge; it is also called the covariant gauge. The Lorentz gauge is convenient, because of the covariance property, especially in the general formulation of classical and quantum electrodynamics. However, even within the Lorentz gauge, there is a certain degree of arbitrariness. For example, if the function $\Lambda(\boldsymbol{r}, t)$ satisfies

$$\nabla^2 \Lambda - \frac{1}{c^2}\frac{\partial^2 \Lambda}{\partial t^2} = 0. \tag{2.8}$$

then $\hat{\boldsymbol{A}}$ and $\hat{\Phi}$ satisfy the Lorentz condition if $\boldsymbol{A}$ and Φ do.

It should be noted that the Lorentz condition (2.5) does not have any physical interpretation, although it looks like some kind of conservation equation. Rather, it is to be regarded only as a mathematical subsidiary condition imposed for convenience in the computations. In fact, other types of mathematical subsidiary relations, corresponding to other "gauges," are also convenient. For example, the condition $\nabla \cdot \boldsymbol{A} = 0$ can be imposed; this condition designates the *Coulomb gauge*.[1] It is particularly convenient in certain radiation problems and is sometimes referred to as the radiation gauge (also as the transverse gauge). Its inconvenience lies in the non-covariant nature of the subsidiary relation.

When describing radiation fields, that is, fields corresponding to propagating electromagnetic waves at large distances from their source in empty space where $\rho = 0$ and $\boldsymbol{j} = 0$, the condition

$$\nabla \cdot \boldsymbol{A} = 0 \tag{2.9}$$

can be imposed together with the condition

$$\Phi = 0. \tag{2.10}$$

For general electromagnetic fields it is not possible to impose both conditions (2.9) and (2.10), but the condition (2.10) follows—again, only for radiation fields—if the condition (2.9) is imposed.[2] Actually, the gauge corresponding to the conditions (2.9) and (2.10) can be considered to be within the Lorentz class, since the Lorentz condition (2.5) is satisfied identically. The gauge (2.9, 2.10) is very convenient in its restricted application to radiation fields, since both $\boldsymbol{E}$ and $\boldsymbol{B}$ are derived from the vector potential alone through the simple relations $\boldsymbol{B} = \nabla \times \boldsymbol{A}$ and $\boldsymbol{E} = -(1/c)\partial \boldsymbol{A}/\partial t$. The subsidiary relation (2.9) also simplifies calculations by requiring the fields (and $\boldsymbol{A}$) to be transverse.

2.1.2 Retarded Potentials in the Lorentz Gauge

The field-source (or potential-source) equations (2.6), for Φ and for each component of $\boldsymbol{A}$ are of the form

$$\nabla^2 \Psi - \frac{1}{c^2}\frac{\partial^2 \Psi}{\partial t^2} = -4\pi s(\boldsymbol{r}, t). \tag{2.11}$$

The equation is linear in its relationship between the field (potential) $\Psi(\boldsymbol{r}, t)$ and the source $s(\boldsymbol{r}, t)$; therefore, a superposition principle will apply, such that the total field

[1] In this gauge the potential Φ satisfies $\nabla^2 \Phi = -4\pi\rho$, which has a solution corresponding to an (instantaneous) Coulomb field. Hence the name *Coulomb gauge*.

[2] This can be seen readily through a substitution of Equation (2.3) into the second of equations (2.1) with $\rho = 0$. The result (2.10) can be considered as a special radiation-field solution of $\nabla^2 \Phi = 0$.

will be a result of a summation over contributions from sources at various spatial points. The solution[3] of Equation (2.11) has a very well-known and particular form in terms of a volume integral over the source:

$$\Psi(\boldsymbol{r}, t) = \int \frac{s(\boldsymbol{r}', t - |\boldsymbol{r} - \boldsymbol{r}'|/c)}{|\boldsymbol{r} - \boldsymbol{r}'|} d^3\boldsymbol{r}'. \tag{2.12}$$

This solution (2.12) exhibits the retardation effect in which the contribution to the field at time t is due to the source characteristics at the time

$$t' = t - |\boldsymbol{r} - \boldsymbol{r}'|/c, \tag{2.13}$$

earlier by an interval $\Delta t = |\boldsymbol{r} - \boldsymbol{r}'|/c$ equal to the time required for propagation at the velocity of light between $\boldsymbol{r}'$ and $\boldsymbol{r}$.

The solution (2.12), in the form of a retarded "Green function," can be derived by various means. The most direct and systematic approach evaluates the Green function in terms of its Fourier transform, with the retardation requirement a consequence of the mathematical analysis. However, a simpler derivation is possible which makes use of a convenient artificial device. We imagine a point charge and current of *variable magnitude*[4] at the origin. The potential $\Phi(\boldsymbol{r}, t)$ due to this charge $q(t)$ will then be a solution of Equation (2.6), which is of the form

$$\nabla^2\Phi - \frac{1}{c^2}\frac{\partial^2\Phi}{\partial t^2} = -4\pi q(t)\delta(\boldsymbol{r}), \tag{2.14}$$

where $\delta(\boldsymbol{r}) = \delta(x)\delta(y)\delta(z)$ is the three-dimensional delta function. From the inherent spherical symmetry in the problem, Φ must be a function of, in addition to the time t, only the magnitude of the radial distance r. Then $\Phi = \Phi(r, t)$ and $\nabla^2\Phi = r^{-2}\partial(r^2\partial\Phi/\partial r)/\partial r$, and if we substitute $\Phi(r, t) = \chi(r, t)/r$, away from the origin χ satisfies

$$\partial^2\chi/\partial r^2 - c^{-2}\partial^2\chi/\partial t^2 = 0. \tag{2.15}$$

This equation has the well-known solution

$$\chi(r, t) = \chi(t \pm r/c), \tag{2.16}$$

and for physical reasons (causality) we choose the minus sign in the argument of χ.

Now we consider the solution to Equation (2.14) in the neighborhood of the origin. As $r \to 0$, the spatial derivatives on the left will be much larger than the time derviative, and Φ will be a solution of the equation

$$\nabla^2\Phi = -4\pi q(t)\delta(\boldsymbol{r}) \qquad (r \to 0). \tag{2.17}$$

The solution to this equation is well known:

$$\Phi(\boldsymbol{r}, t) = q(t)/r \qquad (r \to 0). \tag{2.18}$$

---

[3]Here we are referring to the "special" or "particular" solution of the inhomogenous equation (2.11). This represents the field due directly to the source $s(\boldsymbol{r}, t)$. To this solution must be added the so-called "general" solution of the corresponding homogeneous equation. The latter could represent fields not associated with the specific source $s(\boldsymbol{r}, t)$ such as that connected with some external field.

[4]It should be emphasized that this is essentially a mathematical assumption, made for convenience. This is permissable, even though we know that, because of other independent considerations, an isolated charge cannot change its magnitude.

For this solution to match the result (2.16) above for arbitrary t, the function χ must be identified with q itself, so that the solution of equation (2.14) for general r and t is

$$\Phi(\boldsymbol{r}, t) = q(t - r/c)/r. \tag{2.19}$$

For a general charge distribution we can simply add contributions, according to the superposition principle:

$$\Phi(\boldsymbol{r}, t) = \int \frac{\rho(\boldsymbol{r}', t - |\boldsymbol{r} - \boldsymbol{r}'|/c)}{|\boldsymbol{r} - \boldsymbol{r}'|} d^3\boldsymbol{r}'. \tag{2.20}$$

By similar arguments we obtain the solution to the Equation (2.6) for the vector potential:

$$\boldsymbol{A}(\boldsymbol{r}, t) = \frac{1}{c} \int \frac{\boldsymbol{j}(\boldsymbol{r}', t - |\boldsymbol{r} - \boldsymbol{r}'|/c)}{|\boldsymbol{r} - \boldsymbol{r}'|} d^3\boldsymbol{r}'. \tag{2.21}$$

The two solutions (2.20) and (2.21) are represented by Equation (2.12), and we see how the retardation effect comes in. The contribution of the source to the potentials is that swept up by a spherical wave[5] with amplitude proportional to $|\boldsymbol{r} - \boldsymbol{r}'|^{-1}$ converging on the field point $\boldsymbol{r}$ at the radial velocity c. This concept is useful in obtaining (see Section 4) the expressions for the potentials associated with a single moving point charge, i.e., the so-called Liénard-Wiechert potentials. In this case, the potential Φ, for example, is *not* given by q divided by the retarded distance.

2.2 MULTIPOLE EXPANSION OF THE RADIATION FIELD

2.2.1 Vector Potential and Retardation Expansion

The electric and magnetic fields associated with radiation can, as we have seen in the previous section, within the Lorentz gauge, be evaluated in terms of only the vector potential, Φ being set identically to zero. The vector potential at the field point $\boldsymbol{r}$ at time t is determined by the characteristics of the source current density $\boldsymbol{j}$ at the source point $\boldsymbol{r}'$ and the retarded time $t' = t - R/c$, where

$$R = |\boldsymbol{R}| = |\boldsymbol{r} - \boldsymbol{r}'|. \tag{2.22}$$

We can write the result (2.21) in the form

$$\begin{aligned} \boldsymbol{A}(\boldsymbol{r}, t) &= \frac{1}{c} \int d^3\boldsymbol{r}' \frac{\boldsymbol{j}(\boldsymbol{r}', t - R/c)}{R} \\ &= \frac{1}{c} \iint d^3\boldsymbol{r}' dt' \frac{\boldsymbol{j}(\boldsymbol{r}', t')}{R} \delta(t' - t + R/c). \end{aligned} \tag{2.23}$$

The field point is at a large distance from the source (see Figure 2.1) and if the source has some localization, $r' \ll r$, in

$$R^2 = r^2 - 2\boldsymbol{r} \cdot \boldsymbol{r}' + r'^2, \tag{2.24}$$

[5]This is to be regarded as a "mathematical" rather than a physical wave.

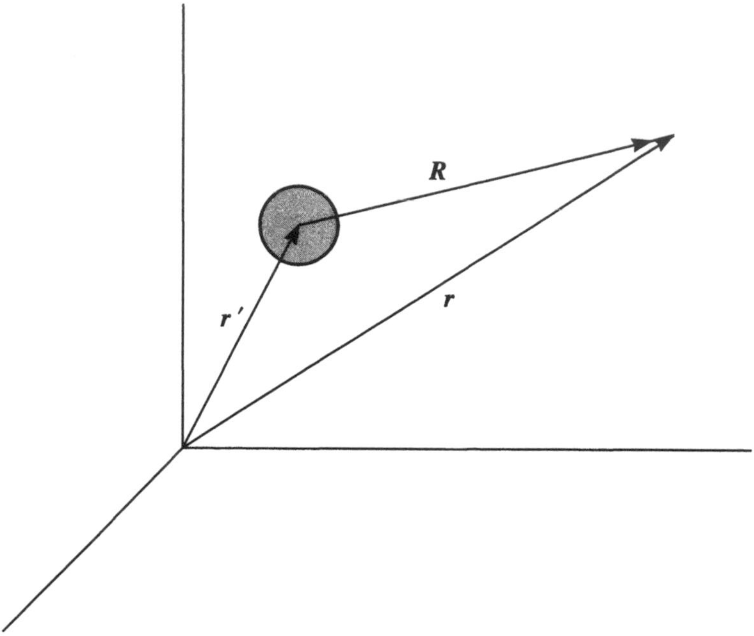

Figure 2.1 Field and source points.

the last term can be neglected. Then

$$R \approx r - \boldsymbol{n} \cdot \boldsymbol{r}', \tag{2.25}$$

where $\boldsymbol{n} = \boldsymbol{r}/r$ is a unit vector in the radial direction. In the current density

$$\boldsymbol{j}(\boldsymbol{r}', t - R/c) \approx \boldsymbol{j}(\boldsymbol{r}', t - r/c + \boldsymbol{n} \cdot \boldsymbol{r}'/c), \tag{2.26}$$

it is convenient to introduce a translated time coordinate

$$t'' = t - r/c, \tag{2.27}$$

which, significantly, does not involve the source coordinate $\boldsymbol{r}'$. Also, the factor $1/R$ in the integrand (2.23) is slowly varying and can be set equal to $1/r$ and taken outside of the integral. We then have, removing the double primes from the translated time coordinate t'',

$$\boldsymbol{A}(\boldsymbol{r}, t) \approx (cr)^{-1} \int d^3\boldsymbol{r}' \boldsymbol{j}(\boldsymbol{r}', t + \boldsymbol{n} \cdot \boldsymbol{r}'/c). \tag{2.28}$$

The retardation term $\boldsymbol{n} \cdot \boldsymbol{r}'/c$ is of the order of the time for propagation at velocity c across the source. If the source motions are non-relativistic, this time is small and the current density can be expanded:

$$\boldsymbol{j}(\boldsymbol{r}', t + \boldsymbol{n} \cdot \boldsymbol{r}'/c) = \boldsymbol{j}(\boldsymbol{r}', t) + \frac{\boldsymbol{n} \cdot \boldsymbol{r}'}{c} \frac{\partial \boldsymbol{j}(\boldsymbol{r}', t)}{\partial t} + \cdots. \tag{2.29}$$

This then yields a *multipole expansion* for the vector potential:

$$\boldsymbol{A}(\boldsymbol{r}, t) = \frac{1}{cr} \int d^3\boldsymbol{r}' \boldsymbol{j}(\boldsymbol{r}', t) + \frac{1}{c^2 r} \frac{d}{dt} \int d^3\boldsymbol{r}' (\boldsymbol{n} \cdot \boldsymbol{r}') \boldsymbol{j}(\boldsymbol{r}', t) + \cdots. \tag{2.30}$$

An expansion of this type is useful when the source motions are non-relativisitic, for which retardation effects are small. When the source consists of charges in relativistic motion, many higher multipoles contribute to the radiation field and the expansion (2.30) is not useful.

It is convenient to replace the formulation in terms of a continuum distribution of source current density by one in terms of a collection of discrete charges. If we set

$$j(\boldsymbol{r}', t) = \sum_\alpha q_\alpha \boldsymbol{v}_\alpha(\boldsymbol{r}')\delta\big(\boldsymbol{r}' - \boldsymbol{r}_\alpha(t)\big), \tag{2.31}$$

where q_α is the charge of particle α and $\boldsymbol{v}_\alpha$ is its velocity, we have

$$\boldsymbol{A}(\boldsymbol{r}, t) = \frac{1}{cr}\sum_\alpha q_\alpha \boldsymbol{v}_\alpha + \frac{1}{c^2 r}\sum_\alpha q_\alpha \frac{d}{dt}\big((\boldsymbol{n}\cdot\boldsymbol{r}_\alpha)\boldsymbol{v}_\alpha\big) + \cdots. \tag{2.32}$$

But (leaving off the subscripts) we can rewrite the second term in the expansion (2.32) using

$$\begin{aligned}(\boldsymbol{n}\cdot\boldsymbol{r})\boldsymbol{v} &= \frac{1}{2}\frac{d}{dt}\big((\boldsymbol{n}\cdot\boldsymbol{r})\boldsymbol{r}\big) + \frac{1}{2}(\boldsymbol{n}\cdot\boldsymbol{r})\boldsymbol{v} - \frac{1}{2}(\boldsymbol{n}\cdot\boldsymbol{v})\boldsymbol{r}\\ &= \frac{1}{2}\frac{d}{dt}\big((\boldsymbol{n}\cdot\boldsymbol{r})\boldsymbol{r}\big) + \frac{1}{2}(\boldsymbol{r}\times\boldsymbol{v})\boldsymbol{\times}\boldsymbol{n}.\end{aligned} \tag{2.33}$$

The vector potential (2.32) can then be written as a sum of (electric) dipole, magnetic dipole, and quadrupole terms:

$$\boldsymbol{A} = \boldsymbol{A}_d + \boldsymbol{A}_m + \boldsymbol{A}_q + \cdots, \tag{2.34}$$

where each term has a time derivative of the corresponding moment. If we define the moments

$$\begin{aligned}\boldsymbol{p} &= \sum_\alpha q_\alpha \boldsymbol{r}_\alpha,\\ \boldsymbol{m} &= \sum_\alpha q_\alpha \boldsymbol{r}_\alpha \boldsymbol{\times} \boldsymbol{v}_\alpha,\\ \boldsymbol{Q} &= 3\sum_\alpha q_\alpha(\boldsymbol{n}\cdot\boldsymbol{r}_\alpha)\boldsymbol{r}_\alpha,\end{aligned} \tag{2.35}$$

the expansion (2.34) can be written

$$\boldsymbol{A} = [\dot{\boldsymbol{p}} + \dot{\boldsymbol{m}}\times\boldsymbol{n} + \ddot{\boldsymbol{Q}}/6c]/cr + \cdots, \tag{2.36}$$

where the dots denote time derivatives.

2.2.2 Multipole Radiated Power

The evaluation of the electric and magnetic fields from the vector potential can be simplified for the case where only the radiation-field components are of interest. These are the fields associated with propagating plane waves and a general superposition of these waves could be represented by a sum over Fourier components;

that is, we could write a corresponding vector potential for a general radiation field in the form[6]

$$\boldsymbol{A}(\boldsymbol{r}, t) = \sum_{\boldsymbol{k}} \boldsymbol{a}_{\boldsymbol{k}} e^{i(\boldsymbol{k}\cdot\boldsymbol{r} - \omega t)}. \tag{2.37}$$

Here the $\boldsymbol{a}_{\boldsymbol{k}}$ are the Fourier amplitudes corresponding to wave vector $\boldsymbol{k}$ with frequency $\omega = c|\boldsymbol{k}|$. For the gauge condition div $\boldsymbol{A} = 0$ to hold identically for all $\boldsymbol{r}$ and t, for each Fourier component the transversality condition

$$\boldsymbol{a}_{\boldsymbol{k}} \cdot \boldsymbol{k} = 0 \tag{2.38}$$

must hold. We can also make use of another general form for the radiation field. Setting coordinate axes with the x-axis along the direction of propagation in the radial direction away from the source, the components of the vector potential associated with the radiation field will have the general form [see Equation (2.16)], in addition to the $1/r$ factor,[7]

$$\boldsymbol{A} \propto f(x - ct) = f(X) \tag{2.39}$$

where $X = X(x, t) = x - ct$ is the argument of the general function f. The magnetic field $\boldsymbol{B}$ = curl $\boldsymbol{A}$ results from differentiation with respect to x, but it is more convenient to express this in terms of a time derivative. Since $\partial f/\partial x = (\partial f/\partial X)(\partial X/\partial x) = \partial f/\partial X$ and $\partial f/\partial t = (\partial f/\partial X)(\partial X/\partial t) = -c(\partial f/\partial X)$, we can write $\partial f/\partial x = -(1/c)\partial f/\partial t$. In terms of the direction of propagation (away from the source) $\boldsymbol{n} = \boldsymbol{k}/k$, the general magentic field can then be written in terms of the time derivative of the vector potential:

$$\boldsymbol{B} = (1/c)\dot{\boldsymbol{A}} \times \boldsymbol{n}. \tag{2.40}$$

This is a general relation for radiation fields and holds for any gauge, including the one imposed to obtain the retarded potentials (2.20) and (2.21) and their expansions, which we are employing. These potentials do not satisfy the (sometimes convenient) radiation gauge conditions div $\boldsymbol{A} = 0$ and $\Phi = 0$, so $\boldsymbol{E} \neq -(1/c)\dot{\boldsymbol{A}}$; instead, again only for radiation fields,

$$\boldsymbol{E} = \boldsymbol{B} \times \boldsymbol{n}. \tag{2.41}$$

From the convenient relation (2.40), the radiation magnetic field can be computed in terms of the time derivatives of the various moments by means of the retardation expansion (2.36). To the various terms in $\dot{\boldsymbol{A}}$ can be added anything proportional to $\boldsymbol{n}$ without changing the (physical) quantity $\boldsymbol{B}$. It is convenient to, in this manner, replace $\boldsymbol{Q}$ in Equation (2.35) by the expression

$$\boldsymbol{Q} = 3 \sum_{\alpha} q_{\alpha} \left[(\boldsymbol{n} \cdot \boldsymbol{r}_{\alpha}) \boldsymbol{r}_{\alpha} - \tfrac{1}{3} \boldsymbol{n} r_{\alpha}^2 \right]. \tag{2.42}$$

The components of this vector can be expressed in terms of a quadrupole tensor through the relation $Q_j = Q_{jk} n_k$ (sum over k), where

$$Q_{jk} = \sum_{\alpha} q_{\alpha} (3x_j x_k - \delta_{jk} r^2)_{\alpha} \tag{2.43}$$

---

[6]Actually, the real part ("Re") of this expression should be taken

[7]In evaluating the radiation fields $\boldsymbol{E}$ and $\boldsymbol{B}$ from $\boldsymbol{A}$, differentiation of the $1/r$ factor is neglected; the radiation fields, like the potential, have a $1/r$ falloff.

is a symmetric traceless tensor. The magnetic field (2.40) is then

$$\begin{aligned} \boldsymbol{B} &= \frac{1}{c^2 r}\left[\ddot{\boldsymbol{p}} \times \boldsymbol{n} + (\ddot{\boldsymbol{m}} \times \boldsymbol{n}) \times \boldsymbol{n} + \frac{1}{6c}\dddot{\boldsymbol{Q}} \times \boldsymbol{n}\right] \\ &= \boldsymbol{B}_d + \boldsymbol{B}_m + \boldsymbol{B}_q. \end{aligned} \tag{2.44}$$

The radiated power from a system of charges is computed from the Poynting vector

$$\boldsymbol{S} = (c/4\pi)\boldsymbol{E} \times \boldsymbol{B} = (c/4\pi)(\boldsymbol{B} \cdot \boldsymbol{B})\boldsymbol{n}, \tag{2.45}$$

which is the energy flux in the direction $\boldsymbol{n}$. Multiplying by $4\pi r^2$, we get the total radiated power in terms of a directional average (of $\boldsymbol{n}$):

$$P = dW/dt = r^2 c\overline{(\boldsymbol{B} \cdot \boldsymbol{B})}, \tag{2.46}$$

where the bar denotes a directional average. In evaluating this average, the cross terms can be shown to vanish; that is,

$$\overline{\boldsymbol{B}_d \cdot \boldsymbol{B}_m} = \overline{\boldsymbol{B}_d \cdot \boldsymbol{B}_q} = \overline{\boldsymbol{B}_m \cdot \boldsymbol{B}_q} = 0. \tag{2.47}$$

To see this, consider the "parity" of the various components of $\boldsymbol{B}$ (behavior under $\boldsymbol{n} \to -\boldsymbol{n}$). From their definitions, $\boldsymbol{B}_d$ is odd while $\boldsymbol{B}_m$ and $\boldsymbol{B}_q$ are even, and this proves the first two of the relations (2.47). The last one involving $\boldsymbol{B}_m$ and $\boldsymbol{B}_q$ is harder to prove. However, from the vector identity $(\boldsymbol{A} \times \boldsymbol{B}) \cdot (\boldsymbol{C} \times \boldsymbol{D}) = (\boldsymbol{A} \cdot \boldsymbol{C})(\boldsymbol{B} \cdot \boldsymbol{D}) - (\boldsymbol{A} \cdot \boldsymbol{D})(\boldsymbol{B} \cdot \boldsymbol{C})$, we have

$$(\dddot{\boldsymbol{Q}} \times \boldsymbol{n}) \cdot \left[(\ddot{\boldsymbol{m}} \times \boldsymbol{n}) \times \boldsymbol{n}\right] = \left[\dddot{\boldsymbol{Q}} \cdot (\ddot{\boldsymbol{m}} \times \boldsymbol{n})\right](\boldsymbol{n} \cdot \boldsymbol{n}) - (\dddot{\boldsymbol{Q}} \cdot \boldsymbol{n})\left[\boldsymbol{n} \cdot (\ddot{\boldsymbol{m}} \times \boldsymbol{n})\right], \tag{2.48}$$

but the last term in brackets on the right is zero. It then remains to prove that the directional average of $\dddot{\boldsymbol{Q}} \cdot (\ddot{\boldsymbol{m}} \times \boldsymbol{n})$ is zero. In terms of the Levi-Civita symbol [see Equation (1.73)], this is

$$\dddot{\boldsymbol{Q}} \cdot (\ddot{\boldsymbol{m}} \times \boldsymbol{n}) = \varepsilon_{jkl} \dddot{Q}_{jn} \ddot{m}_k n_n n_l, \tag{2.49}$$

and since

$$\overline{n_n n_l} = \tfrac{1}{3}\delta_{nl}, \tag{2.50}$$

the directional average is

$$\overline{\dddot{\boldsymbol{Q}} \cdot (\boldsymbol{m} \times \boldsymbol{n})} = \tfrac{1}{3}\varepsilon_{jkl} \dddot{Q}_{jl} \ddot{m}_k. \tag{2.51}$$

But we can relabel the dummy indices j and l, taking half the sum, making use of the symmetric nature of $\dddot{Q}_{jl}$ and the antisymmetric nature of ε_{jkl}, and we see easily that the result (2.51) is identically zero. Thus,

$$P = r^2 c(\overline{\boldsymbol{B}_d^2} + \overline{\boldsymbol{B}_m^2} + \overline{\boldsymbol{B}_q^2}) = P_d + P_m + P_q. \tag{2.52}$$

The terms P_d, P_m, and P_q can be expressed in terms of the individual (time derivatives of) moments through the insertion of the expressions (2.44) and (2.52). In P_d and P_m the angular averages are trivial and we obtain

$$\begin{aligned} P_d &= 2\ddot{\boldsymbol{p}}^2/3c^3, \\ P_m &= 2\ddot{\boldsymbol{m}}^2/3c^3. \end{aligned} \tag{2.53}$$

The quadrupole expression is not so easy to evaluate. Employing for now the notation $q_j = \dddot{Q}_j = \dddot{Q}_{jk}n_k = q_{jk}n_k$, we have

$$(\boldsymbol{q} \times \boldsymbol{n})^2 = \boldsymbol{q}^2 - (\boldsymbol{q} \cdot \boldsymbol{n})^2. \tag{2.54}$$

But

$$\boldsymbol{q}^2 = \boldsymbol{q} \cdot \boldsymbol{q} = q_j q_j = q_{jk} q_{jl} n_k n_l, \tag{2.55}$$

$$(\boldsymbol{q} \cdot \boldsymbol{n})^2 = (q_j n_j)^2 = q_j q_k n_j n_k = q_{jl} q_{kn} n_j n_k n_l n_n. \tag{2.56}$$

The directional average of the product $n_k n_l$ in Equation (2.55) was given in the result (2.50). The average of the four-index product in Equation (2.56) must be symmetric in the indices and expressed in terms of the "fundamental tensor" δ_{jk}. The most general such symmetric function is

$$\overline{n_j n_k n_l n_n} = a(\delta_{jk}\delta_{ln} + \delta_{jl}\delta_{kn} + \delta_{jn}\delta_{kl}), \tag{2.57}$$

where a is a constant. The constant can be evaluated by either contracting on two pairs of indices or by contracting on all four. Doing the latter, letting all four refer to the "z" or "polar" direction for a spherical polar coordinate description, from

$$\overline{n_z^4} = \frac{1}{2}\int_0^\pi \cos^4\theta \sin\theta \, d\theta = \frac{1}{5}, \tag{2.58}$$

we obtain $a = 1/15$. The quadrupole power radiated is then, by Equations (2.52), (2.44), (2.54)–(2.58),

$$P_q = \frac{1}{180c^5} \sum_{jk} \dddot{Q}_{jk}^2. \tag{2.59}$$

It should be noted that in P_d, P_m, and P_q the contribution results from a sum over squares of individual moment ($\ddot{p}_j$, $\ddot{m}_j$, $\dddot{Q}_{jk}$) contributions. Further, remember that these results refer to the total energy radiated, integrated over photon energies and summed over polarization states.

The multipole expansion is useful, as we have stated earlier, in the limit where the source motions are non-relativistic. In this case, the retardation effects are small and the principal contribution comes from the dipole term. Sometimes, however, because of symmetries in the source characteristics,[8] there is no dipole contribution. Compared with the (electric) dipole term, the magnetic dipole and quadrupole terms are small and of the same order of magnitude. If the characteristic velocities of the source particles are $\sim v$, from the definitions of $\boldsymbol{p}$, $\boldsymbol{m}$, and Q_{jk}, we see that

$$P_m \sim P_q \sim (v^2/c^2) P_d. \tag{2.60}$$

2.3 FOURIER SPECTRA

Depending on the details of the source characteristics, the radiation field contains a spectrum of photon energies or frequencies. If $E(t)$ and $B(t)$ represent the magnitude of the electric and magnetic fields, respectively, associated with a particular

[8] An example of this case—one that will be considered later—is that in which the source consists of two identical particles. In the scattering of two electrons, the lowest-order radiation term is quadrupole.

component[9] of polarization, the corresponding contribution to the energy flux in the magnitude of the Poynting vector in the radial direction away from the source is

$$S(t) = dW/dA\,dt = (c/4\pi)E^2(t) = (c/4\pi)B^2(t). \tag{2.61}$$

The fields can be written in terms of Fourier amplitudes:

$$E(t) = \int E_\omega e^{-i\omega t} d\omega, \quad B(t) = \int B_\omega e^{-i\omega t} d\omega, \tag{2.62}$$

where the Fourier components are determined by the specific time dependence of the fields:

$$E_\omega = (2\pi)^{-1} \int E(t) e^{i\omega t} dt, \quad B_\omega = (2\pi)^{-1} \int B(t) e^{i\omega t} dt. \tag{2.63}$$

If the source motions are periodic, instead of a continuum of frequencies, there would be a sum over a discrete spectrum:

$$E(t) = \sum_k E_k e^{-i\omega_k t}, \quad B(t) = \sum_k B_k e^{-i\omega_k t}. \tag{2.64}$$

Let us take the continuum form (2.62) and consider only the magnetic field, since it is equal in magnitude to the associated electric field (although, of course, the corresponding components are mutually perpendicular). Employing the complex forms (2.61) we should write the Poynting vector as $S(t) = (c/4\pi)|B^2(t)|$. We can also introduce the energy flux per unit frequency, integrated over time:

$$I(\omega) = dW/dA\,d\omega, \tag{2.65}$$

so that we can write for the total energy flow per unit area:

$$dW/dA = \int S(t)dt = \int I(\omega)d\omega, \tag{2.66}$$

with

$$S(t) = (c/4\pi)|B(t)|^2 = (c/4\pi) \iint B^*_{\omega'} B_\omega e^{i(\omega'-\omega)t} d\omega' d\omega. \tag{2.67}$$

When this form is substituted into the first integral on the right side of Equation (2.66) and use is made of the identity

$$\int e^{i(\omega'-\omega)t} dt = 2\pi\delta(\omega'-\omega), \tag{2.68}$$

we have

$$dW/dA = \frac{1}{2} c \int |B_\omega|^2 d\omega. \tag{2.69}$$

In the above expressions, the frequency variable ω extends from $-\infty$ to ∞, but the positive and negative frequencies in the radiation field spectrum are physically equivalent and can be lumped together. Thus, we can write

$$I_\omega \equiv I(|\omega|) = 2I(\omega) = c|E_\omega|^2 = c|B_\omega|^2. \tag{2.70}$$

[9] It could be one of the mutually perpendicular linear components transverse to the direction of propagation.

If the element of area through which radiation is flowing is at a distance r from the source and subtends a solid angle $d\Omega$, then $dA = r^2 d\Omega$, and the differential energy radiated within frequency $d\omega$ is

$$dW_\omega = I_\omega r^2 d\omega\, d\Omega. \tag{2.71}$$

In this relation, the intensity I_ω will be proportional to $1/r^2$, so that dW_ω is independent of r.

The photon concept can be introduced into our classical formulation by writing

$$dW_\omega = \hbar\omega\, dw_\omega, \tag{2.72}$$

where dw_ω is the *probability* of photon emission within $d\omega$. Then we have

$$dw_\omega = (c/\hbar) r^2 (d\omega/\omega)\big[|E_\omega|^2 \text{ or } |B_\omega|^2\big] d\Omega \tag{2.73}$$

for the differential probability of emitting a photon within frequency $d\omega$ and within solid angle $d\Omega$. As mentioned earlier, E_ω and B_ω refer to the components associated with a particular polarization, and the corresponding probability dw_ω given by Equation (2.73) would then refer to this photon polarization state. If in dw_ω we are not interested in polarization, and want the total dw_ω summed over polarizations, the terms $|E_\omega|^2$ or $|B_\omega|^2$ in Equation (2.73) would be the sum of the squares of their values associated with each polarization.

The expression (2.73), although it contains $\hbar$, is essentially a classical one, since the only quantum mechanics introduced is the photon concept by means of the relation (2.72). For example, we have not provided the necessary quantum mechanics to calculate E_ω or B_ω from the source motions. In this chapter, we employ only classical theory to relate the fields to source characteristics. Nevertheless, the classical theory, together with the subsequent injection of the photon concept by means of the relation (2.72), does provide a description of certain phenomena often considered to be quantum mechanical in nature. One phenomenon is the so-called infrared divergence or infrared "catastrophe," which is the infinite probability of emitting infinitesimally soft photons in charged-partical processes. The simple semi-classical formulation described above is sufficient to yield this effect as well as, in fact, its explanation. We return to discuss this phenomenon in more detail later in this chapter.

Once again, it may be well to emphasize that if we are not interested in photon polarizations, the E_ω or B_ω in the emission probability (2.73) can refer to the (Fourier transform of the) magnitude of the fields. Alternatively, if the total field is expressed in terms of its two mutually perpendicular polarizations ($\boldsymbol{B} = \boldsymbol{B}_1 + \boldsymbol{B}_2$, with $\boldsymbol{B}_1 \cdot \boldsymbol{B}_2 = 0$; similarly with $\boldsymbol{E}$), the emission probability can be expressed in terms of the sum of the individual polarization components. That is, in the result (2.73) we can write

$$|B_\omega|^2 = |B_{\omega 1}|^2 + |B_{\omega 2}|^2 \tag{2.74}$$

(similarly with $|E_\omega|^2$). For the case of dipole radiation, there is a simple and convenient relation for the angular and spectral distribution of radiation summed over polarizations. By equations (2.36), (2.40), and (2.70), we have

$$(dW/d\omega\, d\Omega)_d = |\ddot{\boldsymbol{p}}_\omega \times \boldsymbol{n}|^2 / c^3, \tag{2.75}$$

in terms of the Fourier transform of $\ddot{\boldsymbol{p}}$. When we integrate over angles of emission, the spectrum of energy radiated is

$$(dW/d\omega)_d = (4\pi/c^3)|\ddot{\boldsymbol{p}}_\omega|^2\overline{\sin^2\theta} = (8\pi/3c^3)|\ddot{\boldsymbol{p}}_\omega|^2; \qquad (2.76)$$

here θ is the angle between $\ddot{\boldsymbol{p}}_\omega$ and $\boldsymbol{n}$, and the spherical average of $\sin^2\theta$ is 2/3. The result (2.76) can be put in slightly different form, since $\ddot{\boldsymbol{p}}_\omega = -\omega^2 \boldsymbol{p}_\omega$, and for a single charge the second time derivative of the dipole moment is directly related to the particle acceleration and thus to the force on it. That is, for a charge q the emission spectrum is directly related to the Fourier transform of its acceleration:

$$(dW/d\omega)_d = (8\pi q^2/3c^3)|\boldsymbol{a}_\omega|^2. \qquad (2.77)$$

We shall return to further general developments concerning radiation field spectra and applications to specific processes later in this chapter and in later chapters.

2.4 FIELDS OF A CHARGE IN RELATIVISTIC MOTION

2.4.1 Liénard-Wiechert Potentials

In the convenient Lorentz gauge, the vector and scalar potentials associated with a distribution of charges and currents are given by the expressions (2.20)–(2.23). An important application of these retarded potentials is to the case of a single particle of charge q moving at an arbitrary velocity $\boldsymbol{v}$. As mentioned earlier, for this problem—even in the case of a point particle—the potentials are *not* given by the non-relativistic values ($R = |\boldsymbol{r} - \boldsymbol{r}'|$)

$$\Phi_{\mathrm{NR}} \to q/R, \quad A_{\mathrm{NR}} \to q\boldsymbol{v}/cR, \qquad (2.78)$$

at the retarded position $\boldsymbol{r}'$ and time t'. This is immediately obvious, since the expressions (2.78) do not form the components of a four-vector. Actually, it is not difficult to construct a covariant form for $A_\mu = (i\Phi, \boldsymbol{A})$ that reduces to the non-relativistic limits (2.78) for $v \ll c$. From $\boldsymbol{R} = \boldsymbol{r} - \boldsymbol{r}'$, the source-to-field point, the four-vector

$$R_\mu = (iR, \boldsymbol{R}) = (ic(t - t'), \boldsymbol{r} - \boldsymbol{r}') \qquad (2.79)$$

can be formed [see Equation (2.22)]; this is a null four-vector:

$$R_\mu R_\mu = 0. \qquad (2.80)$$

Employing the four-vector velocity $v_\mu = \gamma(ic, \boldsymbol{v})$, a scalar

$$R_\nu v_\nu = -\gamma(Rc - \boldsymbol{R}\cdot\boldsymbol{v}) \qquad (2.81)$$

can be constructed that reduces to $-Rc$ in the limit $v \ll c$. The expression

$$A_\mu = -q[v_\mu/R_\nu v_\nu], \qquad (2.82)$$

where the brackets are meant to imply the imposition of the retardation condition, satisfies the covariance requirement and reduces to the required limit in a Lorentz frame where the motion is non-relativistic. It is, therefore, a correct general formula valid in a frame where the particle velocity is arbitrarily large.

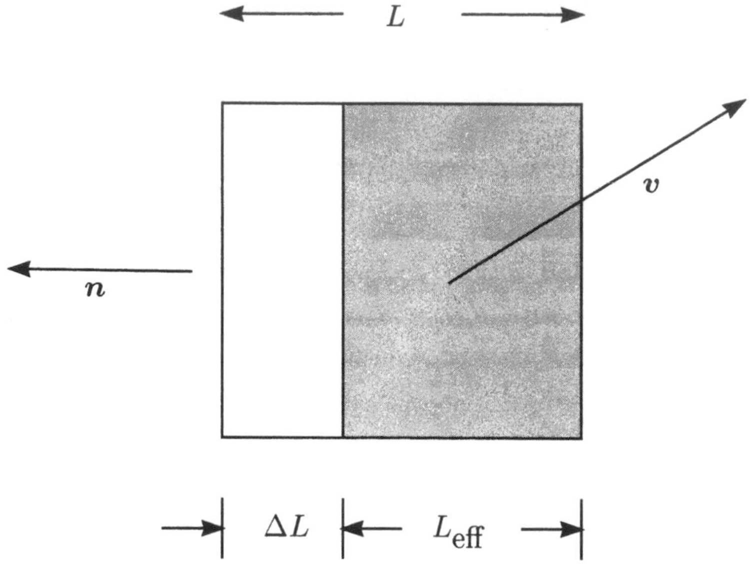

Figure 2.2 "Effective volume" associated with the field of a moving charge.

In terms of the source-to-field point unit vector $\boldsymbol{n} = \boldsymbol{R}/R$, $\boldsymbol{\beta} = \boldsymbol{v}/c$, and

$$\varkappa \equiv 1 - \boldsymbol{n} \cdot \boldsymbol{\beta}, \tag{2.83}$$

the result (2.82) is

$$A_\mu = (i\Phi, \boldsymbol{A}) = [iq/\varkappa R, q\boldsymbol{v}/c\varkappa R]. \tag{2.84}$$

The magnitude of the retardation effect is expressed in the factor $\varkappa^{-1}$. Although we have made use of covariance considerations in deriving this result, this correction factor is not relativistic in nature, the modification being linear in v. The correction is due totally to effects of retardation in the relationship between source motions and field (potential) amplitudes.

The expressions (2.84) are called the *Liénard-Wiechert potentials*, and their derivation as given above, although simple, perhaps obscures the meaning of, in particular, the correction factor $\varkappa^{-1}$. One way of interpreting the factor $q/\varkappa$ is as an "effective charge" (q_{eff}). To see this, refer back to the expressions (2.20)–(2.23) for the potentials that exhibit the manner in which an imaginary spherical wave converging on the field point $\boldsymbol{r}$ at time t sweeps past the charge in motion at the source point $(\boldsymbol{r}', t')$. We assume, for convenience, that the charge has finite spatial extent although, in the end, we can let it approach the point-particle limit. Also for convenience, imagine the field point at the origin, so that $\boldsymbol{n}$ points toward the origin. Then if $\boldsymbol{v}$ is pictured as having a positive component away from the origin, $\boldsymbol{n} \cdot \boldsymbol{v}$ is negative. Now consider the contribution to A and Φ from a cubic (or cylindrical) element of the charge with flat front and back perpendicular to $\boldsymbol{n}$ (see Figure 2.2). During the time Δt that the wave front passes the charge element, the back end of the charge has moved a distance ΔL, so that the *effective* amount of charge that the wave sees is

$$q_{\text{eff}} = q(L_{\text{eff}}/L), \tag{2.85}$$

where[10] L is the length of the element in the radial $(-\boldsymbol{n})$ direction, and

$$L_{\text{eff}} = L + \Delta L \tag{2.86}$$

(remember that ΔL is negative if $\boldsymbol{n} \cdot \boldsymbol{v}$ is negative). But $\Delta L = \boldsymbol{n} \cdot \boldsymbol{v} \Delta t$, $\Delta t = L_{\text{eff}}/c$ and, by the relations (2.85) and (2.86), we have

$$q_{\text{eff}} = q(1 - \boldsymbol{n} \cdot \boldsymbol{\beta})^{-1} = q/\varkappa. \tag{2.87}$$

The spatial extent of the charge can now be made infinitesimally small with no effect on the correction factor $\varkappa^{-1}$.

2.4.2 Charge in Uniform Motion

The fields associated with a charge moving at constant velocity can be computed easily, and the results are of interest in connection with, for example, a topic (Weizsäcker-Williams Method) that is treated later in this chapter. For the case of uniform motion the particle retarded position (P_r) and present position (P_p) are related in a simple way to the field point (P_f), as is indicated in Figure 2.3. In terms of the distance

$$s \equiv \varkappa R = R - \boldsymbol{R} \cdot \boldsymbol{\beta}, \tag{2.88}$$

the vector and scalar potentials are given by

$$\boldsymbol{A} = q\boldsymbol{\beta}/s, \quad \Phi = q/s, \tag{2.89}$$

where, for convenience in the notation, the brackets indicating retardation have been omitted. The fields $\boldsymbol{E}$ and $\boldsymbol{B}$ are computed from $\boldsymbol{A}$ and Φ by differentiation with respect to the field-point coordinates. To do this, it is convenient to employ a coordinate system (K_0) with the origin at the instantaneous present position of the particle. Then, by Equations (2.2) and (2.3), we have $E_x = -\partial\Phi/\partial x_0 - c^{-1}\partial A_x/\partial t_0$, etc. However, since the fields are carried by the particle (moving in the x- or x_0-direction), the time derivatives can be computed from

$$\frac{\partial}{\partial t_0} = -v\frac{\partial}{\partial x_0}. \tag{2.90}$$

Also (see Figure 2.3),

$$\begin{aligned} s^2 &= r_0^2 - (R\beta \sin\theta)^2 = r_0^2(1 - \beta^2 \sin^2\psi) \\ &= x_0^2 + y_0^2 + z_0^2 - \beta^2(y_0^2 + z_0^2). \end{aligned} \tag{2.91}$$

The vector potential has only an x-component (direction of $\boldsymbol{v}$) and we obtain for the fields:

$$\begin{aligned} \boldsymbol{E} &= qs^{-3}(1 - \beta^2)\boldsymbol{r}_0, \\ \boldsymbol{B} &= \boldsymbol{\beta} \times \boldsymbol{E}. \end{aligned} \tag{2.92}$$

---

[10] The lengths are, of course, lab-frame values, which are different from those in the particle rest frame.

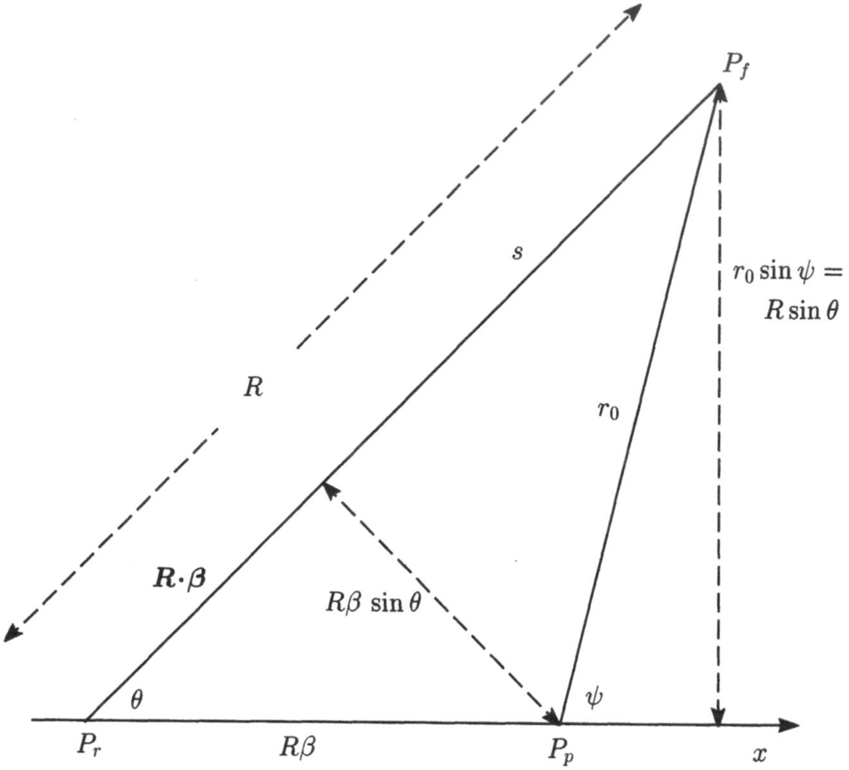

Figure 2.3 Present and retarded positions for a charge in uniform motion.

Note that $\boldsymbol{r}_0$ is the vector from the *present* position of the charge to the field point (rather than from the retarded position). The longitudinal and transverse components of the fields should also be noted, especially in the extreme relativistic limit. The magnitude of the electric field in the transverse direction (perpendicular to the x-axis) can be expressed as a function of ψ and r_0 or as a function of ψ and the "impact" parameter $b = r_0 \sin\psi$:

$$E_t = \frac{q}{r_0^2}\frac{(1-\beta^2)\sin\psi}{(1-\beta^2\sin^2\psi)^{3/2}} = \frac{q}{b^2}\frac{(1-\beta^2)\sin^3\psi}{(1-\beta^2\sin^2\psi)^{3/2}}. \tag{2.93}$$

The longitudinal component is

$$E_l = \frac{q}{r_0^2}\frac{(1-\beta^2)\cos\psi}{(1-\beta^2\sin^2\psi)^{3/2}} = \frac{q}{b^2}\frac{(1-\beta^2)\sin^2\psi\cos\psi}{(1-\beta^2\sin^2\psi)^{3/2}}. \tag{2.94}$$

The magnetic field has only a transverse component, which, at the field point, is perpendicular to both $\boldsymbol{E}_t$ and $\boldsymbol{\beta}$ [see Equation (2.92)]:

$$|\boldsymbol{B}| = \beta E_t. \tag{2.95}$$

The dependence of the fields on ψ provides a description of their time dependence as seen by an observer at the fixed field point P_f as the particle (at P_p) passes. The largest electric field component is the transverse field E_t, which attains a value $E_t \to (q/b^2)(1-\beta^2)^{-1/2}$ as $\psi \to \pi/2$. The longitudinal component E_l does not attain this value and, in fact, changes sign at $\psi = \pi/2$. As $\beta \to 1$, the electric field E_t and its accompanying magnetic field (2.95) become large at $\psi = \pi/2$, and the observer at P_f sees a strong *transverse pulse*. The idea behind the Weizsäcker-Williams method is that this pulse can be regarded as a flux of photons (see Section 8).

2.4.3 Fields of an Accelerated Charge

For a charge in non-uniform motion the calculation of the fields is a more complicated problem. The case of finite acceleration is of great importance, however, because there are now radiation-field components; that is, there are electric and magentic fields that fall off as $1/r$ at a large distance from the charge. The existence of these fields is due to, as in the non-relativistic problem, the effects of retardation. In the computation of the fields from the spatial and time derivatives of the potentials, the retardation results in a certain amount of complexity. This is because, in evaluating the spatial gradients with respect to the field-point coordinate ($\boldsymbol{r}$), it is necessary to include the explicit functional dependence of the potentials on $\boldsymbol{r}$ and dependence that is contained in the retardation condition.

We consider the motion of the source particle to be known; that is, $\boldsymbol{r}' = \boldsymbol{r}'(t')$ is specified, as is its velocity $\boldsymbol{v} = \partial \boldsymbol{r}'/\partial t'$ and acceleration $\dot{\boldsymbol{v}} = \partial \boldsymbol{v}/\partial t'$ (we leave the primes off $\boldsymbol{v}$ asnd $\dot{\boldsymbol{v}}$). The retarded time coordinate t' is the independent variable in the problem. To compute the fields $\boldsymbol{E} = -\nabla\Phi - (1/c)\partial \boldsymbol{A}/\partial t$ and $\boldsymbol{B} = \nabla \times \boldsymbol{A}$ from the Liénard-Wiechert potentials (2.84), it is necessary to express the derivatives with respect to the retarded time t'. This is easily accomplished through consideration of the mathematical statement of the retardation condition, that is, Equation (2.80) or

$$R = \left[(\boldsymbol{r}-\boldsymbol{r}')\cdot(\boldsymbol{r}-\boldsymbol{r}')\right]^{1/2} = c(t-t'). \tag{2.96}$$

Differentiating with respect to t, we have

$$\frac{\partial R}{\partial t} = c\left(1 - \frac{\partial t'}{\partial t}\right) = \frac{\partial R}{\partial t'}\frac{\partial t'}{\partial t} = -\boldsymbol{n}\cdot\boldsymbol{v}\frac{\partial t'}{\partial t}. \tag{2.97}$$

Then

$$\frac{\partial}{\partial t} = \frac{\partial t'}{\partial t}\frac{\partial}{\partial t'} = \frac{R}{s}\frac{\partial}{\partial t'} = \frac{1}{\varkappa}\frac{\partial}{\partial t'}. \tag{2.98}$$

We also need a convenient expression for the gradient operator ∇, which can be written

$$\nabla = \nabla_r + \nabla_{t'} = \nabla_r + (\nabla t')\partial/\partial t'. \tag{2.99}$$

Here ∇_r means differentiation with respect to the field-point coordinate $\boldsymbol{r}$, ignoring the dependence contained within the retardation condition, and $\nabla_{t'}$ accounts for the latter contribution. The factor $\nabla t'$ can be found from the retardation condition (2.96), recognizing that varying $\boldsymbol{r}$ implies a variation in t':

$$\begin{aligned}\nabla t' = -\frac{1}{c}\nabla R &= = -\frac{1}{c}\left(\frac{\boldsymbol{R}}{R} + \frac{\partial R}{\partial t'}\nabla t'\right)\\ &= -\frac{1}{c}\left(\frac{\boldsymbol{R}}{R} - \boldsymbol{n}\cdot\boldsymbol{v}\nabla t'\right).\end{aligned} \tag{2.100}$$

Solving for $\nabla t'$, we have $\nabla t' = -\boldsymbol{R}/sc$, and

$$\nabla = \nabla_r - \frac{\boldsymbol{R}}{sc}\frac{\partial}{\partial t'}. \tag{2.101}$$

The fields $\boldsymbol{E}$ and $\boldsymbol{B}$ are derived from the potentials $\boldsymbol{A}(= q\boldsymbol{v}/cs)$ and $\Phi(= q/s)$, employing the operators (2.98) and (2.101). Carrying through these operations and rearranging the terms, we find

$$\frac{\boldsymbol{E}}{q} = \frac{\boldsymbol{R}_v}{s^3}(1-\beta^2) + \frac{1}{c^2s^3}\boldsymbol{R}\times(\boldsymbol{R}_v\times\dot{\boldsymbol{v}}), \tag{2.102}$$

$$\boldsymbol{B} = \boldsymbol{n}\times\boldsymbol{E}, \tag{2.103}$$

where

$$\boldsymbol{R}_v = \boldsymbol{R} - R\boldsymbol{\beta} \tag{2.104}$$

is the "virtual present source-to-field-point vector." That is, if the source (charge) at the retarded position and velocity were to move at constant velocity, it would be at a certain "virtual" position, and $\boldsymbol{R}_v$ is the vector from that position to the field point. Compare this result with that for a charge in uniform motion (see Figure 2.3). Again, the brackets [ ] designating retardation have been left off the terms on the right-hand side.

For considerations of radiation effects, the important term in the field (2.102) is the second, since it falls off as $1/r$ at large distances. It then yields a Poynting vector component $S \propto 1/r^2$ and an energy flow rate $dW/dt = S\,dA \propto d\Omega$ through an element of area $dA = r^2 d\Omega$ and solid angle $d\Omega$. The radiation term is finite only when there is an acceleration of the particle, and it is a consequence of the effects of retardation.

2.5 RADIATION FROM A RELATIVISTIC CHARGE

It is convenient to express the radiation fields in terms of the dimensionless quantities $\boldsymbol{n} = \boldsymbol{R}/R$, $\boldsymbol{\beta} = \boldsymbol{v}/c$, and $\varkappa = 1 - \boldsymbol{n}\cdot\boldsymbol{\beta}$. Then the second term on the right of Equation (2.102) can be written

$$\boldsymbol{E}_{\text{rad}} = (q/cR\varkappa^3)\boldsymbol{n}\times\big((\boldsymbol{n}-\boldsymbol{\beta})\times\dot{\boldsymbol{\beta}}\big). \tag{2.105}$$

The Poynting vector at the field point $\boldsymbol{r}, t$ is then

$$\begin{aligned} \boldsymbol{S}(\boldsymbol{r},t) &= \boldsymbol{n}(c/4\pi)|\boldsymbol{E}_{\text{rad}}|^2 \\ &= \boldsymbol{n}\frac{q^2}{4\pi c}\left[\frac{\boldsymbol{n}\times\big((\boldsymbol{n}-\boldsymbol{\beta})\times\dot{\boldsymbol{\beta}}\big)}{\varkappa^3 R}\right]^2, \end{aligned} \tag{2.106}$$

where the brackets have now been reinserted to emphasize that quantities therein are to be evaluated at the retarded coordinates. That is, for example, $\boldsymbol{\beta}c$ and $\dot{\boldsymbol{\beta}}c$ are the velocity and acceleration at $\boldsymbol{r}'$ and $t' = t - R/c$.

The energy flow rate per unit solid angle $d\Omega = dA/R^2$ at $\boldsymbol{r}, t$ would be computed from

$$dW/dt\,d\Omega = R^2 S, \tag{2.107}$$

but this is *not* equal to charge's energy radiation rate. If dE' is the change in the energy of the charged particle as a result of the emission of radiation, we can write $dE' = -dW$. However, the interval of time dt associated with the passage of photons through the element of area dA at the field point $\boldsymbol{r}$ is different from the interval dt' corresponding to the emission of these photons by the charge at $\boldsymbol{r}'$. The relationship between these intervals has already been derived and is given by the result (2.98): $dt'/dt = R/s = 1/\varkappa$. Then the rate of radiation of energy by the charge is

$$-dE'/dt' = (dW/dt)(dt/dt') = \varkappa\, dW/dt. \tag{2.108}$$

The times t and t' both refer to events in the lab frame; that is, the prime does not refer to a different Lorentz frame. The $\varkappa$ factor is associated with retardation effects, being first order in β. Since, also, $d\Omega = d\Omega'$, the angular distribution of emitted radiation is given by

$$-\frac{dE'}{dt'd\Omega'} = \frac{q^2}{4\pi c}\frac{\left[\boldsymbol{n}\times\left((\boldsymbol{n}-\boldsymbol{\beta})\times\dot{\boldsymbol{\beta}}\right)\right]^2}{\varkappa^5}. \tag{2.109}$$

Employing the well-known identity for the triple vector product, we can write the quantity in brackets in Equation (2.109) as

$$\begin{aligned}\boldsymbol{n}\times\left((\boldsymbol{n}-\boldsymbol{\beta})\times\dot{\boldsymbol{\beta}}\right) &= (\boldsymbol{n}-\boldsymbol{\beta})(\boldsymbol{n}\cdot\dot{\boldsymbol{\beta}}) - \dot{\boldsymbol{\beta}}\left(\boldsymbol{n}\cdot(\boldsymbol{n}-\boldsymbol{\beta})\right)\\ &= (\boldsymbol{n}-\boldsymbol{\beta})(\boldsymbol{n}\cdot\dot{\boldsymbol{\beta}}) - \varkappa\dot{\boldsymbol{\beta}}.\end{aligned} \tag{2.110}$$

Squaring this expression then yields the squared bracket in Equation (2.109):

$$\left[\boldsymbol{n}\times\left((\boldsymbol{n}-\boldsymbol{\beta})\times\dot{\boldsymbol{\beta}}\right)\right]^2 = \varkappa^2\dot{\beta}^2 + 2\varkappa(\boldsymbol{n}\cdot\dot{\boldsymbol{\beta}})(\boldsymbol{\beta}\cdot\dot{\boldsymbol{\beta}}) - (1-\beta^2)(\boldsymbol{n}\cdot\dot{\boldsymbol{\beta}})^2. \tag{2.111}$$

This relation is convenient for an evaluation of the angular integration over $d\Omega'$ to obtain the total radiative energy loss rate. The form (2.111), together with the factor $\varkappa^{-5}$, describes the angular distribution of the emitted radiation. In the non-relativistic limit, the result is the familiar dipole pattern of the form $\sin^2\vartheta$, where ϑ if the angle between $\boldsymbol{n}$ and $\dot{\boldsymbol{\beta}}$. The result in the extreme relativistic limit ($\beta \to 1$) is more interesting, since in this limit $\varkappa$ can get very small when $\boldsymbol{n}$ is along $\boldsymbol{\beta}$. In this case, the distribution is peaked in the direction of the instantaneous velocity $\boldsymbol{\beta}$; if θ is the angle between $\boldsymbol{n}$ and $\boldsymbol{\beta}$, it is easy to show that the angular width is of the order $\Delta\theta \sim \gamma^{-1} = (1-\beta^2)^{1/2} \ll 1$. The effect is basically kinematic in nature and could also be derived through considerations of transformations between the lab frame and one in which the motion is non-relativistic.

Another characteristic of the angular distribution in the general case should be mentioned. From the form (2.109) we see that the intensity is zero in two directions: when $\boldsymbol{n} - \boldsymbol{\beta}$ is along and opposite to the direction of $\dot{\boldsymbol{\beta}}$.

To compute the total rate of radiation of energy, an integration over $d\Omega'$ (directions of $\boldsymbol{n}$) can be performed. It is convenient to employ coordinate axes in which one axis (say, the y-axis) is instantaneously (at the retarded time) aligned with the velocity $\boldsymbol{\beta}$. Choosing this axis as, in addition, the polar axis of a spherical polar coordinate system describing $\boldsymbol{n} = (\sin\theta\cos\varphi, \cos\theta, \sin\theta\sin\varphi)$, we have $\varkappa = 1 - \beta\cos\theta$. In the terms on the right-hand side of Equation (2.111) the integrations over the azimuthal angle φ are trivial. The subsequent integrations over $d\theta$ are elementary, although the algebra is a little tedious. The result can be written

$$-dE'/dt' = (2q^2/3c)\gamma^6\big[\dot{\beta}^2 - (\boldsymbol{\beta}\times\dot{\boldsymbol{\beta}})^2\big]. \tag{2.112}$$

The derivation of the form (2.112) as outlined above requires a considerable amount of work, especially if we include that necessary to obtain the angular distribution (2.109). However, there is much simpler method of obtaining the energy radiation rate (2.112) that makes use of ideas of covariance. The differential energy radiated can be written $dE_{\rm rad} = E\,dN$, where E is the photon energy and dN is the differential number of photons[11] emitted. As we have seen in Section 1.2.3, the ratio dt/E associated with photon coordinates is an invariant, where the photons may be moving in any direction. Then the radiated power $dE_{\rm rad}/dt$ must be an invariant or scalar. But we have already derived the form for this scalar in the non-relativistic limit; this is the dipole or lowest-order term (2.53), which is, of course, contained within the rate (2.112) when $\beta \ll 1$. It is always possible to consider a Lorentz frame where the motion is, at a particular time, non-relativistic. In this limit, the radiative energy loss is

$$-dE/dt = (2q^2/3m^2c^3)\dot{p}_j\dot{p}_j, \tag{2.113}$$

with a sum over the three spatial indices j. Since dE/dt (for this particular problem) is an invariant, the exact relativistic expression must be a scalar function of the particle's kinematic quantities. It is easy to guess the appropriate generalization, the validity of which is established by its agreement with the result (2.113) in a Lorentz frame where the motion is non-relativistic. The relativistic generalization is obtained through the replacement

$$\dot{p}_j\dot{p}_j \rightarrow (dp_\mu/d\tau)(dp_\mu/d\tau), \tag{2.114}$$

and it remains to show that the covariant form is identical to the non-covariant expression (2.112). Since $d\tau = dt/\gamma$, and $p_\mu = \gamma mc(i, \boldsymbol{\beta})$, we have

$$(dp_\mu/d\tau)(dp_\mu/d\tau) = m^2c^2\gamma^2\big[(\dot{\gamma}\boldsymbol{\beta} + \gamma\dot{\boldsymbol{\beta}})^2 - \dot{\gamma}^2\big]. \tag{2.115}$$

But $\dot{\gamma} = \gamma^3\boldsymbol{\beta}\cdot\dot{\boldsymbol{\beta}}$, $\dot{\beta}^2 - (\boldsymbol{\beta}\cdot\dot{\boldsymbol{\beta}})^2 = (\boldsymbol{\beta}\times\dot{\boldsymbol{\beta}})^2$, and elementary algebraic manipulations yield the form (2.112).

This very simple derivation illustrates again the very considerable power of covariance considerations. For example, the non-relativistic expression (2.113) contains only the dipole contribution, while the relativistic result includes contributions from many multipoles. Yet, the exact formula has been obtained from its limiting form without having to employ the formulation of radiation theory for relativistic particles. The relativistic result (2.112) is, in addition, of great significance in exhibiting important features of the radiation phenomenon. In particular, the factor γ^6 indicates the increased *efficiency* of the radiation phenomenon for relativistic particles. This is especially so for the case where the accelerating force is perpendicular to the particle velocity. The force is related to the acceleration by

$$\boldsymbol{F} = mc\big[\gamma^3(\boldsymbol{\beta}\cdot\dot{\boldsymbol{\beta}})\boldsymbol{\beta} + \gamma\dot{\boldsymbol{\beta}}\big]. \tag{2.116}$$

---

[11]It is really not necessary to introduce the photon concept here, but it is convenient.

When $\boldsymbol{F}$ is along $\boldsymbol{\beta}$ and $\gamma \gg 1$, $\dot{\boldsymbol{\beta}} \to F/mc\gamma^3$ and the radiated power approaches

$$P_{\parallel} \to (2q^2/3m^2c^3)F^2, \tag{2.117}$$

which is independent of energy. On the other hand, if $\boldsymbol{F}$ is perpendicular to $\boldsymbol{\beta}$, we see from Equation (2.116) that $\dot{\beta} \to F/mc\gamma$, and

$$P_{\perp} \to (2q^2/3m^2c^3)F^2\gamma^2, \tag{2.118}$$

increasing as the square of the energy. Comparison of the two results shows why, for example, in the field of high-energy physics, radiative losses are far more important in synchrotrons than in linear accelerators (linacs). The comparison is relevant for astrophysics, where nature has provided both high-energy particles and magnetic fields, which provide the (perpendicular) deflecting force. The associated "synchrotron radiation" from relativistic electrons in cosmic radio sources is produced, as we see, by a very efficient mechanism.

2.6 RADIATION REACTION

The radiation of energy by an accelerated charge has an effect on the motion of the particle that can be described in terms of a radiation reaction force. The form of the expression for this force can be established through considerations of energy conservation. However, a more revealing derivation of the result involves a calculation of the force from various elements of the charge acting on one another. Because of the effects of retardation, the total force is not zero and there results a total "self-force." Treatment of the phenomenon also involves the evaluation of the charge's "self-energy," and, in this simple classical problem, we can introduce the ideas of "renormalization." A full treatment of the problem is clearly beyond the scope of this book. We only touch on the subject, first in the non-relativistic limit—or in a reference frame where the motion is such that $\beta \ll 1$.

2.6.1 Non-Relativistic Limit

Radiation reaction was considered by Lorentz, who first introduced the electron into the subject of electromagnetic theory. For definiteness in the formulation, our charged particle is referred to as an electron, although the classical description would be the same for any particle. We consider a slowly moving electron of finite size and compute the reaction force that results from the interaction of the radiation field from different parts of the electron acting on other parts (see Figure 2.4). Let de and de' be two elements of charge on the electron, located at $\boldsymbol{r}$ and $\boldsymbol{r}'$, respectively. The reaction force is computed from

$$\boldsymbol{F} = \iint de\, d\boldsymbol{E}, \tag{2.119}$$

where $d\boldsymbol{E}$ is the differential electric field at $\boldsymbol{r}$ due to the charge element de' at the source point $\boldsymbol{r}'$. The double integral is then over the elements de and de'.

It is convenient to evaluate the radiation reaction force in an inertial frame in which the electron is instantaneously at rest. The radiation-reaction force terms are

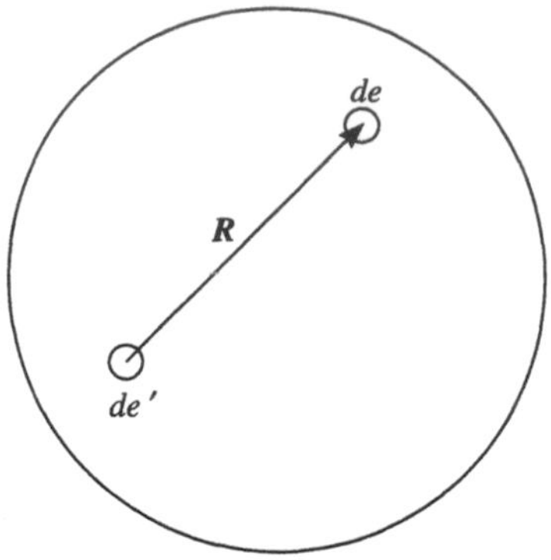

Figure 2.4 Infinitesimal charge elements associated with the same charge distribution.

proportional to $\dot{\boldsymbol{v}}$ and to higher derivatives of the velocity, and these are the same in the lab frame as in the instantaneous rest frame. The integration over de in the double integral (2.119) is instantaneous, but the differential field $d\boldsymbol{E}$ at de is due to the motion of the element de' at the retarded time at which the velocity is not zero. In terms of the source-to-field point distance $\boldsymbol{R} = \boldsymbol{r} - \boldsymbol{r}'$, $\boldsymbol{n} = \boldsymbol{R}/R$, and $\varkappa = 1 - \boldsymbol{n} \cdot \boldsymbol{\beta}$, the differential field $d\boldsymbol{E}$ due to de' would be given by the expression (2.102):

$$d\boldsymbol{E} = \frac{de'}{R^3 \varkappa^3}\left(1 - \frac{v^2(t')}{c^2}\right)\left(\boldsymbol{R} - \frac{R}{c}\boldsymbol{v}(t')\right) + \frac{1}{c^2}\boldsymbol{R} \times \left[\left(\boldsymbol{R} - \frac{R}{c}\boldsymbol{v}(t')\right) \times \dot{\boldsymbol{v}}(t')\right]. \tag{2.120}$$

Note that all quantities on the right are retarded values; that is, $\dot{\boldsymbol{v}}(t') = d\boldsymbol{v}(t')/dt'$, and $\varkappa = 1 - \boldsymbol{n} \cdot \boldsymbol{v}(t')/c$. The retarded time is $t' = t - R/c$, where at the present time the velocity $\boldsymbol{v}(t)$ is zero. We then express the (retarded) velocity and acceleration as expansions in terms of quantities at the time t:

$$\begin{aligned} \boldsymbol{v}(t') &= -(R/c)\dot{\boldsymbol{v}}(t) + \frac{1}{2}(R/c)^2\ddot{\boldsymbol{v}}(t) + \cdots, \\ \dot{\boldsymbol{v}}(t') &= \dot{\boldsymbol{v}}(t) - (R/c)\ddot{\boldsymbol{v}}(t) + \cdots. \end{aligned} \tag{2.121}$$

In the expression for $d\boldsymbol{E}$ we neglect higher-derivative terms ($\dddot{\boldsymbol{v}}$, etc.) and consider only terms linear in $\boldsymbol{v}$, $\dot{\boldsymbol{v}}$, and $\ddot{\boldsymbol{v}}$, neglecting terms in v^2, $v\dot{v}$, etc.; terms having coefficients c^{-p} with $p > 3$ are also ignored. Then

$$\begin{aligned} \varkappa &= 1 + \boldsymbol{R} \cdot \dot{\boldsymbol{v}}/c^2 - (R/2c^3)\boldsymbol{R} \cdot \ddot{\boldsymbol{v}} + \cdots, \\ \varkappa^{-3} &= 1 - 3\boldsymbol{R} \cdot \dot{\boldsymbol{v}}/c^2 + (3R/2c^3)\boldsymbol{R} \cdot \ddot{\boldsymbol{v}} + \cdots, \end{aligned} \tag{2.122}$$

where $\dot{\boldsymbol{v}}$ and $\ddot{\boldsymbol{v}}$ are the values at time t. The differential electric field is, neglecting the higher-order terms in Equation (2.120),

$$d\boldsymbol{E} = de'\left[-\frac{2\boldsymbol{R}(\dot{\boldsymbol{v}} \cdot \boldsymbol{R})}{R^3 c^2} + \frac{\boldsymbol{R}(\boldsymbol{R} \cdot \ddot{\boldsymbol{v}})}{2R^2 c^3} + \frac{\boldsymbol{R}}{R^3} + \frac{\ddot{\boldsymbol{v}}}{2c^3} + \cdots\right]. \tag{2.123}$$

For component k of the field, summing over dummy index j, the result can be written

$$dE_k = de'\left[-\frac{2R_k\dot{v}_jR_j}{R^3c^2} + \frac{R_kR_j\ddot{v}_j}{2R^2c^3} + \frac{R_k}{R^3} + \frac{\ddot{v}_k}{2c^3} + \cdots\right]. \tag{2.124}$$

The total force on the electron is found from Equation (2.119) by integrating over de' and de. For a spherically symmetric distribution,

$$\begin{aligned} &\iint de\,de'\,f(R)R_k = 0, \\ &\iint de\,de'\,f(R)R_kR_j = \frac{1}{3}\delta_{kj}\iint de\,de'\,f(R)R^2, \end{aligned} \tag{2.125}$$

and we have

$$\boldsymbol{F} = \frac{2e^2}{3c^3}\ddot{\boldsymbol{v}} - \frac{4W_{\rm el}}{3c^2}\dot{\boldsymbol{v}} + \cdots, \tag{2.126}$$

where

$$W_{\rm el} = \frac{1}{2}\iint de\,de'/R \tag{2.127}$$

is the electrostatic self-energy. The result can be written

$$\boldsymbol{F} = \boldsymbol{F}_{\rm rr} - m_{\rm el}\dot{\boldsymbol{v}}, \tag{2.128}$$

where the radiation reaction $\boldsymbol{F}_{\rm rr}$ is given by the first term on the right-hand side of Equation (2.126), and $m_{\rm el} = 4W_{\rm el}/3c^2$ is the "electromagentic mass." Note the factor $4/3$; it was suggested long ago by Poincaré that there should be non-electromagnetic binding forces providing a "glue" to stabilize the electron, and that the associated binding energy is $\frac{1}{3}W_{\rm el}$. In fact, if there are external forces acting on the electron, it is appropriate to write the equation of motion in the form

$$m_0\dot{\boldsymbol{v}} - (2e^2/3c^3)\ddot{\boldsymbol{v}} = \boldsymbol{F}_{\rm ext}, \tag{2.129}$$

where the electromagnetic and mechanical inertia terms are combined into an "observed" mass m_0, which can be determined experimentally. Moreover, with m_0 determined empirically, it is then not necessary to know what percentage of it is electromagnetic in origin. In fact, for a point particle, $W_{\rm el}$ diverges. There are higher terms in $\boldsymbol{F}$ that would result from, for example, the next-order terms in the expansions (2.121), and these would be of the order $F_{\rm rr}(\dddot{\boldsymbol{v}}/\ddot{\boldsymbol{v}})r_0/c$, where r_0 is the electron radius. For a point particle these terms go to zero, but, of course, $W_{\rm el}$ is infinite as $r_0 \to 0$. This is what classical electron theory in this elementary formulation has to live with. On the other hand, the expression for $\boldsymbol{F}_{\rm rr}$ is independent of the structure of the electron.

The combining of electromagnetic mass into a total observed mass is an essential part of what is known as "renormalization" in modern quantum electrodynamics. The idea goes back to J. J. Thomson in classical electron theory. It is clear from elementary considerations that quantum mechanics must be brought into the problem. There are still divergences in quantum electrodynamics, although they are not as severe (logarithmic). Nevertheless, quantum electrodynamics does treat the electron as a point particle and these divergences are there, so there is still is a certain amount of, as Feynman used to say, "sweeping the dirt under the rug." We seem to be able to endure these difficulties, however.

There is an extensive literature on electron theory—even on the classical theory and in recent decades. Suggested references are listed at the end of this chapter.[12]

2.6.2 Relativistic Theory: Lorentz-Dirac Equation

The radiation-reaction force is a real effect whose magnitude is increased when charged particles experience rapid changes in their acceleration. It is then of interest to obtain a relativistic generalization of the non-relativistic equation of motion including radiation reaction. If "mass renormalization" has already been carried out, the relativistic generalization of Equation (2.129) should be of the form

$$m_0\, dv_\mu/d\tau = K_\mu + \Gamma_\mu, \tag{2.130}$$

where m_0 is the observed mass, K_μ is some external four-vector force, and Γ_μ is the term due to radiation reaction. The equation of motion (2.130) is sometimes called the Lorentz-Dirac equation and Γ_μ is referred to as the Abraham radiation-reaction four-vector.

It is not hard to obtain the expression for Γ_μ. The most obvious try would be to set it equal to $(2e^2/3c^3)d^2v_\mu/d\tau^2$, whose space component does reduce to non-relativistic form $\boldsymbol{F}_{\rm rr}$ is that limit. However, such an expression for Γ_μ does not satisfy the relation [see Equation (1.65)]

$$\Gamma_\mu v_\mu = 0, \tag{2.131}$$

which is a conservation equation that all forces must satisfy; that is, in general, $v_\mu d^2v_\mu/d\tau^2$ is not identically zero. The next step would be to try the form[13]

$$\Gamma_\mu = (2e^2/3c^3)(d^2v_\mu/d\tau^2 + Sv_\mu), \tag{2.132}$$

where S is a scalar function. This scalar can be established by imposing the condition (2.131), which leads immediately to the result

$$S = \frac{1}{c^2}v_\mu\frac{d^2v_\mu}{d\tau^2} = \frac{1}{c^2}\left[\frac{d}{d\tau}\left(v_\mu\frac{dv_\mu}{d\tau}\right) - \frac{dv_\mu}{d\tau}\frac{dv_\mu}{d\tau}\right] = -\frac{1}{c^2}a_\mu a_\mu, \tag{2.133}$$

since $v_\mu v_\mu = -c^2 =$ constant, and $a_\mu = dv_\mu/d\tau$. The solutions to and properties of the Lorentz-Dirac equation have been studied extensively (see Footnote 12). We shall not go further into this topic, since it is somewhat outside the principal program of this text.

---

[12]A thorough discussion of, in particular, classical electron theory is given in the scholarly book *Classical Charged Particles* by F. Rohrlich (Reading, MA: Addison-Wesley Publ. Co., Inc., 1965). Formulations that can avoid divergence problems are described and an extensive collection of relevant papers are cited, including those published early in the last century.

[13]We do not try an expression with a term proportional to $dv_\mu/d\tau$ because this would be contained within the left-hand side of Equation (2.130).

2.7 SOFT-PHOTON EMISSION

Photons are produced whenever charged particles suffer a change in velocity. There is also photon production accompanying the production of a charged particle, as in, for example, β-decay ($n \to p + e^- + \overline{\nu}_e$ for bound or free neutrons or for protons bound in a nucleus $p \to n + e^+ + \nu_e$).[14] In this case, the electron or positron can be considered to be accelerated instantaneously from rest to a velocity v, and the overall radiative process corresponds to the production of a photon accompanying the other particles in the final state. When the photon has a very low energy, the description of the process is simplified in that usually (but not always) we can derive a photon-emission probability that does not depend on the details of the associated radiationless process. This is because the photons are "soft," and their emission does not disturb the accompanying (radiationless) process that causes the particle acceleration. The formulas for the soft-photon emission probability can be derived by classical electrodynamics, as is done in this section. In Chapter 3, the same formulas will be derived by quantum electrodynamics. By employing the two different approaches, we learn more about the range of validity of the resulting expressions. The formulas are of great value because of their generality, and they allow a convenient calculation of certain important processes like, in particular, bremsstrahlung. Also, although they are restricted to the soft-photon limit, the formulas allow simple estimates for photon-producing processes at general photon energies.

We consider emission by non-relativistic and relativistic particles, and derive some useful expressions by means of the purely classical formulation of the chapter. The formulas will be derived again in the following chapter wherein their range of validity will be discussed further. Chapter 3 will also consider the effects of photon production connected with interactions with particles' intrinsic magnetic moments.

2.7.1 Multipole Formulation

For a system of particles in non-relativistic motion, the total rate of radiation of energy is given by the result (2.52), which can be written

$$dW/dt = \sum_M C_M |M(t)|^2, \tag{2.134}$$

where $M(t)$ is some moment (or, rather, its time derivative), and C_M is a numerical coefficient. The values of C_M and $M(t)$ are given in Equations (2.53) and (2.59) for the first few terms of the multipole expansion (2.134). Further, we can introduce the moment's Fourier transform M_ω by writing

$$M(t) = \int M_\omega e^{-i\omega t} d\omega, \tag{2.135}$$

with

$$M_\omega = (2\pi)^{-1} \int M(t) e^{i\omega t} dt. \tag{2.136}$$

[14]Here the photon "coupling" (see Chapter 3) or emission is associated with the production of the e^- or e^+ rather than with the proton. The e^- and e^+, together with the neutrino, carry away most of the energy and their velocities are much larger than that of the more massive proton.

Then, including only positive frequencies in the emission spectrum $dW/d\omega$ and introducing the photon concept by means of the relation (2.72), we have

$$dw_\omega = (4\pi/\hbar)(d\omega/\omega)\sum_M C_M|M(\omega)|^2 \tag{2.137}$$

for the photon-emission probability (within $d\omega$).

If the individual moment can be written (as it can) as a time derivative

$$M(t) = \dot{\mu}_M \tag{2.138}$$

in terms of some related moment $\mu(t)$, in the soft-photon limit $M_\omega \to \Delta\mu_M/2\pi$, where $\Delta\mu_M$ is the change in μ_M. Then, in this limit, the result (2.137) becomes

$$dw_\omega \to (\pi\hbar)^{-1}(d\omega/\omega)\sum_M C_M|\Delta\mu_M|^2. \tag{2.139}$$

This is a very general formula and certain features of it should be emphasized. First, note that the quantities $\Delta\mu_M$ are independent of ω and so the factor $d\omega/\omega$ exhibits the infrared divergence effect. That is, there is always a factor of this precise form, independent of whether the lowest-order emission is electric or magnetic dipole or quadrupole radiation or from higher multipole radiation. Basically, the fundamental assumption that yields this result is that the system of charges can exist in a continuum of states rather than in a quantized spectrum. This will be true for a system of free particles but not for a system in bound states. It should also be noted how the contributions from the various moments contribute additively with the appropriate coefficients C_M; this is the case when dw_ω represents the probability integrated over angles for the outgoing photon. The coefficients are the same for the x, y, and z components of the dipole contributions and for each contribution from the quadrupole tensor. The probability dw_ω is then determined by the combination of quantities $\Delta\mu_M$.

2.7.2 Dipole Formula

Let us now obtain an important result for soft-photon emission by a single charge ze in non-relativistic motion. In this case, the emission is dipole radiation, and, instead of employing the expressions given above, we refer back to the more general formula (2.73) to exhibit the angular distribution as well. In Equation (2.73) we employ[15] the magnetic-field term with

$$\begin{aligned} B_\omega &= (2\pi)^{-1}\int |\text{curl } \boldsymbol{A}|e^{i\omega t}\,dt \\ &= (2\pi c)^{-1}\int |\dot{\boldsymbol{A}}\times\boldsymbol{n}|e^{i\omega t}\,dt. \end{aligned} \tag{2.140}$$

[15]Although we are computing a radiation field, we do not evaluate its intensity from the electric-field magnitude derived from $-\dot{A}/c$; that is, we are not employing the gauge (2.9) and (2.10). This is because we are deriving the fields from the Liénard-Wiechert potentials, which do not satisfy that gauge condition. With the fields derived from the L-W potentials, it is simpler to employ the relation (2.10) for the magnetic component of the radiation field, since no gauge is specified therein. These remarks were also made in Section 2.2 and are relevant to the upcoming Section 2.7.3.

In the low-frequency limit this becomes

$$B_\omega \to (2\pi c)^{-1}|\Delta \boldsymbol{A} \times \boldsymbol{n}|, \tag{2.141}$$

where ΔA is the change in the vector potential associated with the charge motion; in the non-relativistic limit, ΔA approaches $(ze/R)\Delta\boldsymbol{\beta}$, where $c\Delta\boldsymbol{\beta}$ is the velocity change. Equation (2.73) then gives

$$dw_\omega = (\alpha/4\pi^2)(d\omega/\omega)z^2|\Delta\boldsymbol{\beta}\times\boldsymbol{n}|^2 d\Omega \tag{2.142}$$

for the soft-photon emission probability within frequency $d\omega$ and in the direction $\boldsymbol{n}$ within the solid angle $d\Omega$. We see that the result is proportional to the fine-structure constant α and is determined by $\Delta\boldsymbol{\beta}$. Integrating over $d\Omega$ we get the total dw_ω (if we are not interested in the direction of the outgoing soft photon):

$$dw_\omega = (2\alpha/3\pi)z^2(\Delta\boldsymbol{\beta})^2 d\omega/\omega. \tag{2.143}$$

The dipole result (2.142) can be expressed in a form that is even more general. Both results (2.142) and (2.143) correspond to the photon production probability summed over polarization states. However, the dw per polarization state can be seen from the result (2.142) by rewriting the factor $|\Delta\boldsymbol{\beta}\times\boldsymbol{n}|^2$. The two photon (linear) polarization states can be described by two unit vectors $\boldsymbol{\varepsilon}_a$ and $\boldsymbol{\varepsilon}_b$ in mutually perpendicular directions perpendicular to $\boldsymbol{n} = \boldsymbol{k}/k$; that is, if $\boldsymbol{\varepsilon}$ is a general polarization state, $\boldsymbol{\varepsilon}\cdot\boldsymbol{n} = 0$. Then, if $\boldsymbol{q}$ is any vector [see Equation (2.54)],

$$(\boldsymbol{q}\times\boldsymbol{n})^2 = q^2 - (\boldsymbol{q}\cdot\boldsymbol{n})^2 = \sum_{\boldsymbol{\varepsilon}}(\boldsymbol{\varepsilon}\cdot\boldsymbol{q})^2. \tag{2.144}$$

Another way of obtaining the result (2.144) makes use of the relation $\boldsymbol{n} = \boldsymbol{\varepsilon}_a\times\boldsymbol{\varepsilon}_b$; since $\boldsymbol{\varepsilon}_a\cdot\boldsymbol{\varepsilon}_b = 0$, the elementary vector identity $\boldsymbol{A}\times(\boldsymbol{B}\times\boldsymbol{C}) = (\boldsymbol{A}\cdot\boldsymbol{C})\boldsymbol{B} - (\boldsymbol{A}\cdot\boldsymbol{B})\boldsymbol{C}$ yields $\left(\boldsymbol{q}\times(\boldsymbol{\varepsilon}_a\times\boldsymbol{\varepsilon}_b)\right)^2 = (\boldsymbol{q}\cdot\boldsymbol{\varepsilon}_a)^2 + (\boldsymbol{q}\cdot\boldsymbol{\varepsilon}_b)^2$. That is, from the formula (2.142) we can identify the expression for the probability per photon state for emission in the direction $\boldsymbol{n}$ desginated by the element $d\Omega$:

$$dw_\omega = (\alpha/4\pi^2)z^2(\boldsymbol{\varepsilon}\cdot\Delta\boldsymbol{\beta})^2(d\omega/\omega)d\Omega. \tag{2.145}$$

2.7.3 Emission from Relativistic Particles

For the more general case of emission by relativistic particles it is also convenient to employ the result (2.73) with B_ω given by Equation (2.141). Now we use the relativistic expression (2.84) for the vector potential, and for a system of particles, readily obtain the general formula

$$dw_\omega = \alpha|\boldsymbol{\Delta}_\beta|^2(d\omega/\omega)d\Omega \tag{2.146}$$

for the soft-photon-emission probability. Here

$$\boldsymbol{\Delta}_\beta = (2\pi)^{-1}\left[\sum_k \Delta(z\boldsymbol{\beta}/\varkappa)_k\right]\times\boldsymbol{n}, \tag{2.147}$$

the sum being over the charged particles, and the soft-photon-emission probability is determined by the particles' velocities and changes in velocities. Further, we again obtain the result that dw_ω is proportional to $d\omega/\omega$ in this very general case.

Even for the case of a single particle, integration over $d\Omega$ does not yield a simple result for arbitrary initial and final velocities. We give expressions for dw_ω only for two special, but important, cases. First, consider the case where a single particle is accelerated from rest to (or suddenly produced at) velocity $\boldsymbol{\beta}c$. Then

$$dw_\omega = \frac{\alpha}{4\pi^2} z^2 \beta^2 \frac{d\omega}{\omega} \frac{\sin^2\theta}{(1-\beta\cos\theta)^2} d\Omega, \tag{2.148}$$

where θ is the angle between $\boldsymbol{\beta}$ and $\boldsymbol{n}$. The integration over $d\Omega$ $(= 2\pi \sin\theta d\theta)$ is elementary and yields the total probability

$$dw_\omega = \frac{\alpha}{\pi} z^2 \left(\frac{1}{\beta} \ln \frac{1+\beta}{1-\beta} - 2 \right) \frac{d\omega}{\omega}, \tag{2.149}$$

which is consistent with the dipole formula (2.143) when $\beta \ll 1$.

The second important example of emission from relativistic particles for which a simple expression for dw_ω results is the case where a single highly relativistic particle suffers small-angle elastic scattering. Then, since we are considering soft-photon emission, β is close to unity before and after scattering, $\Delta\beta$ is small and $\boldsymbol{\beta}\cdot\Delta\boldsymbol{\beta}$ is negligible. Then, with $\boldsymbol{f}(\boldsymbol{\beta}) = (1-\boldsymbol{\beta}\cdot\boldsymbol{n})^{-1}\boldsymbol{\beta}$, and $\Delta\boldsymbol{f} = (\partial\boldsymbol{f}/\partial\boldsymbol{\beta})\cdot\Delta\boldsymbol{\beta}$,

$$dw_\omega = (4\pi^2)^{-1}\alpha z^2 (d\omega/\omega) \int (\Delta\boldsymbol{f} \times \boldsymbol{n})^2 d\Omega. \tag{2.150}$$

The evaluation of the integrals[16] of the various terms here is elementary. The algebra is a little tedious, but the result is simple:

$$dw_\omega = (2\alpha/3\pi) z^2 \gamma^2 (\Delta\boldsymbol{\beta})^2 (d\omega/\omega), \tag{2.151}$$

that is, just $\gamma^2 = (1-\beta^2)^{-1}$ times the non-relativistic expression (2.143). It should be emphasized again that this formula holds only in the limit of small-angle scattering and for $\gamma \gg 1$. The formula is useful, however, and will be applied in a later chapter on bremsstrahlung.

Formula (2.151) can be derived in a much simpler way by making use of the non-relativistic expression (2.143). We consider the process in the lab frame (K) where the initial velocity is along, say, the x-axis and in a frame (K') moving in the same direction such that in this frame the particle motion is non-relativistic. Since dw_ω is a probability (a number), it must be an invariant:

$$dw_\omega(\boldsymbol{v}, \Delta\boldsymbol{v}) = dw'_{\omega'}(\boldsymbol{v}', \Delta\boldsymbol{v}'). \tag{2.152}$$

For the right-hand side we can employ the result (2.143). Since $d\omega'/\omega' = d\omega/\omega$, and $\Delta\boldsymbol{v}'$ is in, say, the y-direction, we have only to perform an elementary Lorentz transformation of $\Delta\beta'_y$:

$$\Delta\beta'_y = \Delta\beta_y/\gamma(1-\beta\beta_x). \tag{2.153}$$

But $\beta_x \to \beta$, so $\Delta\beta'_y \to \gamma\Delta\beta_y$ and the result (2.151) is readily obtained.

Finally, let us rewrite and generalize some of the relativistic expressions given here, and again make use of covariance arguments to show the most general formula can be obtained easily from the corresponding expression in the non-relativistic

[16] They are similar to those involved in the derivation of the result (2.112).

limit. We express the formulas in terms of the photon momentum $\boldsymbol{k}$ rather than ω ($= |\boldsymbol{k}|c$), being part of the four-vector $k_\mu = (ik, \boldsymbol{k})$. Also, we let p_μ and p'_μ be, respectively, the initial and final four-momenta of the charged particle. A photon polarization four-vector $\varepsilon_\mu = (\varepsilon_0, \boldsymbol{\varepsilon})$ is introduced, with a gauge chosen such that $\varepsilon_0 = 0$. Then the invariant[17]

$$k \cdot p \equiv k_\mu p_\mu = -\gamma mck\varkappa \tag{2.154}$$

can be used to write

$$\frac{\boldsymbol{\beta}}{\varkappa} = -k\frac{\boldsymbol{p}}{k \cdot p}. \tag{2.155}$$

The factor $(d\omega/\omega)d\Omega$ is expressed in terms of the invariant $d^3\boldsymbol{k}/k$:

$$\frac{dk}{k}d\Omega = \frac{1}{k^2}\frac{d^3\boldsymbol{k}}{k} \tag{2.156}$$

In the gauge with $\varepsilon_0 = 0$,

$$\varepsilon_\mu p_\mu = \varepsilon \cdot p = \boldsymbol{\varepsilon} \cdot \boldsymbol{p}, \tag{2.157}$$

and the formula (2.146) becomes [see Equation (2.144)]

$$dw_\omega = \frac{z^2\alpha}{4\pi^2}\sum_{\boldsymbol{\varepsilon}}\left|\Delta\left(\frac{\varepsilon \cdot p}{k \cdot p}\right)\right|^2\frac{d^3\boldsymbol{k}}{k}. \tag{2.158}$$

That is, per polarization state, the generalization of the non-relativistic expression (2.145) is

$$dw_\omega = \frac{z^2\alpha}{4\pi^2}\left|\frac{\varepsilon \cdot p}{k \cdot p} - \frac{\varepsilon \cdot p'}{k \cdot p'}\right|^2\frac{d^3\boldsymbol{k}}{k}. \tag{2.159}$$

The expression is manifestly covariant, involving factors that are Lorentz invariants. This is to be expected, since dw_ω should be invariant, as has already been noted [see Equation (2.152)]. Actually, the general expression (2.159) could be obtained *directly* from the formula (2.145) by rewriting the latter in terms of factors that are manifestly covariant and that reduce to the non-relativistic factors in that limit. It is not difficult to do this; the $\boldsymbol{\varepsilon} \cdot \boldsymbol{\beta}$ term is replaced by the invariant $\varepsilon \cdot p$ and the $1/k$ inside the square is replaced by the form $(k \cdot p)^{-1}$ [see Equation (2.155)]. In the next chapter, in Section 3.5, all of the soft-photon formulas derived here will be derived in a quantum-mechanical treatment.

2.8 WEIZSÄCKER-WILLIAMS METHOD

The idea for this method was first introduced by Fermi[18] in 1924 and was developed more fully ten years later by von Weizsäcker and especially by Williams. Sometimes the procedure is referred to, in a more descriptive way, as the "Method of Virtual

[17]For a four-dimensional dot product we do not use boldface symbols, to distinguish it from the three-dimensional case.

[18]*Z. Phys.* **29**, 315 (1924).

Quanta," and, more recently, the designation "Equivalent Photon Method" (e.p.m.) has been used. We stick with the older name, employing, for brevity, simply "W-W."

The W-W method is really quite powerful. It allows the calculation, by very simple means, of certain proceses that otherwise would be extremely difficult to evaluate. Generally, it is the cross section for some process that is computed, and the method allows its evaluation to a relative accuracy $\sim (\ln N)^{-1}$, where N is some large number. The actual application of the method for various specific processes will be deferred to later chapters. Applications will show, for example, how it allows an alternative derivation for a process and how it always provides more insight into the nature of the process. Here the foundations for the development of the procedure will be given and the basic assumptions involved will be discussed, along with the limitations of the method. We also try to indicate basically why it works so well in general, although this will only be fully clarified later when the specific applications are outlined.

2.8.1 Fields of a Moving Charge

In the W-W method the effects of the fields on a moving charge q on a target system T are described in terms of an "equivalent" flux of photons. The charge is incident on T at an impact parameter b with a velocity $\boldsymbol{v}$, and to describe the fields it is convenient to introduce two reference frames K and K' with x- and x'-axes oriented along $\boldsymbol{v}$, with q moving at the origin of K' (see Figure 2.5). It is assumed that the charge is moving fast enough that the interaction with T does not cause an appreciable deviation from a straight line path. The system T is located at a distance b along the y-axis of K, and experiences fields from q that are time variable. The equivalent flux of photons incident on T then has a frequency spectrum that is determined by the details of this time dependence.

In the frame K' there is only an electric field with components at T equal to $\boldsymbol{E}' = (q/r'^3)(x'_1, b, 0)$; here x'_1 is the coordinate of T in K', and $r'^2 = b^2 + x'^2_1$. With K and K' coinciding at $t' = 0$, $x'_1 = -vt'$. What are needed are the fields in K at the point $(0, b, 0)$ where the target is located. The time coordinate t at this point is gotten from t' by the elementary Lorentz transformation $t' = \gamma(t - vx_1/c^2) = \gamma t$ (since $x_1 = 0$), so that we can write $x'_1 = -\gamma vt$. The electromagnetic fields in K are found from the tensor transformation [see Equations (1.28), (1.73)]

$$F'_{\rho\lambda} = a_{\rho\mu} a_{\lambda\nu} F_{\mu\nu}, \tag{2.160}$$

where

$$F_{\mu\nu} = \begin{pmatrix} 0 & iE_1 & iE_2 & iE_3 \\ -iE_1 & 0 & B_3 & -B_2 \\ -iE_2 & -B_3 & 0 & B_1 \\ -iE_3 & B_2 & -B_1 & 0 \end{pmatrix} \tag{2.161}$$

is the electromagnetic field tensor. As in this problem, when the relative motion is along the x- and x'-axes, the transformation coefficients are given by Equations (1.23)–(1.25). Then, the transformations (2.152) for individual components

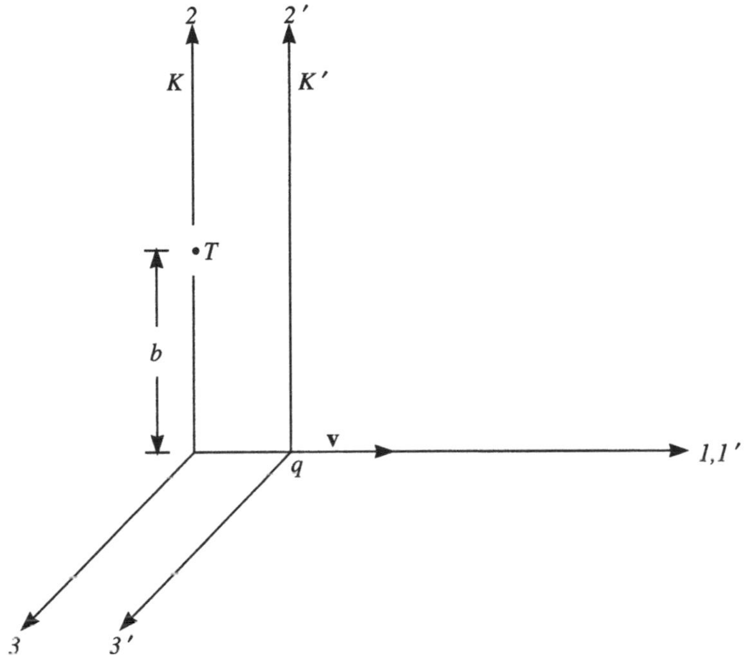

Figure 2.5 Moving charge and "target" reference frames.

are found to be

$$\begin{aligned}
E_1' &= E_1, \\
E_2' &= \gamma(E_2 - \beta B_3), \\
E_3' &= \gamma(E_3 + \beta B_2), \\
B_1' &= B_1, \\
B_2' &= \gamma(B_2 + \beta E_3), \\
B_3' &= \gamma(B_3 - \beta E_2).
\end{aligned} \tag{2.162}$$

Actually, in our problem we have to go from the frame K' to K, but those transformations are the same as the ones (2.162) with the sign of β changed and with primes transferred to the unprimed fields. Of course, there are only the fields E_1' and E_2' in K'. The only fields in K are then

$$\begin{aligned}
E_1 = E_1' &= -\frac{q\gamma vt}{(b^2 + \gamma^2 v^2 t^2)^{3/2}}, \\
E_2 = \gamma E_2' &= \frac{q\gamma b}{(b^2 + \gamma^2 v^2 t^2)^{3/2}}, \\
B_3 = \beta\gamma E_2' &= \beta E_2,
\end{aligned} \tag{2.163}$$

now expressed in terms of the K-frame variable t. These are the fields experienced by the target system.

2.8.2 Equivalent Photon Fluxes

If the target system consists of charges in non-relativistic motion,[19] magnetic forces are negligible and the principal perturbation is from the electric fields E_1 and E_2. Then, if there were, in addition to E_1 and E_2, two fictitious[20] magnetic fields $(B_3)_a = E_2$ and $(B_3)_b = -E_1$, the perturbation would be essentially the same. However, now the perturbation can be described in terms of two radiation pulses with Poynting vectors $S_1 = (c/4\pi)E_2^2$ and $S_2 = (c/4\pi)E_1^2$ striking the target in two directions. These pulses can be considered to consist of photons of linear polarization, with a spectrum of frequencies determined by the time dependence of E_2 and E_1. In terms of the Fourier transforms of these fields, the spectral distributions of the equivalent photon fluxes will be [see Equation (2.70)]

$$\begin{aligned} dJ_1/d\omega &= dN_1/dA\,d\omega = (c/\hbar\omega)|E_{2\omega}|^2, \\ dJ_2/d\omega &= dN_2/dA\,d\omega = (c/\hbar\omega)|E_{1\omega}|^2. \end{aligned} \tag{2.164}$$

Specifically, the Fourier amplitudes are given by the integrals

$$E_{2\omega} = \frac{q}{2\pi bv}\int_{-\infty}^{\infty} \frac{e^{ia\xi}}{(1+\xi^2)^{3/2}}d\xi, \tag{2.165}$$

$$E_{1\omega} = -\frac{q}{2\pi bv\gamma}\int_{-\infty}^{\infty} \frac{\xi e^{ia\xi}}{(1+\xi^2)^{3/2}}d\xi, \tag{2.166}$$

where the obvious variable change $\xi = \gamma vt/b$ has been made, and the dimensionless parameter a is given by

$$a = \omega b/\gamma v. \tag{2.167}$$

All of the frequency dependence is contained in a. Moreover, the main contributions to the integrals (2.165) and (2.166) come from $|\xi| \lesssim 1$, so that for $a \gg 1$ the integrands are oscillatory and the integrals are small. The characteristic frequency of the equivalent photon fluxes is then

$$\omega_c \sim \gamma b/v, \tag{2.168}$$

and this is also the effective maximum frequency of both of the distributions (2.164). The integrals (2.165) and (2.166) are actually representations of modified Bessel functions, so that they can be evaluated from tables for any value of the parameter a. However, only the asymptotic forms will be important for our considerations, both as to the foundations of the method as well as for most (but not all) the applications.

It is the asymptotic form at low frequency (small a) that is most relevant. The Fourier amplitudes in this limit may be obtained through consideration of the integrals (2.165) and (2.166) for $a \ll 1$. The first integral is very simple and we have

$$E_{2\omega} \to q/\pi bv. \tag{2.169}$$

[19] It will be shown below that this assumption can be relaxed, so that the W-W method can be employed when there are relativistic motions in the target system.

[20] Actually, as $\beta \to 1$, the field $B_3 = E_2$ really *is* present [see Equations (2.163)], so that in this limit $(B_3)_a$ is not fictitious.

In the second integral the asymptotic form can be obtained in good approximation by setting $e^{ia\xi} = 1 + ia\xi + \cdots$ and by applying effective cutoffs to the integral, giving the approximate result[21]

$$2\int_1^{1/a} d\xi/\xi = 2\ln(1/a), \tag{2.170}$$

$$E_{1\omega} \to -(iqa/\pi bv\gamma)\ln(1/a). \tag{2.171}$$

The important ratio is then

$$|E_{1\omega}|^2/|E_{2\omega}|^2 \to \left[(a/\gamma)\ln(1/a)\right]^2, \tag{2.172}$$

which is very small in the limit $a \ll 1$, so that in the limit of very soft "virtual" photons, $dJ_1/d\omega \propto \omega^{-1}$ while $dJ_2/d\omega \propto \omega$. It is because of this result and, therefore, in applications (treated later) that the main contribution to the evaluation of cross sections by the W-W method comes from these virtual photons, and that the pulse J_1 always gives the principal result, J_2 being unimportant. The method always gives a final formula involving a logarithmic factor whose argument (N) is very large. Since N is evaluated to an accuracy up to an undetermined multiplying constant of order unity, use of the asymptotic forms given above is sufficient in the general formulation. That is, it is really unnecessary to introduce the precise forms for the distributions (2.164) in terms of modified Bessel functions. In the end, the asymptotic forms are always taken and the argument N is not precisely determined.

It is, however, appropriate to exhibit here a general formula for the cross section $d\sigma$ for a process evaluated by means of the W-W method. If we describe the process in terms of the interaction of the equivalent virtual photons accompanying the fast charge, we can write

$$d\sigma = dN\, d\sigma_v \tag{2.173}$$

where dN is the differential number of virtual photons and $d\sigma_v$ is their interaction cross section. For example, in an important application of the W-W method, bremsstrahlung can be considered as Compton scattering of the virtual photons of the Coulomb field of the scattering center by the incoming fast particle. In this case, $d\sigma_v$ would be the Compton cross section and $d\sigma$ the bremsstrahlung cross section. The differential dN is obtained by multiplying the virtual photon flux $dJ_1/d\omega$ by $d\omega$ and the differential area $2\pi b\, db$ associated with charged particles incident on the target at impact parameters within db. This summary over azimuthal angles means that $d\sigma_v$ should be the cross section for unpolarized (virtual) photons. Employing the asymptotic form (2.169) for $E_{2\omega}$, we then have, for $q = ze$, the result

$$dN = (2\alpha/\pi)(z/\beta)^2(d\omega/\omega)(db/b), \tag{2.174}$$

where α is, again, the fine structure constant and $\beta = v/c$. There will, in the end, be an integration over db, yielding a factor $\ln(b_{\max}/b_{\min})$ and, depending on the

[21]The exact asymptotic form of this integral is obtained by replacing $\ln(1/a)$ by $\ln(2/\Gamma_E a) = \ln(1.123/a)$, where $\ln \Gamma_E = 0.5772$ is Euler's constant. An outline of an elementary derivation of this more precise result, without resort to general identities on Bessel funcitons, may be found in R. J. Gould, *Am. J. Phys.* **38**, 189 (1970).

process considered, there may be an integration over $d\omega$ (or a transformation of $d\omega$ into, say, the differential energy of the particle produced in the process).

Finally, we might note another reason for neglecting the effects of the pulse associated with J_2, at least for the case of the highly relativistic incident charge. This is exhibited in the factor $1/\gamma$ multiplying the integral (2.166) and means that, for $\gamma \gg 1$, the flux $dJ_2/d\omega$ is small compared with $dJ_1/d\omega$ for *all* frequencies of virtual photons. Since, for $\gamma \gg 1$, the flux $dJ_2/d\omega$ is truly fictitious for a relativistic target (since there is no magnetic field to accompany E_1), and since magnetic interactions may be important, the equivalent-photon description for this "transverse" pulse has its limitations. On the other hand, as has been remarked earlier, the field E_2 really does have a magnetic counterpart (B_3) of equal magnitude as $\beta \to 1$, so that the "longitudinal" pulse J_1 does have a valid equivalent-photon description, even when the target is relativistic and magnetic forces are important.

2.9 ABSORPTION AND STIMULATED EMISSION

A radiation field, incident on a system of charges, can, itself, provide the perturbation to cause the system to undergo a transition. This "external" radiation field can thereby induce a transition resulting in the production of a new photon (stimulated emission) or a transition in which the system absorbs energy from the radiation field, removing a photon from it (absorption). The two processes, stimulated emission and absorption, are a result of the same interaction or perturbation and can be considered as just the time reverse of one another. This is indicated pictorially in Figure 2.6 where a photon beam (wavy lines) is shown incident on the charge system s. The beam is considered to have a definite direction specified by, say, a solid angle element $d\Omega$, and we consider a class of its photons having a particular polarization and a frequency within $d\omega$. Thus, the photon states of the beam are completely specified and $\bar{n}$ is the associated photon occupation number. Also, we consider a transition in s between two specified states ("1" and "2"). There can occur, in addition to the processes induced by the beam, spontaneous emission, in general in any direction; Figure 2.6 indicates the case of emission in the beam direction.[22]

A most important result is the relation between the rates for stimulated emission and absorption and that for spontaneous emission. The relation is extremely simple and is sometimes referred to in terms of the "Einstein A and B coefficients," although that old-fashioned terminology and notation will not be employed here. Also, it should be noted that, to derive the fundamental result, the only quantum-mechanical concept that will be introduced is that of the photon. That is, it is not necessary to make use of the detailed formalism of quantum field theory. This fact has dictated the inclusion of the topic in a chapter on classical radiation theory rather than in the following one on quantum electrodynamics.

[22]Stimulated emission, being the exact reverse of absorption, always takes place in the direction of the incident radiation beam.

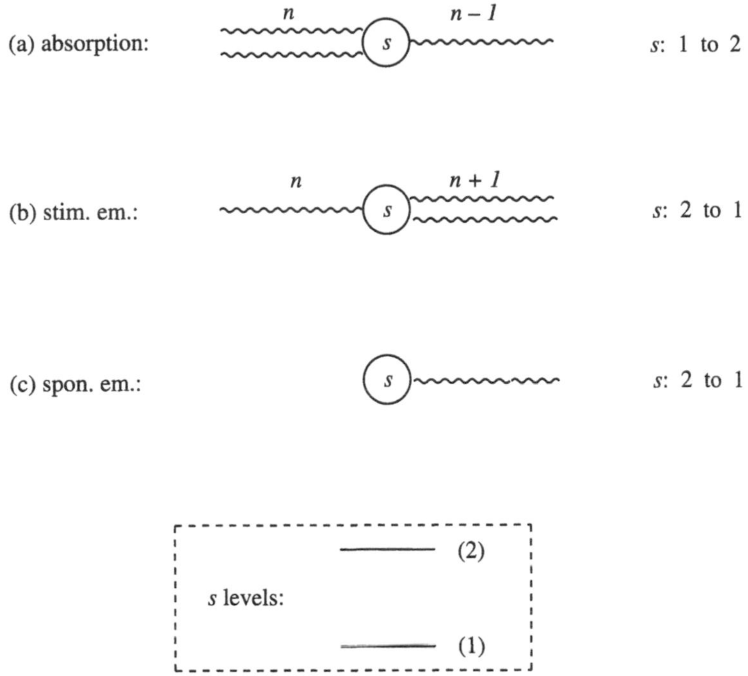

Figure 2.6 Representations of absorption, stimulated emission, and spontaneous emission.

2.9.1 Relation to Spontaneous Emission

Consider again the three processes indicated in Figure 2.6 with the photon state completely specified in terms of its polarization, frequency (or energy), and direction of motion. As a result of the interaction with the photon beam, the system s undergoes transitions between states 1 and 2. The system can also undergo spontaneous radiative transitions that can deposit photons into the beam. The rates for the absorption (a) and stimulated-emission (b) processes will be proportional to the occupation number ($\overline{n}$) of the photon beam and to the numbers N_1 and N_2 of the system in the lower and upper state, respectively. That is,

$$\begin{aligned} R_{\mathrm{a}} &= R_{1\to 2} = R_{\mathrm{abs}} \propto \overline{n}N_1, \\ R_{\mathrm{b}} &= R_{2\to 1} = R_{\mathrm{stim}} \propto \overline{n}N_2. \end{aligned} \tag{2.175}$$

The proportionality constants (w_{abs} and w_{stim}) in the rates (2.175) must be identical, since the two processes are just the time reverse of one another. The spontaneous rate will, of course, be proportional to N_2:

$$R_{\mathrm{c}} = R_{\mathrm{spon}} \propto N_2, \tag{2.176}$$

and we designate the associated proportionality constant as w_{spon}. The rate constants w_{abs} ($= w_{\mathrm{stim}}$) and w_{spon} are determined by the characteristics of the charge system s; that is, they are "atomic" parameters. A convenient way of determining the relationship between them is to consider the case where the system s is in thermal equilibrium with the surrounding photon gas. Then, balancing the $1 \to 2$ and $2 \to 1$ rates involving photons produced or absorbed within the beam direction and with particular polarization, we have

$$\bar{n} N_1 w_{\text{abs}} = \bar{n} N_2 w_{\text{stim}} + N_2 w_{\text{spon}}. \tag{2.177}$$

With[23]

$$\begin{aligned} \bar{n} &= (e^{\hbar\omega/kT} - 1)^{-1}, \\ N_2/N_1 &= e^{-\hbar\omega/kT}, \end{aligned} \tag{2.178}$$

and the identity $w_{\text{abs}} = w_{\text{stim}}$, we obtain the simple and fundamental result

$$w_{\text{spon}} = w_{\text{abs}} = w_{\text{stim}} \equiv w. \tag{2.179}$$

In other words, there is only *one* (w) fundamental radiative transition constant. The coefficient w will, of course, be different for each type of system and transition—and, in general, the w's are difficult to compute—but the identity (2.179) must hold for each given transition. This identity corresponds to the relation between the "Einstein A and B coefficients."

2.9.2 General Multiphoton Formula

Because of the result (2.179), for any process for which there is a finite rate coefficient w for the spontaneous production of a photon in some polarization state and in some direction of motion and with frequency ω, the total rate will be given by

$$R_{\text{spon}} + R_{\text{stim}} = w N_2 (1 + \bar{n}). \tag{2.180}$$

The factor $1 + \bar{n}$ simply corrects for simulated emission. The result (2.180) holds for any background photon gas (not just for that of a blackbody, of course), and $\bar{n}$ is the occupation number of the gas causing the stimulated process. Again, we must remember that stimulated emission takes place only in the direction of the (incident) photons that cause it. Note, further, that $\bar{n}$ is the occupation number for the characteristics of the outgoing photon. The process is stimulated only by these same kind of photons.

There are some radiative processes that take place by the spontaneous production of two photons. For example, the neutral pion decays spontaneously into two γ-rays: $\pi^0 \to \gamma_1 + \gamma_2$. The total decay rate for photons emitted in specific directions is obtained from the spontaneous rate by multiplying by $(1 + \bar{n}_1)(1 + \bar{n}_2)$, where $\bar{n}_1$ and $\bar{n}_2$ are the occupation numbers of the surrounding photon gas at the energies $\hbar\omega_1$ and $\hbar\omega_2$ (for a pion at rest these energies are equal) and corresponding directions and polarizations. This corrects for stimulated emission. That precisely the factor $(1 + \bar{n}_1)(1 + \bar{n}_2)$ is required can be seen if we consider the pion to be in thermal equilibrium with a (blackbody) photon gas. Then in a steady state condition

$$N(\pi^0) w (1 + \bar{n}_1)(1 + \bar{n}_2) = N(2\gamma) w \bar{n}_1 \bar{n}_2 \tag{2.181}$$

[23]The blackbody occupation number [first of Equation (2.178)] is a result of (only) the assumption that the photon is a massless boson.

where w is the transition rate. The "2γ" two-photon state of the π^0 will have relative numbers $N(2\gamma)/N(\pi^0) = \exp[(\hbar\omega_1 + \hbar\omega_2)/kT]$ and $1 + \overline{n} = \overline{n}e^{\hbar\omega/kT}$ for a blackbody distribution. Thus, we see that the condition (2.181) is satisfied.

Another example of a two-photon radiative process is the decay of the $2s$ state of a hydrogenic system:

$$a_{2s} \rightarrow a_{1s} + \gamma_1 + \gamma_2, \tag{2.182}$$

with $\hbar\omega_1 + \hbar\omega_2 = \frac{3}{4}Z^2\mathrm{Ry}$. Corrected for stimulated emission, the total decay rate is obtained from the spontaneous rate by multiplying by $(1 + \overline{n}_1)(1 + \overline{n}_2)$; there is an integration over photon energies and directions. Again, we see that this factor is needed to satisfy the detailed balance relation

$$N_{2s}w(1 + \overline{n}_1)(1 + \overline{n}_2) = N_{1s}w\overline{n}_1\overline{n}_2, \tag{2.183}$$

with $N_{2s}/N_{1s} = \exp[-(\hbar\omega_1 + \hbar\omega_2)/kT]$ and $\overline{n}_1$ and $\overline{n}_2$ the blackbody occupation numbers.

The generalization to the case of a process in which there are any number of photons in the final state is clear. If $R_{\mathrm{spon}}(\omega_1, \omega_2, \ldots)$ is the spontaneous rate, the rate corrected for stimulation by an external radiation field is

$$R = R_{\mathrm{spon}}(\omega_1, \omega_2, \ldots)(1 + \overline{n}_1)(1 + \overline{n}_2)\cdots. \tag{2.184}$$

2.9.3 Stimulated Scattering

In addition to stimulated photon-emission process, there can be stimulation of photon scattering by an external radiation field. The rate is enhanced by a factor $1+\overline{n'}$, where $\overline{n'}$ is the occupation number of the photon gas for the state (energy, polarization, direction of motion) of the scattered photon. The scattering can be by a free electron (Compton scattering) or by an atomic or molecular system. For a system with internal degrees of freedom the scattering can either leave the system unchanged (Rayleigh scattering) or cause an excitation in the system at the same time (Raman scattering). If, in the scattering, the photon frequency changes from ω to ω' (and its polarization and direction changes) and the scattering system changes from s to s', we can see the necessity of precisely the factor $1 + \overline{n'}$ to correct for stimulated scattering by considering detailed balance. With the scattering system in equilibrium with a surrounding blackbody radiation field, the condition

$$\overline{n}Nw_{\mathrm{sc}}(1 + \overline{n'}) = \overline{n'}N'w_{\mathrm{sc}}(1 + \overline{n}) \tag{2.185}$$

is satisfied identically with $\overline{n}$ and $\overline{n'}$ the blackbody occupation numbers and $N/N' = \exp[-\hbar(\omega - \omega')/kT]$. Of course, the stimulation correction $1 + \overline{n'}$ (or $1 + \overline{n}$) applies whether the external radiation field is or is not in thermal equilibrium with the scattering system.

Bibliographical Notes

For general classical electrodynamics and classical radiation theory the three books by Jackson, Panofsky and Phillips, and Landau and Lifshitz are excellent. These texts, listed as References 4, 5, and 6 at the end of Chapter 1, will be referred to here, for brevity, as J, P^2, and L^2. Each is especially good on certain topics covered in this chapter, as we indicate below. Another scholarly work is that of Rohrlich (R), mentioned in Footnote 12. Also useful is the book by W. Heitler, *The Quantum Theory of Radiation*, 3rd ed., Oxford, UK: Oxford University Press, 1954, referred to as H; the first chapter of Heitler's book is on classical electrodynamics. For the various topics in this chapter, the suggested references are as follows:

Gauge invariance: H.
Retarded potentials: L^2, J, P^2.
Multipole expansion: L^2.
Fourier spectra: L^2.
Fields of relativistic charge: P^2.
Radiation from relativistic charges: L^2, J, P^2.
Radiation reaction: P^2, R, H.
Soft-photon emission: L^2, J, P^2.
Weizsäcker-Williams method: J.
Absorption and stimulated emission (quantum mechanical approach): H.

Chapter Three

Quantum Electrodynamics

A comprehensive treatment of quantum electrodynamics (QED) cannot be given here in a single chapter. In any case, that would be beyond the scope of this book, and, as with the classical theory, the general subject has already been covered in a number of very fine textbooks. The full theory is, in fact, quite complicated and requires a lengthy formulation if it is to be laid out in a systematic way. Actually, the final result of this theory is a fairly simple set of prescriptions (Feynman rules) for the calculation of rates of electromagnetic processes. Instead of trying to "derive" the rules in a rigorous and pedestrian manner, we shall try to take an intuitive approach—like Feynman did—to shortcut the path leading to the rules themselves. In an attempt to make the reasoning seem more natural, we start first with a restricted limiting form of the theory. This limiting form is the non-relativistic theory restricted to the case where the photons are radiation-field particles. That is, the photons are considered to exist in given observable states described by their wave vector ($\boldsymbol{k}$) and polarization unit vector ($\boldsymbol{\varepsilon}$). The photons are considered to couple to non-relativistic charges and magnetic moments, and the rates (cross sections) for the various processes are computed by elementary perturbation theory. To calculate these rates we need the perturbation Hamiltonian terms (H') associated with these couplings, and we can obtain these by elementary, but rigorous, methods. Having formulated this restricted corner of the subject, we then try to make the jump to the relativistic theory. In later chapters, when the theory is applied to treat various processes, the same approach will be taken. That is, the non-relativistic limit will be treated first, to be followed by the more general case where the particles can have relativistic energies.

The non-relativistic theory is, to be sure, an important subject in itself, with many useful applications. Some processes are sufficiently complicated that analytic results cannot be derived in general covariant formulations, although exact formulas can be derived in the non-relativistic theory with its restricted energy domain. This chapter treats only the developments of the theory for its use in later applications, and it differs from the way the subject is presented in the standard textbooks. Modern books generally formulate the covariant theory from the very beginning, a result of the great success of this general and preferred approach. The whole subject has an interesting history, and we begin with a short outline of it.

3.1 BRIEF HISTORICAL SKETCH

A lot of people have the impression that QED started in 1948 with the work of Feynman, Schwinger, and others. The real beginning came in 1905 with the introduction of the concept of the photon. This concept was not readily accepted by twentieth-century physicists; only by about, say, 1917 did the particle nature of electromagnetic radiation begin to be considered seriously.[1] In 1922, the particle character of the photon was seen clearly in the observation of the Compton effect which demonstrated the existence of momentum and energy of individual photons. With the subsequent developments in quantum mechanics in 1925 and the description of processes in terms of a general time-dependent Schrödinger equation

$$H\psi = i\hbar\,\partial\psi/\partial t, \tag{3.1}$$

the formulation of a quantum theory of electromagnetism followed, principally in the work of Dirac, Heisenberg, Pauli, and Fermi.

The essence of QED—a quantum theory of the electromagnetic field—is the introduction of the photon and the description of the fields in terms of a collection of these particles in various one-particle states. That is, the field is specified in terms of the values of the occupation numbers of the particle (photon) states. The classical Maxwell theory serves as a guide for QED, since in the limit of large occupation numbers the theory must approach that of the classical continuum. In a quantum-mechanical formulation describing many charged particles and photons, the wave function ψ in Equation (3.1) would contain information on the charges and photons, and the Hamiltonian would consist of parts associated with each plus an interaction part:

$$H = H_{\text{ch}} + H_{\text{ph}} + H_{\text{int}}, \tag{3.2}$$

just as in classical theory. Dirac gave the first such treatment of charged particles and photons in 1927 for the special case of radiation-field photons. As in classical electrodynamics, the quantum mechanics of the radiation field is simpler than that for general electromagnetic fields.

General QED, formulated to describe electromagnetic effects not confined to radiation fields, presents more formidable problems, especially in connection with the requirements of gauge invariance. The first formulations of the general theory was by Heisenberg and Pauli in 1929 and Fermi in 1930. The theory was then successfully applied in the early 1930s to a number of problems involving the interaction of charged particles and photons. At the same time, however, it was found that there were difficulties with the theory when applied to certain problems. In particular, it was found that the theory seemed incapable of calculating effects to higher order in the electromagnetic coupling constant α. That is, quantities like cross sections, calculated by perturbation theory in a power series in α, came out in satisfactory form in lowest order, but yielded infinities in the higher-order corrections. The divergence problems, which we discuss briefly later in this chapter, show up in the evaluation of these "radiative corrections" and, in particular, in the

[1] Even as late as 1917, only Einstein was actively advocating acceptance of the idea of the photon (or "quantum") [see A. Pais, *Rev. Mod. Phys.* **51**, 861 (1979)].

attempt to calculate the electromagnetic self-energy of an electron. As first shown by Weisskopf, the divergence in the self-energy problem is not as serious as in classical electrodynamics, when treated by relativistic QED; the divergence, instead of being linear, is now of a logarithmic type.

Quantum electrodynamics remained in this state until the late 1940s when there began a new phase in theoretical and experimental developments on the subject. The experiments were partly a result of advances in microwave techniques from research carried out during World War II. The first important measurement was the (1947) determination of the *Lamb shift*, a displacement of the $2s_{1/2}$ level with respect to the $2p_{1/2}$ level in atomic hydrogen. The small level shift, suspected from earlier spectroscopic measurements, provided a challenge for theory, and an approximate calculation of the effect was carried out by Bethe soon after the experimental discovery. This was an important discovery, and the method used in the theoretical calculation of the effect was also of great significance, for the level shift corresponds to a "radiative correction," and the measurement showed that the effect is real and finite. Bethe's was the first calculation making use of "renormalization" techniques, thereby eliminating the necessity of dealing with the ("ultraviolet") divergences that occur in these higher-order calculations. The idea, first introduced into QED by Kramers[2] in 1937, removes the divergences by noting that the observed mass of the electron includes the electromagnetic self-energy. Then, when the higher-order radiative effects are expressed in terms of the observed mass, the divergent integrals subtract out.

In the years 1948–1950 there was a great deal of work on QED, especially on the theoretical side. Independently, Feynman and Schwinger reformulated the theory, providing a description that was relativistically covariant at every stage. Actually, it turned out that work on covariant QED had already begun in Japan in 1943 by Tomonaga; however, Tomonaga's papers, first published in Japanese, were unknown to almost everyone. Feynman's approach first seemed hard to comprehend, and Schwinger's papers were not easy to read. The work of Feynman was highly intuitive and, although no one else really knew why, with his formulation he was apparently able to perform calculations of a number of processes with great ease. The relationship between the work of Tomonaga, Feynman, and Schwinger with field theory was shown by Dyson. It is Dyson's formulation of covariant QED and covariant perturbation theory that is found in all subsequent textbooks on the subject.

The advantage of a fully covariant formulation of QED became evident, and the theory, with the help of renormalization procedures, was shown to be capable of performing calculations to very high accuracy. In particular, radiative corrections to atomic energy levels (Lamb shift) and to the magnetic moment of the electron were evaluated and found to be in excellent agreement with new very precise measurements. In its ability to calculate quantities that can be measured—to an accuracy of one part in 10^8 in atomic energy levels—the theory is highly successful. Perhaps it should be emphasized, nevertheless, that modern QED is fundamentally no different

[2]See, for example, the note on p. 453 in H. A. Kramers, *Quantum Mechanics*, New York: Dover Publ. Inc., 1964.

from the Dirac-Heisenberg-Pauli-Fermi theory described in 1930. The computational capabilities are much more advanced in the fully covariant theory, however, and this again demonstrates the advantages of a covariant formulation. For example, using the modern methods, the derivation of a formula for, say, some cross section might take a couple of hours, while months of work would be required using the older non-covariant procedures. For some calculations, even the methods of modern QED involve very lengthy mathematical manipulations requiring the evaluation of a great many integrals, etc. These higher-order calculations, which employ the techniques of renormalization, are beyond the capability of the older formulations of QED. Still, the theory is not in a totally satisfactory state in that the divergences have not been eliminated. For example, we cannot compute the electromagnetic self-energy of the electron; indeed, the theory still gives infinity for this quantity. If we can forget about difficulties like that, the theory can be applied with confidence to the calculation of virtually any observable electromagnetic phenomenon.

3.2 RELATIONSHIP WITH CLASSICAL ELECTRODYNAMICS

Purely classical electrodynamics (CED) is a continuum theory where the concept of a field particle (photon) is foreign. It is also a limiting case of the more general theory (QED), so that it is contained within the latter. This relationship can be exploited, as it is in the following section (3.3), to employ CED as a guide in formulating QED. There is, of course, a limit to what can be done along these lines, since the more general theory is greater in content and describes phenomena that are beyond the capability of the classical theory. Nevertheless, like all fundamental theories, QED is simple in its foundations, and it is not difficult to formulate it with the help of our knowledge of its limiting form.

The classical domain of a radiation field corresponds to the condition

$$\bar{n} \gg 1, \tag{3.3}$$

where $\bar{n}$ is the photon occupation number; this is essentially a specification of the applicability of a continuum-description theory of the photon field. For a particular process involving the interaction of these photons, the classical limit requires that the effects having to do with individual photons be negligible. That is, the particle (photon) characteristics must not play a role in this limit. It would seem that the condition that the photons are "soft" would be sufficient to satisfy this requirement. The soft-photon limit would confine the photon energy and momentum (that is, its kinematic properties) to be small compared with the energies and momenta (and their changes) of the charged particles involved in the particular process. For most processes this criterion would be sufficient for the validity of a classical treatment. However, for a particular process[3] it may be that the photon kinematic properties play a key role even in the soft-photon limit. In such cases, a semi-classical calculation is not completely valid if its only quantum-mechanical aspect is the introduction

[3]An example, treated later, would be electron-electron bremsstrahlung in the Born-approximation limit. This process must be given a quantum-mechanical treatment even for the case where the photons are soft.

of the photon concept. Some processes, such as photon-photon scattering, simply do not exist in a purely classical theory. This process, which can be thought of as involving virtual electron-positron pairs in an intermediate state, requires the fully relativistic QED for its treatment. However, immediately upon the introduction of quantum-mechanical ideas, it becomes fairly clear that such processes must be possible. In fact, even without a detailed quantum-mechanical calculation, we can have some idea of the order of magnitude of the associated cross section for this higher-order process. Moreover, for some processes, the relativistic effects are such that they can be included in a calculation in an approximate way by modifying the (large) argument of some logarithmic factor.

For a process that does have an essentially classical description, it may be that only certain kinematic properties of the photon need be limited for the domain of applicability. For example, it may be that only the photon momentum is important in kinematic considerations while the photon energy is not limited. Or it may be that a classical description is valid even though the photon momentum is comparable to that of the charged particles involved. Bremsstrahlung and Compton scattering in the non-relativistic limit are examples of, respectively, each of these cases.

Sometimes it is an accident that a certain classical formula is identical to a particular quantum-mechanical one. If this happens to be so, it is important to understand the reason for the coincidence. There is certainly one aspect of QED that is highly significant, and this has to do with the fundamental coupling constant associated with electromagnetic processes. The results (2.143) and (2.149), derived classically and obtained later in this chapter in a quantum-mechanical formulation, indicate how α $(= e^2/\hbar c \cong 1/137)$ determines the probability of photon emission. The electromagnetic coupling is weak (but not *very* weak), so that the calculation of processes can be carried out through the use of perturbation theory. Also, in the comparison of classical and perturbation-theory formulas, because α is small, we can understand why identical results are obtained. In the calculation of, say, photon-emission probability, the classical formula really gives the probability of *one or more* photons being produced, while the perturbation-theory result is for the probability of production of *one* photon. Because multiple photon production is improbable,[4] the "one or more" of the classical formula essentially means one.

The degree to which quantum mechanics must be included in a problem involving the electromagnetic interaction depends on the details of the process. For example, in bremsstrahlung it may be that "photon emission" can be described classically while the rest of the process (scattering, say) is treated quantum mechanically. The experimental or observational conditions can also dictate how some phenomenon is handled theoretically. If individual photons are not being detected and quantum mechanics does not play a role in the process, then classical electrodynamics is an appropriate description. Quantum electrodynamics can also be employed in such a problem, but it simply makes no sense to take the more general approach. In fact, the exact quantum-mechanical treatment might be so complicated that the problem cannot be handled this way. The relationship between CED and QED will become clearer as we treat the various electromagnetic processes in later chapters.

---

[4]However, this is not so at ultrahigh energies [see Equation (3.191)].

3.3 NON-RELATIVISTIC FORMULATION

3.3.1 Introductory Remarks

By confining our treatment to describe only processes involving radiation-field photons, we can simplify the subject. Thus, our photons are always real[5] and their parameters (polarization and wave vector) are numbers that are fixed in the initial and/or final states. This relaxes, to a certain degree, the extent to which quantum mechanics must be employed in the description of the photon field and its interaction. However, it is inappropriate to refer to our treatment as "semi-classical," since all the necessary quantum mechanics is included, and the formulation—with, however, its limited applicability—is completely rigorous. We do include the particle aspects of the photon field, that is, the effects of the individual-particle momentum ($\hbar \boldsymbol{k}$) and energy ($\hbar c k = \hbar c|\boldsymbol{k}|$). Also, we give a complete quantum mechanical description of the charged particles that interact with the photons. The only restriction placed on the charge motion is that $\beta = v/c \ll 1$.

A wide variety of problems can be worked out with the help of the formulation outlined here. It can even be applied to problems involving pair production and annihilation, as long as the charged particles involved have non-relativistic kinetic energies. For some problems or applications, the non-covariant development is actually more convenient, even for the case where the charges are in relativistic motion. However, nowadays everyone learns QED in the modern fully covariant form. There are, as a result, few treatments of non-relativistic, non-covariant QED in modern textbooks. Perhaps, therefore, the simplified approach taken in this chapter will be helpful to people not inclined to delve into the details of the full theory.

3.3.2 Classical Interaction Hamiltonian

Consider first the classical motion of a charge in an electromagnetic field described in terms of a vector potential $\boldsymbol{A}(\boldsymbol{r}, t)$ and scalar potential $\Phi(\boldsymbol{r}, t)$. If there is present, in addition, a velocity-independent potential $V(\boldsymbol{r}, t)$, the total Lagrangian function would be (see Chapter 1)

$$L = \tfrac{1}{2}mv^2 + (q/c)\boldsymbol{A} \cdot \boldsymbol{v} - q\Phi - V. \tag{3.4}$$

As is well known, this Lagrangian yields the correct equation of motion including the Lorentz force. The corresponding Hamiltonian is

$$H = \frac{1}{2m}\left(\boldsymbol{p} - \frac{q}{c}\boldsymbol{A}\right)^2 + q\Phi + V. \tag{3.5}$$

It should be noted that

$$\boldsymbol{p} = \partial L/\partial \boldsymbol{v} = m\boldsymbol{v} + (q/c)\boldsymbol{A} \tag{3.6}$$

---

[5]That is, they are not "virtual." A virtual photon is one that is emitted and reabsorbed and exists only in intermediate states. If the photons in intermediate states are described by plane wave states, the amplitude and coupling derived here for radiation photons can still be used to describe these states, however.

is the canonical momentum and *not* the particle momentum; it is this quantity that is replaced by the operator $-i\hbar\nabla$ in going to a Schrödinger equation. The particular form of the Hamiltonian should also be noted, in particular, the first term in Equation (3.5); it is the same as that for a free particle with the replacement

$$\boldsymbol{p} \to \boldsymbol{p} - (q/c)\boldsymbol{A}. \qquad (3.7)$$

The prescription (3.7) for the formation of a Hamiltonian function including electromagnetic interactions is sometimes referred to as a "Principle of Minimal Electromagnetic Coupling." The coupling is "minimal" in the same sense that it yields a particular combination of terms involving the potentials Φ and $\boldsymbol{A}$ such that the gauge invariance of the theory is satisfied.[6] In covariant form, in terms of the gradient operator, the relation (3.7) would become, for quantum-mechanical wave equations,

$$\partial_\mu \to \partial_\mu - (iq/\hbar c)A_\mu. \qquad (3.8)$$

The Hamiltonian (3.5), when written as

$$H = p^2/2m + V + H_{\text{int}}, \qquad (3.9)$$

yields an interaction part

$$H_{\text{int}} = -(q/2mc)(\boldsymbol{p}\cdot\boldsymbol{A} + \boldsymbol{A}\cdot\boldsymbol{p}) + (q^2/2mc^2)\boldsymbol{A}\cdot\boldsymbol{A} + q\Phi \qquad (3.10)$$

associated with the coupling to the electromagnetic field. In this expression, since we want to make a transition to a quantum-mechanical formulation, the terms $\boldsymbol{p}\cdot\boldsymbol{A}$ and $\boldsymbol{A}\cdot\boldsymbol{p}$ are not equated. When $\boldsymbol{p}$ is replaced by the gradient operator and the term $\nabla\cdot\boldsymbol{A}$ operates on a function ψ to the right, the identity

$$\nabla\cdot\boldsymbol{A}\psi = (\nabla\cdot\boldsymbol{A})\psi + \boldsymbol{A}\cdot\nabla\psi \qquad (3.11)$$

is employed. As we have seen in Chapter 2, for radiation fields a gauge can be chosen such that $\nabla\cdot\boldsymbol{A} = 0$ *and* $\Phi = 0$. In this case, for a system of charges interacting with an electromagnetic radiation field, the part of the Hamiltonian associated with this coupling is

$$H_{\text{int}} = \frac{i\hbar}{c}\sum_\alpha \frac{q_\alpha}{m_\alpha}\boldsymbol{A}(\boldsymbol{r}_\alpha, t)\cdot\nabla_\alpha + \frac{1}{2c^2}\sum_\alpha \frac{q_\alpha^2}{m_\alpha}\boldsymbol{A}(\boldsymbol{r}_\alpha, t)\cdot\boldsymbol{A}(\boldsymbol{r}_\alpha, t). \qquad (3.12)$$

This expression is a generalization of Equation (3.10), employing the identity (3.11) and summing over particles (α), replacing the momenta by the operators $-i\hbar\nabla_\alpha$.

The formulation given above is adequate to calculate the rate at which the system of charges undergoes transitions as a result of the action of the perturbation (3.12). The state of the charge system is described in terms of a wave function $\psi(\boldsymbol{r}_\alpha, t)$ with the index α running over 1 to N (= number of charges). The charges can be in free-particle states or bound in atoms, molecules, or nuclei. However, the electromagnetic field, expressed in terms of the vector potential $\boldsymbol{A}$ in Equation (3.12), is treated as an "external" potential, that is, as a specified function of $\boldsymbol{r}$ and t. This corresponds to a classical description of the photon field, and it is important to understand the limitations of the treatment and why it is adequate for a

[6] See, for example, F. Rohrlich, *Classical Charged Particles*, Reading, MA: Addison-Wesley Publ. Co., 1965; W. Pauli, *General Principles of Quantum Mechanics*, Berlin: Springer-Verlag, 1980.

certain class of problems. The formulation works only for radiation-field photons, that is, for photons in free-particle (plane-wave) states. These photons are thus in fixed states, either generated prior to incidence on the charge system or as detected after interaction with the system. The simplified treatment cannot handle problems that require consideration of electromagentic perturbations corresponding to purely intermediate-state photons[7] ("photons" that are emitted and reabsorbed during the process). Although the photons must be either incoming or outgoing on the charge system, they can be created (or annihilated) as a result of the perturbation (3.12). The elementary theory given here is capable of calculating these processes without introducing the formalism of quantum field theory and annihilation and creation operators, etc.

Photons existing only in intermediate states do not have observable characteristics and, in fact, have a spectrum of kinematic properties. A full quantum-mechanical treatment would be required to evaluate processes involving such states, and this apparatus is provided by conventional quantum field theory. However, the formulation outlined here is adequate to treat problems in which the charged particles exist in (unobserved) intermediate states "in between" the action of perturbations. The elementary theory is also capable of treating problems involving pair production and annihilation as long as the charged-particle kinematic energies are non-relativistic. However, the non-relativistic theory cannot treat problems in which there are "virtual" charged pairs in intermediate states, since the characteristic energies of these pairs is such as to force a relativistic treatment. An example of a process of this type is photon-photon scattering—a phenomenon that does not exist in purely classical theory. As mentioned earlier, photons can scatter off one another because of the possibility of virtual electron-positron pairs in an intermediate state. The incoming and outgoing photons couple to these "particles" and this allows the process to take place. For center-of-mass photon energies such that $\varepsilon_1 = \varepsilon_2 \ll mc^2$, one might think that a non-relativistic theory would be capable of computing the cross section for the process. However, if pairs are produced in an intermediate state, they will have characteristic kinetic energies $\sim mc^2$ even for the scattering of low-energy photons.

Finally, on the subject of the interaction Hamiltonian, we should introduce an additional term associated with coupling of the electromagnetic field to a particle's intrinsic magnetic moment. Charge coupling to the "orbital" motion is contained in the Hamiltonian (3.12), but (permanent) magnetic moments in a magnetic field have an energy $-\boldsymbol{\mu}\cdot\boldsymbol{B}$, and a corresponding term would have to be added to the Hamiltonian. In terms of the vector potential of the radiation field, for a collection of moments, the expression

$$H_{\text{int}(\mu)} = -\sum_{\alpha} \boldsymbol{\mu}_\alpha \cdot \text{curl}\, \boldsymbol{A}(\boldsymbol{r}_\alpha, t) \tag{3.13}$$

should be added to H_{int} in Equation (3.12). This term, like that for the $\boldsymbol{A}\cdot\boldsymbol{p}$ coupling, is linear in the electromagnetic field amplitude.

[7]This is a rather loose terminology; the designation "photon" should perhaps be reserved for free-particle states. On the other hand, all photons are eventually absorbed and in that sense could even be regarded as virtual.

3.3.3 Quantum-Mechanical Interaction Hamiltonian

For the calculation of the rates for various processes involving coupling to photons, we need the forms for the Hamiltonian terms associated with these couplings. Basically, the couplings are the forms H_{int} for which we must now substitute in the appropriate expression for the vector potential corresponding to a single propagating photon. In terms of the photon wave vector $\boldsymbol{k}$ and unit polarization vector $\boldsymbol{\varepsilon}$, the vector potential is[8]

$$\boldsymbol{a}(\boldsymbol{r}, t) = a_0 \boldsymbol{\varepsilon} \cos(\boldsymbol{k} \cdot \boldsymbol{r} - \omega t). \tag{3.14}$$

In the oscillatory term with the phase $\phi = \boldsymbol{k} \cdot \boldsymbol{r} - \omega t$, if we write

$$\cos\phi = \tfrac{1}{2}(e^{i\phi} + e^{-i\phi}), \tag{3.15}$$

in evaluating the rate for the process by perturbation theory, only one of the terms $(e^{\pm i\phi})$ would be picked out, depending on whether the photon is incoming or outgoing. The task, then, is to determine the amplitude factor a_0; this would then establish the precise forms for each of the three kinds of terms in H_{int}. That is, we would have the $\boldsymbol{a}\cdot\boldsymbol{p}$, $\boldsymbol{a}\cdot\boldsymbol{a}$, and $\boldsymbol{\mu}\cdot$curl $\boldsymbol{a}$ couplings that, for various processes, determine the associated rate or cross section.

It is easy to fix the parameter a_0, and there are at least two simple procedures for doing this.[9] The most direct method fixes a_0 by relating the photon energy flux computed with the vector potential (3.14) to what it should be for a simple propagating photon of energy $\hbar\omega = \hbar ck$. If for photon states we take a unit normalization volume, the number density of photons in the state described by $\boldsymbol{\varepsilon}$ and $\boldsymbol{k}$ would be $n_\gamma = 1$. In terms of the photon energy density u_γ, the magnitude of the photon energy flux (Poynting vector) in the direction of $\boldsymbol{k}$ would be

$$S = c\hbar\omega = cu_\gamma. \tag{3.16}$$

If E and B are the magnitude of the electric and magnetic fields carried by the photon, then

$$u_\gamma = (\langle E^2\rangle + \langle B^2\rangle)/8\pi = \langle E^2\rangle/4\pi, \tag{3.17}$$

where the brackets denote a time average. With the electric field $\boldsymbol{E}$ derived from $-(1/c)\partial\boldsymbol{a}/\partial t$, and the time average $\langle\cos^2\phi\rangle = \langle\sin^2\phi\rangle = \frac{1}{2}$, we combine the above two equations to give

$$a_0 = (2\pi\hbar c^2/\omega)^{1/2}, \tag{3.18}$$

which should be then substituted into Equation (3.14) and employed to fix the forms for H_{int}.

In the other method (see Footnote 9) used to evaluate a_0, some process is computed classically and quantum mechanically and the results are then compared. The simplest process to consider is soft-photon production when a charged particle is accelerated suddenly to a velocity v, say. The probability can be computed for the acceleration to be accompanied by the production of a photon of frequency within $d\omega$.

[8]We adopt notation here in which the lower case letter ($\boldsymbol{a}$) is used for the vector potential associated with the state of a single photon.

[9]See R. J. Gould, *Astrophys. J.* **362**, 284 (1990).

This can be done by employing the soft-photon formulas of classical electrodynamics that were derived in Chapter 2, and the comparison with a quantum-mechanical calculation then yields precisely the result (3.18).

The treatment referred to in Footnote 9 uses the notation R, T, and S for the $\boldsymbol{a} \cdot \boldsymbol{p}$, $\boldsymbol{a} \cdot \boldsymbol{a}$, and $\boldsymbol{\mu} \cdot \operatorname{curl} \boldsymbol{a}$ perturbations, standing for "radiation," "two-photon," and "spin." The $\boldsymbol{a} \cdot \boldsymbol{a}$ coupling is quadratic in the electromagnetic field; that is, it involves two fields which can represent two photons. Then, if we write $\boldsymbol{a} = \boldsymbol{a}_1 + \boldsymbol{a}_2$ for the total field,

$$\boldsymbol{a} \cdot \boldsymbol{a} = a_1^2 + a_2^2 + 2\boldsymbol{a}_1 \cdot \boldsymbol{a}_2. \tag{3.19}$$

If, for example, there is one incoming and one outgoing photon, the factor 2 multiplying $\boldsymbol{a}_1 \cdot \boldsymbol{a}_2$ would account for the possibility that labels 1 and 2 (fields 1 and 2) could describe either photon. If, on the other hand, we had two outgoing photons, the factor 2 would be regarded as accounting for a direct and exchange amplitude. We represent the coupling T as a "two-photon vertex" for which the photon states are described by $\boldsymbol{\varepsilon}, \boldsymbol{k}$ and $\boldsymbol{\varepsilon}', \boldsymbol{k}'$, and to remind the reader of the origin of the factor of 2, we enclose it in parentheses in the formula for T. In the expression for S (coupling to the "spin" moment), the curl operator on the plane-wave photon state can be written as a cross product involving $\boldsymbol{k}$ and $\boldsymbol{\varepsilon}$. For coupling to a charge $q = ze$ and magnetic moment $\boldsymbol{\mu}$, the formulas for the three types of H_{int} perturbations can be summarized as

$$R = \frac{ze}{m}\left(\frac{2\pi\hbar}{\omega}\right)^{1/2} \boldsymbol{\varepsilon} \cdot \boldsymbol{p}, \tag{3.20}$$

$$T = \frac{(2)\pi z^2 e^2 \hbar}{m} \frac{\boldsymbol{\varepsilon} \cdot \boldsymbol{\varepsilon}'}{(\omega\omega')^{1/2}}, \tag{3.21}$$

$$S = -ic\left(\frac{2\pi\hbar}{\omega}\right)^{1/2} \boldsymbol{\mu} \cdot (\boldsymbol{k} \times \boldsymbol{\varepsilon}). \tag{3.22}$$

3.3.4 Perturbation Theory

The standard time-dependent perturbation theory employed for the calculation of rates of processes in quantum-mechanical systems was first developed by Dirac. The procedure, often referred to as the "method of variation of constants," is very general in that it is formulated for an arbitrary system described by a Schrödinger equation

$$(H_0 + H')\psi = i\hbar\, \partial\psi/\partial t. \tag{3.23}$$

Here H_0 is the unperturbed Hamiltonian, which satisfies

$$H_0\psi^{(0)} = i\hbar\, \partial\psi^{(0)}/\partial t, \tag{3.24}$$

for which there is a spectrum of stationary-state solutions

$$\psi_m^{(0)} = u_m^{(0)} \exp(-iE_m^{(0)} t/\hbar), \tag{3.25}$$

the $u_m^{(0)}$ being independent of time but functions of the coordinates of the particles. The general state of the unperturbed system is

$$\psi^{(0)} = \sum_m a_m^{(0)} \psi_m^{(0)}, \tag{3.26}$$

a superposition of possible states.

The solution to the "perturbed" Schrödinger equation (3.23) is written in the form

$$\psi = \sum_m a_m(t) \psi_m^{(0)}, \tag{3.27}$$

that is, as a linear combination of the unperturbed wave functions. This is a convenient procedure since often we have a problem in which, before the perturbation acts, the system is in some unperturbed state $\psi^{(0)} = \psi_0^{(0)}$ at, say, $t = -\infty$. Thus,

$$\begin{aligned} a_0^{(0)} &= 1, \\ a_m^{(0)} &= 0 \quad (m \neq 0), \end{aligned} \tag{3.28}$$

the subscript 0 referring to the initial state. As a result of the perturbation H', the system makes transitions to different states. After the action of the perturbation, at $t = +\infty$, the system will be in the final state

$$\psi_f(t = +\infty) = \sum_m a_m(t = +\infty) \psi_m^{(0)}, \tag{3.29}$$

and the probability of a particular state k will be given by

$$W_k = |a_k(t = +\infty)|^2. \tag{3.30}$$

Perturbation theory works when H' is small and $W_k \ll 1$, except for W_0. Since we are interested in transitions of various types, the case $k \neq 0$ is important. In general, if the perturbation H' is sufficiently weak,

$$a_m(t) = \delta_{0m} + a'_m(t), \tag{3.31}$$

with $\left|a'_m(t)\right|^2 \ll 1$. If the general solution (3.27) is substituted into the perturbed equation (3.23), an equation for the a's is obtained:

$$(H_0 + H') \sum_m a_m \psi_m^{(0)} = i\hbar \sum_m (a_m \partial \psi_m^{(0)} / \partial t + \dot{a}_m \psi_m^{(0)}). \tag{3.32}$$

Multiplying from the left by $\psi_k^{(0)}$ and integrating over the spatial volume dV (which may be multidimensional), employing the forms (3.25), we have an equation for a_k:

$$\sum_m a_m e^{i\omega_{km} t} H'_{km} = i\hbar \dot{a}_k. \tag{3.33}$$

The orthogonality of the $\psi_m^{(0)}$ has been employed to get a single term on the right, and

$$\omega_{km} = \left(E_k^{(0)} - E_m^{(0)}\right)/\hbar, \tag{3.34}$$

$$H'_{km} = \int \left(u_k^{(0)}\right)^* H' u_m^{(0)}\, dV. \tag{3.35}$$

Note that in the definition (3.35) the matrix element is computed from time-independent wave functions $u_k^{(0)}$ and $u_m^{(0)}$. The perturbation Hamiltonian H' may still be a function of time, so that H'_{km} may be as well. We consider the important special cases where H' is time independent and later where H' is an harmonic function of t.

For the condition (3.28) in which the system started in the initial state $\psi_0^{(0)}$, the coefficients are given by the expression (3.31). When this is subtituted into the (exact) equation (3.33) the main contribution on the left side is from $m = 0$ and we have, in a first iteration,

$$a'_k(t) = -(i/\hbar)\int_{-\infty}^{t} e^{i\omega_{k0}t} H'_{k0}\, dt. \tag{3.36}$$

With H'_{k0} independent of time[10] the expressions (3.36) can be integrated; writing the lower limit as $-T$, we have

$$a'_k(t) = \frac{H'_{k0}}{E_0 - E_k}(e^{i\omega_{k0}t} - e^{-i\omega_{k0}T}). \tag{3.37}$$

This result can then be employed in Equation (3.31) and then substituted back into the exact equation (3.33) to give a second-iteration result:

$$i\hbar\dot{a}'_k = e^{i\omega_{k0}t} H'_{k0} + \sum_m \frac{H'_{m0}}{E_0 - E_m} e^{i\omega_{m0}t} e^{i\omega_{km}} H'_{km}. \tag{3.38}$$

In this step, the second term in parentheses in Equation (3.37) has been ignored; this term, for $T \to \infty$, does not contribute, being highly oscillatory and having the essential value zero. Since $\omega_{km} + \omega_{m0} = \omega_{k0}$, the result (3.38) is the same as that for the first iteration in Equation (3.36) with

$$H'_{k0} \to H'_{k0} + \sum_m H'_{km}\frac{1}{E_0 - E_m}H'_{m0}. \tag{3.39}$$

For reasons to be discussed later, the factors in the above sum have been written in this special way.

To simplify the equations, we express them with only the first term in the perturbation series (3.39). By means of the Equations (3.30) and (3.36), the probability for the transition $0 \to k$ occurring due to the action of the perturbation H' can be expressed explicitly in terms of an integral[11] involving the matrix element H'_{k0}:

$$W_k = \frac{1}{\hbar^2}\left|\int_{-\infty}^{\infty} e^{i\omega_{k0}t} H'_{k0}(t)\, dt\right|^2. \tag{3.40}$$

This is an important formula that exhibits a number of significant features. In particular, when the perturbation is a slowly varying function of time, the oscillatory exponential factor makes the value of the integral small. That is, when the perturbation is "adiabatic" (slow) the probability of a transition is small and, instead, the

[10] We shall see shortly how to modify our formulas for the case where $H'_{k0} \propto e^{i\omega t}$, as in electromagentic perturbations.

[11] In fact, this integral is 2π times the Fourier transform of $H'_{k0}(t)$.

system adjusts itself gradually to the instantaneous effect of the perturbation. This effect is known as the *Adiabatic Theorem.*

Another very important result can be derived from Equations (3.40) or (3.30) and (3.36). For transitions to continuum states, the transition probability per unit time is of interest. For a particular one (k) of these states, this is given by

$$\frac{\Delta W_k}{\Delta t} = \lim_{\tau\to\infty} \frac{1}{2\tau}\frac{1}{\hbar^2}\left|\int_{-\tau}^{\tau} e^{i\omega_{k0}t} H'_{k0}\,dt\right|^2 . \tag{3.41}$$

When H'_{k0} is independent[12] of time, it can be taken out of the integral, which is then equal to $2i(H'_{k0}/\omega_{k0})\sin\omega_{k0}\tau$. But since

$$\lim_{\tau\to\infty}\frac{\sin^2\omega_{k0}\tau}{\omega_{k0}^2\tau} = \pi\delta(\omega_{k0}) = \pi\hbar\delta(E_k - E_0), \tag{3.42}$$

we have

$$\frac{\Delta W_k}{\Delta t} = \frac{2\pi}{\hbar}\left|H'_{k0}\right|^2 \delta(E_k - E_0). \tag{3.43}$$

Here the δ-function manifests energy conservation in the overall process, and the result (3.43) is often written with a "density of final states" $\rho(E_k) = dN_k/dE_k$ instead of the δ-function. Since a summation over final states is always performed in the applications of the result for $\Delta W_k/\Delta t$, the δ-function is employed then, and the effect is the same as with the inclusion of $\rho(E_k)$. The formula (3.43) has many applications and was called "Golden Rule Number Two" by Fermi. It is well to emphasize its general applicability in that it is based essentially on the perturbed general Schrödinger equation (3.23), which does not specify the form of the Hamiltonian. The Golden Rule has assumed that H' is not an explicit function of time; however, we can also employ the result when H' is a harmonic function of time for our special applications.

Another aspect of the Golden Rule formula ought to be emphasized. Although it was derived in a perturbation theory formulation, the formula itself is more general in that it holds *even if the perturbation is not weak.* To understand this, we should recognize that the formulation was a description of "flow of probability," and as long as the expired time is not long, the final-state coefficients $a'_k(t)$ will be small even if the perturbation is not weak. Actually, if H' is strong, the effective matrix element H'_{k0} can be regarded as the result of the multiple action of some coupling evaluated to higher order, assuming that the perturbation series converges.

In one special application of the Golden Rule formula this more general validity is inherently assumed. The application is to the determination of the ratio of cross sections for processes in the forward and reverse directions. The cross-section ratio yields the corresponding ratio of the associated phase-space factors for the forward and reverse processes. The squared matrix elements are the same because the effective coupling Hamiltonian is Hermitian. Then, for example, if the processes

---

[12]Here we mean that H' is not an *explicit* function of time. The results can still be applied, for example, in the calculation of scattering cross sections when the scattering potential (which would act as H') is time independent. In such scattering, the particle experiences a variable perturbation, but as a result of the dependence on position coordinate and not as a result of explicit time dependence.

themselves involve complicated multiple actions of a strong coupling, the ratio of the forward and reverse cross sections can be obtained from the Golden Rule formula.

3.3.5 Processes, Vertices, and Diagrams

The possible electromagnetic processes are determined by the interaction Hamiltonians (3.20), (3.21), and (3.22). For an overall process there may be other perturbations involved and, correct to second order in the perturbations, the rate would be determined by the effective perturbation matrix element (3.35). We are restricting our treatment to processes involving only radiation-field photons, since this allows simplification. At the same time, it limits the number of processes that our elementary theory can treat. However, the techniques employed in the restricted theory are very similar to those in general QED, thereby providing a useful introduction to the subject.

Of special interest are processes that occur as a result of the action of two perturbations, one or both of which are electromagnetic (that is, involving production or destruction of photons). If U and V are the interaction Hamiltonians associated with the perturbations, the effective perturbation matrix element for the combined process would be, generalizing[13] the result (3.39),

$$H'_{f0} = \sum_I U_{fI} \frac{1}{E_0 - E_I} V_{I0} + \sum_{I'} V_{fI'} \frac{1}{E_0 - E_{I'}} U_{I'0}. \qquad (3.44)$$

Here I and I' are "intermediate" states and these states are summed over. The probability of the process between the initial state 0 and the final state f would be proportional to $|H_{f0}|^2$ and would determine a cross section or transition probability. The states 0 and f are observed or specified in the sense that the system is considered to be measured or detected in these states. The intermediate states are, of course, not observed, and it is essential in the evaluation of the overall $0 \to f$ process that all of these accessible intermediate states be included in the total amplitude for the process.

The probability of a process resulting from two perturbations can be written in the form

$$W_{f0} = \left| \sum_I A_{fI} A_I A_{I0} \right|^2. \qquad (3.45)$$

An individual amplitude is the product of these factors, reading from right to left: (i) an amplitude (A_{I0}) associated with the action of a perturbation acting as to cause the system to make a transition $0 \to I$, (ii) an amplitude (A_I) that is only a function of the intermediate state, and (iii) an amplitude (A_{fI}) from a perturbation acting to cause a transition $I \to f$. If a process can take place through the action of a single perturbation, there is no intermediate state and there would be only the amplitude A_{f0} in lowest order. On the other hand, for a given process, even in lowest order in a

[13] This can be done by simply replacing H' with $U + V$, yielding the four types of second-order terms (UU, VV, UV, VU), each with an energy denominator.

coupling constant,[14] the total amplitude may involve both a simple direct amplitude A_{f0} and a combination of amplitudes involving intermediate states as in the form (3.45). That is, in general, the total amplitude for a process can involve a number of perturbations and intermediate states.

The "intermediate-state amplitude" A_I is just the factor $(E_0 - E_I)^{-1}$ involving the energy denominators. It corresponds to, in modern covariant perturbation theory, the Feynman *propagator* or propagation factor. Although the idea has limited physical meaning, being simply a factor in a perturbation-theory development, it is a convenient notion to introduce. As is indicated in the notation in Equation (3.45), we can think of the total amplitude for the process involving two perturbations as if A_I were an amplitude for propagation in the intermediate state between the two interactions. Basically, this is the reason why the factor with the energy denominator in Equation (3.39) was written in between the two matrix elements. It is also convenient (but not necessary) to introduce a pictorial representation of the process whereby a particle undergoes the transition $0 \to f$ by means of the two perturbations U and V and the intermediate states I and I' [see Equation (3.44)]. The diagrams represent terms in a perturbation series and are nothing more than a bookkeeping device. Note that time runs vertically in these pictures, with the initial state indicated at the bottom and the final state at the top. The horizontal scale can represent position in a rough sense, so that the picture is like that of a "world line," taking some terminology from relativity. An actual path or line for the particle is, of course, not implied, since this would be a classical notion foreign to our quantum-mechanical formulation. These diagrams are not to be taken "literally"; rather, they serve as a guide or reminder in writing down terms in a perturbation series.

Two diagrams are indicated in Figure 3.1, corresponding to the second-order perturbation for the case where the perturbations U and V are different and able to act in either order to cause the transition $0 \to I$ or $I' \to f$. An important example of a problem of this type would be that of bremsstrahlung, which takes place through the combined process of scattering (U, say) and photon production (V). Here, by "photon production" is meant the interaction (3.20) between the charge and the field of the outgoing photon. Charged particles are always subjected to the photon-emission Hamiltonian [Equations (3.20) as well as (3.21) and (3.22)]. However, energy conservation does not permit an isolated charge to produce a photon. An additional perturbation is required, such as a scattering potential, and through the combined action of both perturbations the photon can be produced. This has a corresponding feature in classical radiation theory in which emission takes place because of the charge's acceleration (as a result of the scattering potential).

The interaction associated with photon emission can be represented in a picture by a *vertex* with a wavy line designating the photon. Both the Hamiltonian (3.20) corresponding to the interaction of the charge with the radiation field and the expression (3.22) from the interaction of the intrinsic magnetic moment can be represented by a vertex of the type in Figure 3.2. Again with time running in the vertical direction, this picture is a designation of a matrix element H'_{ba} in which a and b refer to the

[14] As noted already, the perturbation (3.20) is first order in the charge while the perturbation (3.21) is second order.

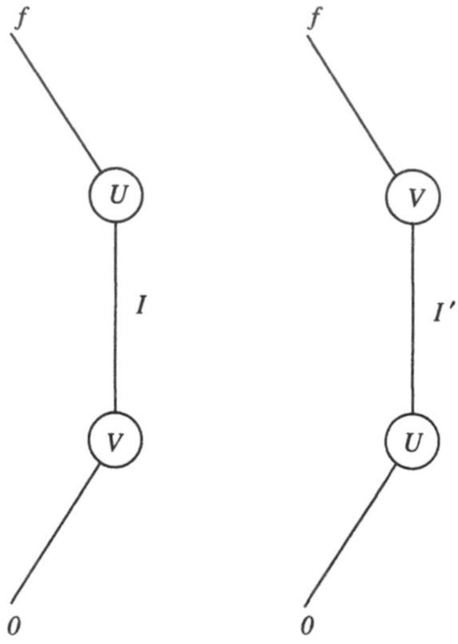

Figure 3.1 Pictorial representation of the action of two perturbations. The vertices represent the matrix elements of the U and V interactions and the line between the vertices represents the energy denominator factor.

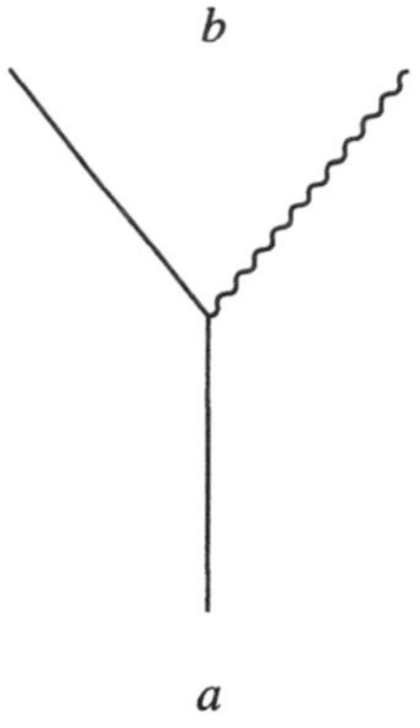

Figure 3.2 Photon-coupling vertex corresponding to photon "production" (outgoing photon).

initial and final states, respectively, with the state b containing a photon. Since an isolated charge or magnetic moment cannot emit a photon, the vertex of Figure 3.2 must be just part of a diagram if a photon process is being represented. As noted above, it could be part of the bremsstrahlung diagrams, and we see that the two diagrams correspond to emission "before and after" the scattering. Again, it should be emphasized that this terminology, and the diagrams, should not be taken literally. "Emission" is associated with the probability for the process, and the probability is the squared total amplitude. The total amplitude itself is not observed nor are its individual components associated with individual diagrams.

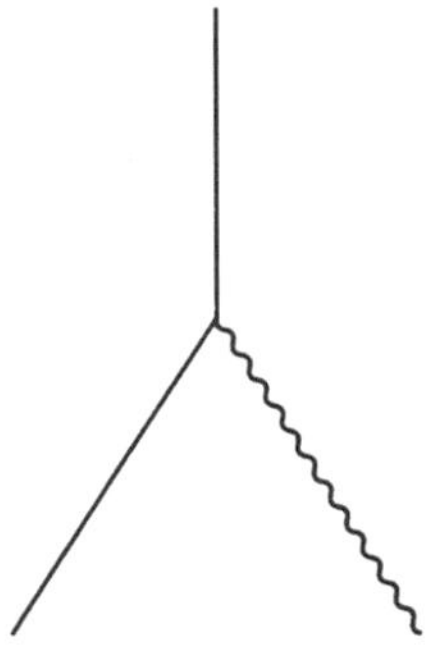

Figure 3.3 Photon-coupling vertex associated with photon "absorption".

Photons can be absorbed as well, and the vertex or matrix element of the Hamiltonian (3.20) associated with interaction of an incoming photon by this coupling would be represented as in Figure 3.3. Basically, the vertices of Figures 3.2 and 3.3 are the same, being the time reverse of one another. However, they do represent distinctly different interactions or perturbations, since one involves an incoming photon and the other an outgoing photon. This means, for example, that in the evaluation of the process of photon scattering by a free charge (Compton scattering), two diagrams occur that involve the vertices of Figures 3.2 and 3.3. That is, if these vertices correspond to the interactions U and V, respectively, of Figure 3.1, the "absorption" and "emission" in the two-step scattering process can occur in either order.[15] Although the basic nature of the interaction is the same (for U and V) in this case, the two photons involved are different and so, therefore, are the associated perturbations. In a later chapter, we give a more complete discussion of Compton scattering, and we shall see that the two diagrams referred to above actually do not contribute to the cross section. The whole contribution comes from a third diagram that is associated with the interaction Hamiltonian (3.21). This diagram, a vertex associated with particle coupling to two photons or two photon fields, would be, for one incoming and one outgoing photon, as shown in Figure 3.4. That is, the coupling involves the point interaction of the fields of four particles: the incoming and outgoing charges and the two photons. The interaction is of higher order than that associated with the coupling (3.20) (and the diagrams in Figure 3.2 or 3.3), being second order in the charge.

The intrinsic magnetic moment coupling (3.22) to the photon field has a vertex looking like that for the coupling (3.20), that is, like Figures 3.2 or 3.3. However, for processes involving non-relativistic charges, this coupling is weaker than the

[15]Again, we do not mean this to be taken literally. Without performing an experiment to do so—and thereby disturbing the process—we cannot know which perturbation acted first. The separate diagrams refer only to mathematical terms in a perturbation-theory formulation.

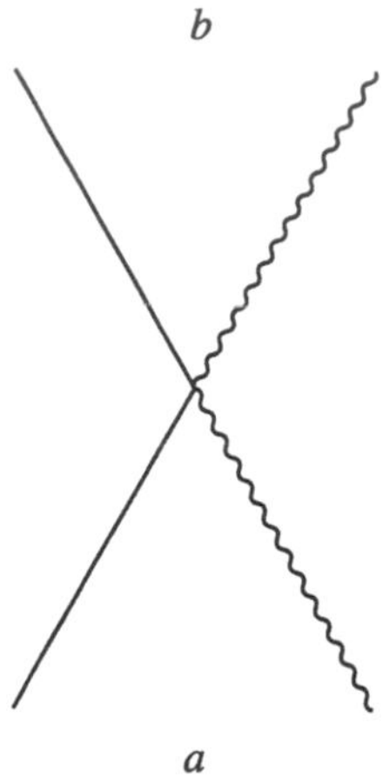

Figure 3.4 Diagram representing the coupling (vertex) of two photon fields, in this case an incoming and an outgoing photon.

lowest-order term R. Since $\mu_\alpha \sim q_\alpha \hbar / m_\alpha c$, we see that

$$R/S \sim k/k_\alpha \sim \hbar\omega/(m_\alpha c^2 E_\alpha)^{1/2}, \tag{3.46}$$

where E_α is the particle kinetic energy. The ratio (3.46) is small, being $\sim (E_\alpha/m_\alpha c^2)^{1/2}$ for $\hbar\omega \sim E_\alpha$. If, on the other hand, a particle had no charge but possessed a magnetic moment, the coupling (3.22) would be the only interaction with the photon field. It is perhaps appropriate to remark at this point that covariant perturbation theory for spin-$\frac{1}{2}$ particles has only one basic interaction or vertex. This vertex is of the type in Figure 3.2 (or 3.3), associated with a linear coupling to the electromagnetic field. The non-relativistic theory, on the other hand, has three basic couplings (3.20)–(3.22), and so the transition to the relativistic theory is somewhat complex. The non-relativistic, non-covariant theory is really quite different from the covariant formulation, and there is not a one-to-one correspondence between the vertices and diagrams in our treatment here and those associated with the relativistic theory.

The "elementary" particles have an important fundamental property—their particle-antiparticle symmetry. That is, in addition to the photon, proton, neutron, electron, pion, muon, and neutrino (electron and muon types), there are the corresponding antiparticles. The antiphoton is absolutely identical to the photon ($\overline{\gamma} = \gamma$). The neutron is also neutral, but it has a magnetic moment and for $\overline{n}$ there is a sign difference in the relation between magnetic moment and spin. The antineutrino has a helicity opposite in sign to that of the neutrino. The antielectron (positron) is identical to the electron except for being oppositely charged (which also has a corresponding effect on its magnetic moment). This is also true of the other charged "elementary" particles[16] (proton, muon, and charged pion). The particle-

[16]On the other hand, the "strange" particles do not possess this symmetry between counterparts of opposite charge.

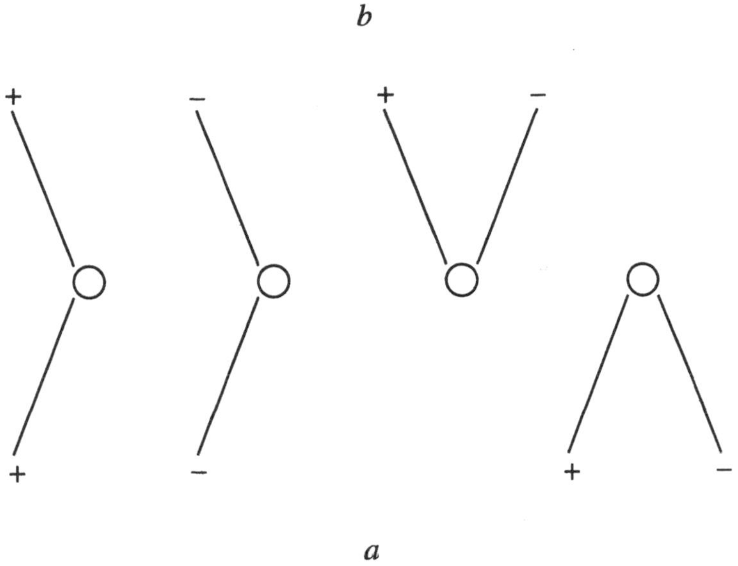

Figure 3.5 Particle and antiparticle diagrams.

antiparticle symmetry means, for example, that in quantum electrodynamics a single unified theory describes electron and positron processes. In particular, it means that, consistent with charge conservation, for any perturbation matrix element

$$V_{ba} = \langle b|V|a\rangle \,, \tag{3.47}$$

the only requirement on the states a and b is that they be consistent with charge conservation. This can be represented pictorially as in Figure 3.5, which gives the four possibilities for interactions involving electrons and/or positrons. In these vertices or diagrams, the circle can represent any kind of interaction. It can correspond to some external potential or one of the three electromagnetic perturbations (3.20)–(3.22). For the latter perturbations, the photons involved can be either incoming or outgoing in all possibilities, so that there are many subcategories[17] for the vertices in Figure 3.5. Some of these specific vertices may have a matrix element that is zero; this is, in fact, the case, as will be seen later.

The two vertices on the right of Figure 3.5 would be part of processes involving pair production and annihilation, respectively. Because of the basic symmetry of the theory, a unified treatment of these processes can be given in terms of a basic interaction that can cause a variety of processes. For example, bremsstrahlung is closely related to pair production (and annihilation), the diagrams for the processes being essentially the same when one set is placed on its side. These processes will be treated in detail in later chapters.

[17]For the interactions (3.20) and (3.22), each of the vertices in Figure 3.5 can involve either an incoming or an outgoing photon. For the interaction (3.21), each vertex could involve an incoming and an outgoing photon, two incoming photons, or two outgoing photons.

3.4 RELATIVISTIC THEORY

Although the treatment of non-relativistic QED in the last section is fairly self-contained, only a very superficial formulation of the relativistic theory can be given here.[18] A substantive exposition of modern QED would require much more space, but perhaps we can provide some understanding of the essence of the theory in this brief outline.

3.4.1 Modifications of the Non-Covariant Formulation

The early ($\sim$1930) formulations of relativistic QED were not manifestly covariant and were somewhat complicated in their derivations of the foundations and in the associated computational techniques in the applications of the theory. The newer ($\sim$1950) methods are preferable and benefit from the more "natural" formulation in terms of covariant equations. Although no new physics is introduced in the covariant formulation, it is quite different in its mathematical layout and in its subsequent computational methods. In the interest of also retaining some simplicity, let us now consider how the non-relativistic theory might be modified.

Some aspects of the theory can be considered to be very general, and we can expect to carry over some of the principles introduced in the non-relativistic development. Again, it is convenient to formulate a perturbation theory, and certain features of the non-covariant formulation in the last section are retained. The non-covariant perturbation theory is still a valid procedure in applications involving even relativistic particles, since it is based on the general Schrödinger equation (3.23) and the superposition principle. Transitions are viewed as the result of the action of perturbations, and more than one perturbation may be required to produce a non-vanishing total amplitude for a certain process. In the mathematical formulation, the perturbations result in matrix elements associated with transitions between two states, and we have found it convenient to introduce the notion of a "vertex" in the pictorial representation of the transition. One or both of these states may be "intermediate" so that the "transition" involving a particular vertex may not be directly related to an observable event; in that case, the associated matrix element is just one of the factors in the expression for the total amplitude for the process. To be consistent with the general literature on the subject, we use the notation, say, M_{lk} for one of these matrix elements (vertices) corresponding to the $k \to l$ transition.

In an amplitude for a process in which intermediate states are involved, it can be expected that there will be a factor corresponding to the energy denominator $(E_0 - E_I)^{-1}$ in the non-covariant formulation. This factor is a function only of the characteristics of the intermediate state, and in covariant perturbation theory it is called a (Feynman) *propagator*. For now we designate it as P_I for an intermediate state I, and we try to give simple arguments for inferring its expected form. For a process corresponding to some (total) transition $0 \to f$ in which an arbitrary number of intermediate states are involved, the probability or rate would be given

[18] There are a large number of excellent textbooks on relativistic QED. A selection of these, chosen for their variety of approach and penetration of the subject, are listed at the end of this chapter.

by an expression of the form

$$W_{f0} = \left| \sum_{I_k \cdots I_n} M_{fI_n} P_{I_n} M_{I_n I_m} \cdots M_{I_l I_k} P_{I_k} M_{I_k 0} \right|^2 . \tag{3.48}$$

The total amplitude in this expression is meant to be read from right to left. Loosely stated, the matrix element M_{I_k0} causes the transition from the initial state 0 to the intermediate state I_k, there is then a "propagation" (P_{I_k}) to the next interaction or vertex $M_{I_lI_k}$, and so forth, to the last interaction M_{fI_n}. The M's are just the perturbation Hamiltonians, and, if there is more than one such perturbation, any one of them can be among the string of factors in Equation (3.48). That is, there can be various combinations of interactions resulting in a variety of processes, and for a given process there may be several combinations yielding the total amplitude for the process.

In covariant perturbation theory, the matrix elements M_{lk} are evaluated from a spacetime integration over the invariant four-dimensional volume d^4x or, in the momentum representation, over d^4p. This results in a fundamental difference from the non-covariant theory in which matrix elements are evaluated from an integration over the three-dimensional spatial volume $d^3\boldsymbol{r}$ (or $d^3\boldsymbol{p}$). The time (as well as space) integration in the matrix elements yields a simplification in the perturbation theory, which is described conveniently in terms of a diagrammatic representation. As noted by Stückelberg in 1942 and fully exploited by Feynman in 1948, positrons—holes in the sea of electrons in negative energy states in Dirac's relativistic electron theory—can be described as electrons moving backward in time.[19] In a process that involves, for example, two interactions, that is, two matrix elements, non-covariant perturbation theory must include the two types of diagrams indicated in Figure 3.6. One involves pair production with the particle of opposite sign to the incoming particle annihilating with the latter; this diagram corresponds to an intermediate state with three particles. On the other hand, in the covariant theory, in which there is an integration over all t for each interaction matrix element, both diagrams on the left in Figure 3.6 are effectively included in a single one. That is, in following the particle's "world line" in covariant theory, we do not distinguish the two diagrams on the left, which are regarded as essentially the same. This can also be seen through consideration of a Lorentz transformation or rotation in the space-time plane; in another Lorentz frame, the pair-production diagram is that on the far left in the figure. The non-covariant theory is, as we have stated earlier, valid—even for relativistic problems. However, it is not as convenient as the covariant theory, which is fundamentally the more natural approach. In non-relativistic problems, the non-covariant theory is more convenient, and the pair-producing diagram is negligible. This is because the corresponding intermediate-state energy is very large, and the factor $(E_0 - E_I)^{-1}$ is very small in the term in the effective perturbation Hamiltonian [see Equation (3.44)].

There is another feature of covariant perturbation theory that may be considered a consequence of the four-dimensional integrations in matrix elements. Matrix

[19] This can be seen in the phase factor $k_\mu x_\mu = \boldsymbol{k} \cdot \boldsymbol{r} - \omega t$ in the wave function of a propagating particle. Replacing ω ($= E/\hbar$) by $-\omega$ corresponds to a description in which the direction of motion is changed unless $t \to -t$.

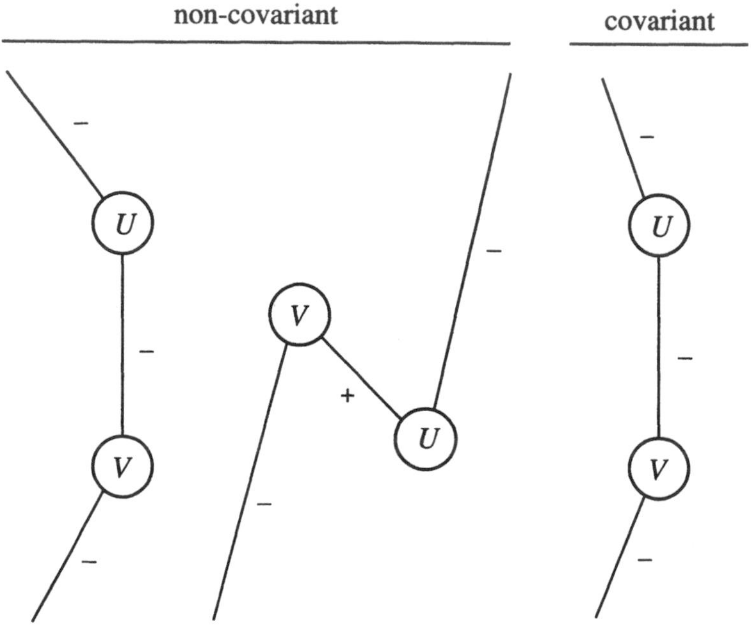

Figure 3.6 Diagrams representing the action of two perturbations in the non-covariant formulation and in the covariant formulation.

elements in the non-covariant theory yield momentum-conservation δ-functions, since the position-space integrals are always of the form

$$\int e^{i\boldsymbol{\kappa}\cdot\boldsymbol{r}}\,d^3\boldsymbol{r} = (2\pi)^3\delta^{(3)}(\boldsymbol{\kappa}), \tag{3.49}$$

[see Equation (2.68)], where $\delta^{(3)}(\boldsymbol{\kappa}) = \delta(\kappa_x)\delta(\kappa_y)\delta(\kappa_z)$, and

$$\boldsymbol{\kappa} = \sum_{\text{in}}\boldsymbol{k} - \sum_{\text{out}}\boldsymbol{k} \tag{3.50}$$

is just the difference between the sum of the wave vectors (or momenta) for the incoming and outgoing particles involved in a vertex. Momentum conservation results because the matrix elements are evaluated from

$$M_{lk} = \int u_l^* M u_k\, d^3\boldsymbol{r} \quad \text{(non-cov.)}, \tag{3.51}$$

with the two spatial wave functions being products of (unperturbed) factors $\exp(i\boldsymbol{k}\cdot\boldsymbol{r})$. In the covariant theory, the matrix elements involve the space- and time-dependent free-particle wave functions (3.25)

$$M_{lk} = \int \psi_l^* M \psi_k\, d^4x \quad \text{(cov.)}, \tag{3.52}$$

with the ψ's of the form $\exp(ik_\mu x_\mu)$. The integrals then yield the result

$$\int e^{i\kappa_\mu x_\mu}\,d^4x = (2\pi)^4\delta^{(4)}(\kappa_\mu) = (2\pi)^4\delta(\kappa_0)\delta^{(3)}(\boldsymbol{\kappa}), \tag{3.53}$$

corresponding to conservation of *both* momentum and energy at each vertex. This then represents a fundamental difference from non-covariant perturbation theory where there was not energy conservation at vertices and in intermediate states. Consequently, diagrams and factors in amplitudes for processes do not have the same precise meaning in the non-covariant and covariant formulations.

The starting point for all relativistic theories is an Action Principle:

$$\delta \int \mathcal{L}\, d^4x = 0, \tag{3.54}$$

where $\mathcal{L}$ is a scalar Lagrangian density. The Lagrangian is a function of the various field components $\phi^\alpha(x_\mu)$ and their derivatives $\partial\phi^\alpha/\partial x_\mu$, where the index α refers to the various types of particles (photons, electrons, etc.). For the integral (3.54) to have a stationary value for independent variations of the field components, the set of field equations must hold:

$$\frac{\partial\mathcal{L}}{\partial\phi^\alpha} - \frac{\partial}{\partial x_\mu}\left(\frac{\partial\mathcal{L}}{\partial(\partial\phi^\alpha/\partial x_\mu)}\right) = 0. \tag{3.55}$$

The Lagrangian is constructed to yield the correct equations of motion, including the effects of interactions, when substituted into the set of relations (3.55). If only electrons, positrons, and photons are involved, $\mathcal{L}$ would be of the form

$$\mathcal{L} = \mathcal{L}_{\mathrm{D}} + \mathcal{L}_{\mathrm{M}} + \mathcal{L}_{\mathrm{I}}, \tag{3.56}$$

consisting of a Dirac (electrons and positrons), Maxwell (photons), and interaction part. Since the Dirac equation is quantum mechanical, the Lagrangian (3.56) is as well, although further steps must be taken to develop the full quantum field theory. Since the theory must allow for production and destruction of photons and particles (in pairs), it is inherently "many-particle" in nature. That is, the ψ in the Dirac Lagrangian is not a one-particle wave function but an operator for the Dirac *field*.

The older formulations of relativistic QED, developed around 1930, were from a "Hamiltonian" approach with commutation relations introduced for the field operators. We now know that this method of field quantization is not very good. It is cumbersome and not fully covariant in its formulation. Feynman, Schwinger, and Dyson showed around 1950 how to reformulate the subject in a much better way. Dyson, in particular, gave a systematic treatment of QED by making the appropriate modifications in the non-covariant field theory to make the formulation covariant at every stage. In addition to demonstrating the connection with the work of Schwinger, Dyson was able to provide a basis for the so-called Feynman rules (and Feynman diagrams) as a logical consequence of covariant perturbation theory. Rather than try to reproduce this formulation, which would have to be lengthy, we attempt to give a shortcut superficial account that may at least provide some understanding of the theory and its applications.

3.4.2 Photon Interactions with Charges without Spin

The simplest, relativistic, one-particle wave equation corresponding to the non-relativistic Schrödinger equation was first written down by Schrödinger himself. However, to avoid confusion with *the* Schrödinger equation, it is referred to as the

Klein-Gordon equation, named after others who first considered relativistic wave equations. If ψ is a scalar function of only $\boldsymbol{r}$ and t (that is, not describing any spin states), a free-particle wave equation could be written as $i\hbar\partial\psi/\partial t = H\psi$ with $H = (p^2c^2 + m^2c^4)^{1/2}$ as the Hamiltonian, replacing p^2 by the operator $-\hbar^2\nabla^2$ as in the non-relativistic theory. This wave equation with the square root is unwieldy, and there is also the problem of the appropriate sign outside the squre root. No one has made progress with the equation and, instead, an equation was formed from the relativistic relation $p_\mu p_\mu = -m^2c^2$, replacing p_μ with $-i\hbar\partial_\mu$ and allowing the terms to operate on $\psi(\boldsymbol{r}, t)$. This is what is now known as the Klein-Gordon (KG) equation[20]:

$$(\partial_\mu\partial_\mu - \kappa^2)\psi = (\Box^2 - \kappa^2)\psi = 0, \tag{3.57}$$

where $\kappa = mc/\hbar$. For particles satisfying such one-particle equations, a field theory can be formulated to describe many-particle states. If there are several types of such particles, characterized by their mass (κ) parameter, the Lagrangian density for the scalar field of each type would be of the form[21]

$$\mathcal{L} = \mathcal{L}_{\text{KG}} = -\tfrac{1}{2}(\partial_\mu\phi\partial_\mu\phi + \kappa^2\phi^2); \tag{3.58}$$

by the Lagrange equation (3.55), the field then satisfies a Klein-Gordon equation of the form (3.57):

$$(\Box^2 - \kappa^2)\phi = 0. \tag{3.59}$$

The sign and factor in front of the Lagrangian function (3.58) are arbitrary for yielding Equation (3.59). However, with this choice the canonical momentum is

$$\pi = \partial\mathcal{L}/\partial\dot{\phi} = \dot{\phi}/c^2, \tag{3.60}$$

and the Hamiltonian

$$\mathcal{H} = \pi\dot{\phi} - \mathcal{L} = \tfrac{1}{2}\pi^2 + \tfrac{1}{2}(\nabla\phi)\cdot(\nabla\phi) + \tfrac{1}{2}\kappa^2\phi^2 \tag{3.61}$$

is positive definite.

One problem with the Klein-Gordon equation is that if the probability current density is given, as in the non-relativistic Schrödinger theory, by

$$\boldsymbol{j} = (\hbar/2mi)(\psi^*\nabla\psi - \psi\nabla\psi^*), \tag{3.62}$$

for the conservation equation

$$\partial\rho/\partial t + \nabla\cdot\boldsymbol{j} = \partial_\mu j_\mu = 0 \tag{3.63}$$

to hold, the probability density must be given by

$$\rho = \frac{i\hbar}{2mc^2}\left(\psi^*\frac{\partial\psi}{\partial t} - \frac{\partial\psi^*}{\partial t}\psi\right). \tag{3.64}$$

Although this expression reduces to the correct non-relativistic limit, it is not positive definite in general. The theory remained in this stage until 1934 when Pauli and

[20] In the various textbooks, the reader will notice sign differences in terms in equations associated with the relativistic theory. This has to do with the somewhat arbitrary choice for the "metric." For example, in many texts $\Box^2$ is meant to designate $(1/c^2)\partial^2/\partial t^2 - \nabla^2$, which is the same as our $\partial_\mu\partial_\mu$ but with a minus sign.

[21] In this function the ∂_μ is meant to operate only on the ϕ to its immediate right.

Weisskopf reinterpreted the density ρ of the Klein-Gordon theory in terms of a charge density. The formulation of a theory of scalar particles of positive and negative charge is conveniently done by introducing a complex scalar field

$$\phi = 2^{-1/2}(\phi^{(1)} + i\phi^{(2)}), \tag{3.65}$$

where $\phi^{(1)}$ and $\phi^{(2)}$ are two real scalar fields. They can be written in terms of ϕ and ϕ^* as

$$\begin{aligned} \phi^{(1)} &= 2^{-1/2}(\phi + \phi^*), \\ \phi^{(2)} &= (1/2^{1/2}i)(\phi - \phi^*). \end{aligned} \tag{3.66}$$

The total Lagrangian $\mathcal{L} = \mathcal{L}^{(1)} + \mathcal{L}^{(2)}$, with $\mathcal{L}^{(1)}$ and $\mathcal{L}^{(2)}$ of the form (3.58), is then, very simply,

$$\mathcal{L} = -\partial_\mu \phi^* \partial_\mu \phi - \kappa^2 \phi^* \phi. \tag{3.67}$$

Here ϕ^* and ϕ are to be considered as independent fields (instead of $\phi^{(1)}$ and $\phi^{(2)}$). Both ϕ and ϕ^* satisfy Equation (3.59), as we see on applying the Lagrange equation (3.55) separately for ϕ and ϕ^* using the Lagrangian (3.67). Further, $\pi = \partial\mathcal{L}/\partial\dot{\phi} = \dot{\phi}^*/c^2$ and $\pi^* = \partial\mathcal{L}/\partial\dot{\phi}^* = \dot{\phi}/c^2$; the Hamiltonian is

$$\begin{aligned} \mathcal{H} &= \pi\dot{\phi} + \pi^*\dot{\phi}^* - \mathcal{L} \\ &= \pi^*\pi + (\nabla\phi^*)\cdot(\nabla\phi) + \kappa^2\phi^*\phi. \end{aligned} \tag{3.68}$$

We can construct a four-current density of the form

$$j_\mu = \text{const}(\phi^{(1)}\partial_\mu\phi^{(2)} - \phi^{(2)}\partial_\mu\phi^{(1)}), \tag{3.69}$$

and since $\Box^2\phi^{(1)} = \kappa^2\phi^{(1)}$ and $\Box^2\phi^{(2)} = \kappa^2\phi^{(2)}$,

$$\partial_\mu j_\mu = 0. \tag{3.70}$$

The current can be written conveniently in terms of ϕ and ϕ^*:

$$j_\mu = \text{const}\left[\phi\frac{\partial\mathcal{L}}{\partial(\partial_\mu\phi)} - \phi^*\frac{\partial\mathcal{L}}{\partial(\partial_\mu\phi^*)}\right], \tag{3.71}$$

with, for example, $\rho = \text{const}(\pi\phi - \pi^*\phi^*)$. This gives some indication of how the theory is applied to describe scalar particles of equal and opposite sign.

This theory, with the Lagrangian (3.67), can be used as a starting point for the description of the interaction of charged pions (π^+, π^-) with photons. From the free-particle Lagrangian (3.67), the interaction Lagrangian can be obtained by replacing ∂_μ with $\partial_\mu - (iq/\hbar c)A_\mu$:

$$\mathcal{L}_{\text{KG}} + \mathcal{L}_{\text{I}} = -\left(\partial_\mu\phi - \frac{iq}{\hbar c}A_\mu\phi\right)^*\left(\partial_\mu\phi - \frac{iq}{\hbar c}A_\mu\phi\right) - \kappa^2\phi^*\phi. \tag{3.72}$$

The interaction Lagrangian is then (see, also, Footnote 21)

$$\mathcal{L}_{\text{I}} = \frac{iq}{\hbar c}(\phi\partial_\mu\phi^* - \phi^*\partial_\mu\phi)A_\mu + \left(\frac{q}{\hbar c}\right)^2\phi^*\phi A_\mu A_\mu. \tag{3.73}$$

To the Lagrangian (3.72) should be added that for a pure photon field. This is usually referred to as the *Maxwell Lagrangian* ($\mathcal{L}_{\mathrm{M}}$) and a convenient form has been introduced by Fermi[22]:

$$\begin{aligned}\mathcal{L}_{\mathrm{M}} &= -\frac{1}{16\pi}F_{\mu\nu}F_{\mu\nu} - \frac{1}{8\pi}(\partial_\mu A_\mu)^2 \\ &= -\frac{1}{16\pi}(\partial_\mu A_\nu - \partial_\nu A_\mu)(\partial_\mu A_\nu - \partial_\nu A_\mu) - \frac{1}{8\pi}(\partial_\mu A_\mu)^2.\end{aligned} \tag{3.74}$$

The total Lagrangian for charged spinless bosons interacting with photons is then

$$\mathcal{L} = \mathcal{L}_{\mathrm{KG}} + \mathcal{L}_{\mathrm{M}} + \mathcal{L}_{\mathrm{I}}. \tag{3.75}$$

The coupling terms in the interaction Lagrangian (3.73) are similar in form to the forms (3.20) and (3.21) that arise in the non-relativistic theory. And, of course, the diagrams that are introduced as a guide in a perturbation-theory description look the same. Because the non-relativistic theory has fundamental differences from the covariant formulation, we have not previously referred to them as "Feynman diagrams," believing that the designation should be reserved for the covariant theory. Although we cannot give a complete description of this theory, we can infer something about it in a simple way, and the form (3.73) can be employed for this purpose.

The first term on the right in the Lagrangian (3.73) has a coupling strength linear in the charge and is associated with, say, the interaction of two charged meson fields with a photon field. It corresponds to a vertex with two meson lines and one photon line like that in Figure 3.2 or 3.3. All lines can be incoming or outgoing. If one meson (field) is incoming and the other is outgoing, the matrix element associated with the interaction or vertex is of the form

$$M_{ba}^{(q)} \propto q\varepsilon_\mu(k_{a\mu} + k_{b\mu}), \tag{3.76}$$

where ε_μ is the photon polarization four-vector arising from the photon field $A_\mu \propto \varepsilon_\mu \exp(\pm i k_\mu x_\mu)$, where the $+$ $(-)$ would be associated with an incoming (outgoing) photon, and $k_{a\mu}$ and $k_{b\mu}$ are the wave vectors for the incoming and outgoing mesons. Since the meson fields are of the form $\exp(ik_{a\mu}x_\mu)$ and $\exp(ik_{b\mu}x_\mu)$, the matrix element is a result of the integration

$$\begin{aligned}M_{ba}^{(q)} &\propto q\varepsilon_\mu(k_{a\mu} + k_{b\mu}) \int d^4x\, e^{i(k_{a\mu} - k_{b\mu} \pm k_\mu)x_\mu} \\ &= (2\pi)^4 q\varepsilon_\mu(k_{a\mu} + k_{b\mu})\delta^{(4)}(k_{a\mu} - k_{b\mu} \pm k_\mu).\end{aligned} \tag{3.77}$$

The δ-function is simply a manifestation of energy and momentum conservation at the vertex.

---

[22] We shall not employ the specific form for $\mathcal{L}_{\mathrm{M}}$ given here, but quote the result for completeness (see references at end of chapter). It is expressed here in c.g.s. or Gaussian units, but most books on QED use the Heaviside-Lorentz units, which have the advantage of eliminating factors of $(4\pi)^{-1}$ in equations for electromagnetic phenomena. In these (HL) units, however, the electronic charge has a different value; for example, the fine-structure constant is given by $(e^2/4\pi\hbar c)_{\mathrm{HL}} = \alpha \simeq 1/137$. The reader should be aware of the use of these different unit systems.

The second term in the interaction Lagrangian (3.73) is second order in the charge and involves two meson fields and two photon fields (see Figure 3.4). It yields a matrix element of the form

$$
\begin{aligned}
M_{ba}^{(q^2)} &\propto q^2 \varepsilon_{1\mu}\varepsilon_{2\mu} \int d^4x \, e^{i(k_{a\mu} - k_{b\mu} \pm k_{1\mu} \pm k_{2\mu})x_\mu} \\
&= (2\pi)^4 q^2 \varepsilon_{1\mu}\varepsilon_{2\mu} \delta^{(4)}(k_{a\mu} - k_{b\mu} \pm k_{1\mu} \pm k_{2\mu}),
\end{aligned}
\tag{3.78}
$$

where the + (−) sign on the photon four-momenta is associated with incoming (outgoing) states.

The form of the couplings can be obtained in another way by making the replacement $\partial_\mu \to \partial_\mu - (iq/\hbar c)A_\mu$ directly in the Klein-Gordon equation $(\partial_\mu \partial_\mu - \kappa^2)\phi = 0$. The "perturbed" equation is then

$$
(\Box^2 - \kappa^2)\phi = (iq/\hbar c)[A_\mu \partial_\mu \phi + \partial_\mu (A_\mu \phi)] + (q/\hbar c)^2 A_\mu A_\mu \phi. \tag{3.79}
$$

The right-hand side of this equation represents source terms $S^{(q)}$ and $S^{(q^2)}$ and is linear in the charged boson field ϕ. Transition matrix elements would be evaluated from integrals of the form $\int \phi_b^* S_a d^4x$. The q-term of the source (3.79), with ∂_μ operating on ϕ_a and A_μ, yields

$$
\begin{aligned}
M_{ba}^{(q)} &\propto q\varepsilon_\mu (k_{a\mu} + k_{b\mu} \pm k_\mu) \int d^4x \, e^{i(-k_{b\mu} + k_{a\mu} \pm k_\mu)} \\
&= (2\pi)^4 q \varepsilon_\mu (k_{a\mu} + k_{b\mu}) \, \delta^{(4)}(k_{a\mu} - k_{b\mu} \pm k_\mu),
\end{aligned}
\tag{3.80}
$$

identical to the result (3.77). In a similar manner, the q^2-term (3.78) is again obtained.

The other fundamental type of quantity in the amplitude for a process is the Feynman propagator P_I. Let us see[23] if the form for this factor can be inferred from simple arguments. We consider again the effects of the action of the two perturbations V and U in a process involving a scalar charged particle (see Figure 3.6 and Section 4.1). As we have already seen, in a covariant formulation, the two diagrams of the non-covariant theory are replaced by one. The essential form of the Feynman propagator for scalar particles can be obtained if we start from the non-covariant amplitude in terms of energy denominators and try to inject some simple ideas of the modern theory. One is a more suitable covariant definition of interaction matrix elements in terms of "entry" and "exit" states: $\langle \text{exit}|M|\text{entry}\rangle$. With this description, we follow the "world line" of a particle even it goes backward in time. For a process involving, say, a π^-, the right-hand, non-covariant diagram in Figure 3.6 would correspond to the initial production (by U) of a π^+, π^- pair with the π^+ then annihilating (V) with the incident π^-. The diagram on its left would be associated with an amplitude

$$
A_1 = U_{fI} P_I V_{I0}, \tag{3.81}
$$

[23]Here we are relying heavily on R. P. Feynman, *Theory of Fundamental Processes*, New York: W. A. Benjamin, Inc., 1962.

with $P_I = (E_0 - E_I)^{-1}$. For the diagram with pair production (or with the π^- going backward in time), taking the exit-entry definition of matrix elements, the amplitude is

$$A_2 = V_{I0} P_I' U_{fI}, \tag{3.82}$$

where $P_I' = [E_0 - (E_I + E_0 + E_f)]^{-1} = -(E_0 + E_I)^{-1}$, since $E_0 = E_f$. Then

$$A_1 + A_2 = U_{fI}(P_I + P_I')V_{I0}, \tag{3.83}$$

where

$$P_I + P_I' = 2E_I/(E_0^2 - E_I^2). \tag{3.84}$$

The result (3.84) can be cast in another form that exhibits its covariant character. In attempting to obtain the form for the Feynman propagator, we should remember that the ultimate expression should be characteristic of the scalar nature of the particle (that is, not of its interactions) and be determined by, for example, its free-particle wave equation. Although we are considering electromagnetic phenomena, we can consider some special process convenient for the determination of the propagator for the intermediate state. We can, for example, consider the interaction V to be some (external) scattering potential (say, produced by some very heavy particle) such that the incident π^- suffers a change in the direction of its momentum but not in its magnitude. The interaction U can be a photon-emission vertex, and let us assume that the energies of the photons are infinitesimally small; then, even for the non-covariant diagram on the right in Figure 3.6, the photon momentum and energy in the intermediate (and final) state are negligible. Dropping, for simplicity, the subscript I for the intermediate state, we have $E_I^2/c^2 = -p^2 + \boldsymbol{p}^2$, where $p^2 \equiv p_\mu p_\mu$. Further, $E_0^2/c^2 = m^2c^2 + \boldsymbol{p}_0^2 = m^2c^2 + \boldsymbol{p}^2$. Thus we obtain[24]

$$(E_0^2 - E_I^2)/c^2 = p^2 + m^2c^2 \quad (p^2 \equiv p_\mu p_\mu). \tag{3.85}$$

We see that the resulting expression is an invariant. Note also that, although $p_\mu = (iE/c, \boldsymbol{p})$ for the intermediate state, $p^2 = p_\mu p_\mu \neq -m^2c^2$. The other factor $(2E_I)$ in the propagator (3.84) is of less significance but has some meaning in terms of the transition to the covariant formulation. In the covariant theory, a different normalization convention is more appropriate. The conservation of probability equation (3.70) requires that j_μ be a four-vector function of x_μ, and this is accomplished by adopting a normalization convention[25] $\int |u|^2 d^3\boldsymbol{r} = 2E$ (rather than unity) for free-particle states. This makes the probability density ρ [see also Equation (3.64)] the 0-component of a four-vector. Since u_I always appears twice in the total amplitude, this explains the occurrence of the factor $2E_I$. The essential factor in the covariant Feynman propagator for scalar particles is then

$$P_F(\text{scalar}) = (p^2 + m^2c^2)^{-1}. \tag{3.86}$$

---

[24] Often in the literature and in texts the propagator for scalar particles is given as $(p^2 - m^2c^2)^{-1}$. The sign difference has to do simply with the convention (metric) chosen for the four-dimensional scalar product. Again, the reader will have to be aware of these various sign conventions that are employed. Also, units with c (and $\hbar$) equal to unity are almost always used in this subject.

[25] Taking $2E$ rather than E just happens to be more convenient.

An alternative simplified approach[26] to obtaining P_F is illuminating in a different way. The coupled field equation (3.79) can be expressed in the form

$$(\Box^2 - \kappa^2)\phi = -S, \tag{3.87}$$

where the right-hand side can be considered a source term. It is written with a minus sign for convenience [compare Equation (2.71)]. Both ϕ and S are functions of x_μ and their Fourier transforms can be introduced. With f standing for either ϕ or S,

$$f(x_\mu) = \int f(k_\mu)e^{ik_\nu x_\nu} d^4k, \tag{3.88}$$

$$f(k_\mu) = (2\pi)^{-4} \int f(x_\mu)e^{-ik_\nu x_\nu} d^4x. \tag{3.89}$$

Substituting into the equation (3.87), we obtain the inhomogeneous solution ($k^2 = k_\mu k_\mu$)

$$\phi(x_\mu) = \int (k^2 + \kappa^2)^{-1} S(k_\mu)e^{ik_\nu x_\nu} d^4k', \tag{3.90}$$

which yields the factor $(k^2 + \kappa^2)^{-1}$ as in the result (3.86). The spacetime form of the propagator can be obtained by writing $S(k_\mu)$ in the terms of $S(x_\mu)$ by means of the relation (3.89). We then have

$$\phi(x_\mu) = \int D(x_\mu - x'_\mu)\, S(x'_\mu)\, d^4x', \tag{3.91}$$

where

$$D(x_\mu - x'_\mu) = \int \frac{e^{ik_\mu(x_\mu - x'_\mu)}}{k^2 + \kappa^2} \frac{d^4k}{(2\pi)^4} \tag{3.92}$$

is the propagator in the position representation. The form (3.91) exhibits how ϕ depends on S at other spacetime points.

3.4.3 Spin-$\frac{1}{2}$ Interactions

The general theory of the interactions of relativistic spin-$\frac{1}{2}$ particles with the electromagnetic field (photons) is what is commonly known as quantum electrodynamics. The spin-$\frac{1}{2}$ case is much more important than that for spin-0 because nature's smallest-mass charged particles, electrons and positrons, are of this type. Unfortunately, the theory is, on the whole, more complicated than that for spin-0 which we have just outlined. Thus, in imposing brevity, we shall have to be quite superficial and cannot really formulate the subject to provide the foundation necessary to perform extensive calculations. Nevertheless, certain basic characteristics of the theory can be seen without getting deeply into a more substantive exposition.

Basically, the content of the theory is determined by the form of the one-particle relativistic wave equation for the electron and positron, that is, the Dirac equation. In seeking this equation, Dirac was guided very much by a feeling that it should

[26] We are, again, relying heavily on R. P. Feynman, *Theory of Fundamental Processes*, New York: W. A. Benjamin, Inc., 1962.

have a form similar to the general time-dependent Schrödinger equation (3.1). Like the non-relativistic equation—and not like the Klein-Gordon equation—it should be first-order in the time derivative $\partial_t = \partial/\partial t$. But, since covariant equations have spatial derivatives ($\partial_j = \partial/\partial x_j$; $j = 1, 2, 3$) appearing in the same manner as time derivatives, it would seem that only the first derivatives ∂_j might be expected. The simplest such equation of this form is

$$(1/c)\partial_t \psi + \alpha_j \partial_j \psi + i\kappa\beta\psi = 0; \tag{3.93}$$

here α_j and β are numerical coefficients and κ ($= mc/\hbar$) is the basic inverse-length parameter characteristic of the particle whose state is described by $\psi(\boldsymbol{r}, t)$. The factor $1/c$ in the first term is natural in order to give the same dimensions as the second and third terms, if β and the three α_j are to be dimensionless. The factor i in the third term is arbitrary (but convenient) and could, alternatively, be incorporated into β.

Equation (3.93) is the free-particle Dirac equation, and it remains to determine the nature of β and the α_j. The wave function ψ is, moreover, allowed to have more than one component in order, for example, to describe particle spin; that is,

$$\psi = \begin{pmatrix} \psi_1 \\ \psi_2 \\ \vdots \end{pmatrix}. \tag{3.94}$$

In fact, the number of components has to be four if (for spin-$\frac{1}{2}$) two spin substates ($s_z = \pm 1/2$) are to be allowed as well as the two types of charge ($e^{\pm}$) or energy or antiparticle (hole) states in the theory. the four components are also called for by the resulting requirements on the coefficients β and α_j; these have to be *matrices* and at least 4×4. The form (3.93) chosen for the relativistic wave equation inherently implies an application to spin-$\frac{1}{2}$ particles with finite mass. The matrices β and α_j are determined by requiring that ψ also satisfy the Klein-Gordon equation (3.57), which is essentially a consequence of the relativistic relation $p_\mu p_\mu = -m^2c^2$ required also for free particles of spin-$\frac{1}{2}$. An equation of this type can be obtained if we operate with $(1/c)\partial_t - \alpha_l \partial_l - i\kappa\beta$ from the left on the Dirac equation (3.93). The coefficients and signs in this operator are fixed by the requirement that there be no terms in ∂_t and $\partial_t \partial_j$. Comparison with Equation (3.57) then yields the following requirements[27] on β and the α_j:

$$\begin{aligned} &\tfrac{1}{2}(\alpha_j \alpha_l + \alpha_l \alpha_j) = \delta_{jl}, \\ &\beta\alpha_j + \alpha_j \beta = 0, \\ &\beta^2 = I \quad (\text{unit matrix } I_{jl} = \delta_{jl}). \end{aligned} \tag{3.95}$$

It is possible to obtain solutions to these equations for β and the α_j in terms of the 2×2 Pauli spin matrices and the 2×2 identity matrix. However, instead of employing such explicit expressions, it is better not to indicate a particular representation, and

[27] The first of the relations (3.95), obtained through consideration of the term in $\partial_j \partial_l$, results in writing the coefficient of $\partial_j \partial_l$ ($= \partial_l \partial_j$) in the most general form to include all terms in the dummy indices j and l.

instead to regard the commutation relations (3.95) as fundamental and make use of identities derived directly from them. The β and α_j must be, in general, non-commuting operators or matrices as we see, in particular, from the second of the identities (3.95). Rather than employing these identities, it is, in fact, preferable to cast them—and the Dirac equation (3.93)—into four-dimensional covariant form; this is easily done.

Because of the last of the identities (3.95), if we operate from the left on the Dirac equation with β, we eliminate the matrix as a factor in the third term. The equation is then

$$\beta(1/c)\partial_t\psi + \beta\alpha_j\partial_j\psi + i\kappa\psi = 0, \tag{3.96}$$

and, in terms of $x_\mu = (ict, x_1, x_2, x_3)$ and ∂_μ, we see that the equation simplifies if we define

$$\begin{aligned} \gamma_0 &= \beta, \\ \gamma_j &= i\alpha_j\beta = -i\beta\alpha_j, \end{aligned} \tag{3.97}$$

and $\gamma_\mu = (\gamma_0, \gamma_1, \gamma_2, \gamma_3)$. The equation (3.96) then becomes

$$(\gamma_\mu\partial_\mu + \kappa)\psi = 0, \tag{3.98}$$

or, in terms of $p_\mu = -i\hbar\partial_\mu$,

$$(\not{p} - imc)\psi = 0, \tag{3.99}$$

where

$$\not{p} = \gamma_\mu p_\mu. \tag{3.100}$$

The covariant form (3.99) of the Dirac equation[28] is extremely simple. Moreover, the new Dirac matrices (3.97) satisfy the very compact identity

$$\gamma_\mu\gamma_\nu + \gamma_\nu\gamma_\mu = 2\delta_{\mu\nu}, \tag{3.101}$$

which replaces the three identities (3.95). The identity (3.101) is easily obtained by operating on the first of the relations (3.95) from the right and left with β.

The notation $\not{p}$, read "p dagger" or "p slash," was introduced by Feynman. For any four-vector, we define

$$\not{B} = \gamma_\mu B_\mu. \tag{3.102}$$

With A_μ and B_μ being four-vectors, the identity (3.101) yields the useful relation

$$\not{A}\not{B} + \not{B}\not{A} = 2A_\mu B_\mu = 2A \cdot B. \tag{3.103}$$

There are many other identities involving Dirac matrices, and often the relation (3.103) is employed in proving the more complicated ones.

[28]The equation appears in this form in some textbooks, but more often it is exhibited without the factor i multiplying mc. The difference has to do with the convention chosen for the metric (see, also, Footnote 24), which we avoid throughout by taking imaginary time components of four-vectors, etc. Our subsequent expressions, such as the Feynman propagator for spin-$\frac{1}{2}$ particles, will, as a result of our notation, also contain the factor i.

To include the effects of electromagnetic interactions, the replacement $\partial_\mu \rightarrow \partial_\mu - (iq/\hbar c)A_\mu$ or

$$\not{p} \rightarrow \not{p} - (q/c)\not{A} \tag{3.104}$$

can be made in the free-particle Dirac equation (3.99). The equation is then

$$(\not{p} - imc)\psi = (q/c)\not{A}\psi = (q/c)\gamma_\mu A_\mu \psi. \tag{3.105}$$

Immediately we see two important characteristics of the covariant theory of electron-photon interactions. The source or coupling is of the γ_μ type—one of the fundamental kinds of coupling in covariant perturbation theory. In covariant QED for spin-$\frac{1}{2}$ particles, there is just this one type of coupling. This is to be compared with the non-relativistic formulation, which must include *three* types of coupling [see Equations (3.12) and (3.13) and Figures 3.2 and 3.4] associated with the $\boldsymbol{A} \cdot \boldsymbol{p}$, $\boldsymbol{A} \cdot \boldsymbol{A}$, and $\boldsymbol{\mu} \cdot \text{curl}\, \boldsymbol{A}$ terms in the interaction Hamiltonian. The single coupling term in the covariant theory accounts for all of these interaction terms in the non-relativistic theory (even the magnetic moment coupling!). As a result, the number of diagrams representing perturbations in a covariant formulation is significantly reduced for the spin-$\frac{1}{2}$ case. The only vertex is one involving two (charged) particle lines and one photon line—as in Figure 3.2. As we have seen, basically the occurrence of only this type of coupling is a result of the inherent simplicity of the Dirac equation, being linear in the momentum (or gradient) operator.

The other important factor in covariant perturbation theory that can be inferred from Equation (3.105) is the Feynman propagator for spin-$\frac{1}{2}$ particles. Comparison of Equation (3.105) with the corresponding field-source equation (3.87) for scalar particles suggests a propagator of the form

$$P_F(\text{spin-}\tfrac{1}{2}) = (\not{p} - imc)^{-1}. \tag{3.106}$$

Making use of the identity (3.103), this can also be written

$$\frac{1}{\not{p} - imc} \cdot \frac{\not{p} + imc}{\not{p} + imc} = \frac{\not{p} + imc}{\not{p}^2 + m^2c^2} = \frac{\not{p} + imc}{p^2 + m^2c^2}, \tag{3.107}$$

where, as in the scalar formulas, $p^2 = p_\mu p_\mu$. Again, the denominator is not zero because $p^2 \neq -m^2c^2$ in intermediate states.

For both the treatment of scalar (spin-0) QED in Section 4.2 and of spinor (spin-$\frac{1}{2}$) QED in Section 4.3, we have given only a sketchy outline. Not only are the derivations not rigorous, for example, those to obtain P_F (spin-0) and P_F (spin-$\frac{1}{2}$), but we have not derived the multiplying constants. This is not important; there are other ways of obtaining the factors. For example, we could compute some formula for which there is also a classical or non-relativistic derivation and compare the results to determine the factor. Of course, there is a detailed systematic method for obtaining the results and prescriptions in covariant perturbation theory. This takes much more time. However, we have given a fairly complete and self-contained treatment of the non-relativistic theory. Perhaps a comparison of this formulation with our more sketchy outline of the relativistic theory will help in understanding the latter. We shall be making some applications of the results derived here and that should also help in the understanding.

3.4.4 Invariant Transition Rate

Although a very general result, the Fermi Golden Rule formula (3.43) is inappropriate in that form for applications involving covariant perturbation theory. We can rewrite it for this purpose, however, with the resulting expression containing factors that are manifestly Lorentz invariant. The most important feature of the relativistic form of the transition rate formula is the invariant nature of the amplitude for the process. To distinguish it from the non-covariant amplitude (M_{f0}) we use the notation $\mathcal{M}_{f0}$ for the quantity. All of the physics of a process is contained in the corresponding $\mathcal{M}_{f0}$; the rest of the factors in the transition rate formula are kinematic in nature.

One characteristic of a relativistic formulation that is different concerns the normalization convention for free-particle wave functions. It is no longer appropriate[29] to take $|u|^2 = 1$; instead we adopt

$$|u|^2 = 2E \tag{3.108}$$

(some textbooks take $2E/m$ or E/m). This type of normalization results naturally from a relativistic formulation, as can be seen from formula (3.64), for example, and has a simple explanantion or interpretation. To provide an invariant probability (volume integral of $|u|^2$), Lorentz-Fitzgerald contraction along the direction of motion then compensates the factor E. Thus, with our modified normalization we make the replacement

$$|M_{f0}|^2 = \frac{|\mathcal{M}_{f0}|^2}{\left(\prod_{(f)} 2E\right)\left(\prod_{(0)} 2E\right)}; \tag{3.109}$$

here the products are over the $2E$ factors for all the incoming (0) and outgoing (f) particles. There would also be normalization factors associated with intermediate states, but these can be assumed to be contained within $\mathcal{M}_{f0}$. These factors could be incorporated within the Feynman propagators, for example, since for every pair of intermediate states, there is a P_F.

There are two other modifications in the non-covariant formula (3.43) that are to be made. The transition matrix element M_{f0} always yields a (total) momentum-conservation δ-function even when multiple perturbations are involved, arising from an integral of the type $\int e^{i\boldsymbol{k}\cdot\boldsymbol{r}} d^3\boldsymbol{r} = (2\pi)^3\delta^{(3)}(\boldsymbol{k})$, where $\boldsymbol{k}$ is the total momentum change. We extract this ubiquitous factor from M_{f0} and combine it with the energy-conservation δ-function:

$$(2\pi)^3\delta^{(3)}\Big(\sum_{(f)} \boldsymbol{p} - \sum_{(0)} \boldsymbol{p}\Big)\delta\Big(\sum_{(f)} E - \sum_{(0)} E\Big) = (2\pi)^3\delta^{(4)}\Big(\sum_{(f)} p_\mu - \sum_{(0)} p_\mu\Big), \tag{3.110}$$

with the four-dimensional covariant δ-function now expressing energy and momentum conservation.

---

[29] In this subsection, we simplify the algebra by setting $c = 1$ and $\hbar = 1$; this provides an easier comparison with treatments in modern textbooks. We also take a unit normalization volume.

The second additional modification of the non-covariant transition rate formula is, like the factor (3.110), essentially kinematic. It concerns the implied factor, eventually integrated, for the differential number of states[30] for the (N) outgoing particles:

$$dN_f = \prod_{(f)}^{N} d^3\boldsymbol{p}/(2\pi)^3. \tag{3.111}$$

Instead of counting the states of N particles, momentum conservation could be applied and $N-1$ factors of $d^3\boldsymbol{p}/(2\pi)^3$ could be taken, and the momentum-space δ-function [times $(2\pi)^3$] could be omitted. However, it is better to express results in this more symmetric way. It allows the employment of the covariant δ-function (3.110) and also allows the N factors $2E$ in the denominator of Equation (3.109) to be combined with the $d^3\boldsymbol{p}$ in product (3.111). This ratio $d^3\boldsymbol{p}/E$ is an invariant [see Equation (1.47)]; it can also be written in another form. With $p^2 \equiv p_\mu p_\mu = \boldsymbol{p}^2 - E^2$ (= inv.), we can write $p^2 + m^2 = (E_p + E)(E_p - E)$, where E_p is the positive square root of $\boldsymbol{p}^2 + m^2$. Then, if it is understood that only the positive values of E will be included in integrations, a factor $\delta(E - E_p)dE$ can be inserted to multiply each $d^3\boldsymbol{p}$. But, with the implied inclusion of only positive energies, $\delta(E - E_p) = 2E\delta(p^2 + m^2)$. Then

$$d^3\boldsymbol{p}/2E = \delta(p^2 + m^2)\, d^4p, \tag{3.112}$$

which is manifestly invariant.

The transition rate can now be expressed in the desired covariant form:

$$\frac{\Delta W}{\Delta t} = (2\pi)^4 \frac{|\mathcal{M}_{f0}|^2}{\prod_{(0)}(2E)}\, \rho, \tag{3.113}$$

where

$$\rho = \Big(\prod_{(f)} 2\pi\delta(p^2 + m^2)\frac{d^4p}{(2\pi)^4}\Big)\delta^{(4)}\Big(\sum_{(f)} p_\mu - \sum_{(0)} p_\mu\Big) \tag{3.114}$$

is the invariant phase space density. The factors of $2E$ for the incident particles in the denominator of Equation (3.113) are kinematic in nature. An important feature of this result is, as already emphasized, the invariant nature of the matrix element $\mathcal{M}_{f0}$. In fact, in some applications, it can even be possible to make an educated guess as to its form from considerations of invariance. There are many applications of the transition rate formula, with the physics of particular processes being contained in $\mathcal{M}_{f0}$. In particular, it is employed to compute lifetimes and cross sections for processes.

---

[30] If the particles have spin and we are not interested in their polarization states, there will also be a sum over these spin substates. Also, if, as is usually the case in the application considered, the incident particles are unpolarized, there would be an average over these polarizations. For simplicity in the notation, we are omitting spin summations and averages from the transition rate formula.

3.5 SOFT-PHOTON EMISSION

When the photon energy and momentum are small, that is, when the photon is "soft," some very general and useful results can be derived. In this section, we give quantum-mechanical derivations of certain formulas applicable in the non-relativistic limit and in the general case. All of the expressions are valid only in the Born approximation, and they will be compared with corresponding results derived through classical electrodynamics. The formulas are identical to the classical ones, but it turns out that the non-relativistic expressions derived quantum mechanically have a more general validity, being applicable even away from the soft-photon limit.

3.5.1 Non-Relativistic Limit

Consider the motion of a charge in the presence of some perturbation V that does not involve the spin of the particle.[31] The charge also feels a photon-emission perturbation, the lowest-order type being the interaction Hamiltonian (3.20) designated H_q'. The charge always experiences the perturbation H_q'; that is, it is always trying to produce photons. However, as mentioned earlier in this chapter, an isolated charge without internal structure cannot do this because of energy conservation. The combined action of the two perturbations V and H_q' does allow the phenomenon of photon emission to take place. The total perturbation Hamiltonian is then

$$H' = V + H_q', \tag{3.115}$$

and this perturbation can account for a variety of processes. The non-electromagnetic part V could represent various mechanisms. For example, it could be some scattering potential, or it could even be some interaction that causes the creation of a charge as in β-decay. In the latter process, a charge of opposite sign must be created (as in neutron decay: $n \to p + e^- + \overline{\nu}_e$), or the charge on a proton must be transferred to a (positive) electron as in positron-producing β-decay. It is always the electromagnetic interactions with the e^+ or e^- that are important for photon production because of their small mass. Further, regarding the application in β-decay, the e^- or e^+ involved must be non-relativistic for the applicability of the formulation in this subsection. There are such β-decays, an example being the decay of tritium (H^3), for which the maximum e^- energy is about 18.7 keV.

We are interested in this section in evaluating the probability that a soft photon of energy within $\hbar\, d\omega$ accompanies some radiationless process, the photon emission being just a small perturbation on the rest of the overall process. Both interactions V and H_q' are treated as perturbations; that is, in lowest order, they are considered to "act once." This is always a valid assumption for the electromagnetic perturbation

---

[31] We ignore spin to simplify the formulation. This assumption is really not necessary, and, if spin interactions are involved, there is no difficulty in modifying the equations by adding the spin coordinate to a total description of the particle state: $\psi_{\text{tot}} = \psi(\boldsymbol{r}, t)\psi_{\text{spin}}$. If there are no spin interactions, the orthogonality of the spin eigenfunctions would simply tell us that the spin coordinate remains unchanged in the overall process and can be ignored. In the following subsection (5.2), we consider a problem where spin plays a major role.

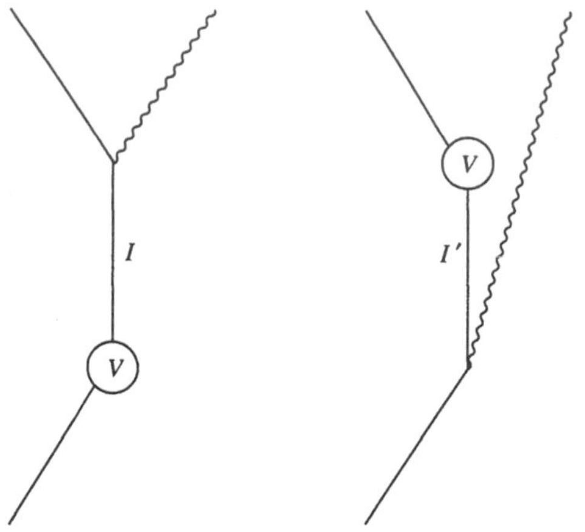

Figure 3.7 Diagrams representing the combined action of two perturbations, one being some coupling or perturbation V and the second representing an electromagnetic coupling, in this case associated with the production of an outgoing photon.

H_q', since the interactions is fundamentally weak. The smallness of the fine structure constant guarantees this: $\alpha \approx 1/137$. Treating V as a perturbation that acts once is a more restricting assumption that generally confines the application to the domain of the Born approximation. The Born approximation would require the energy of the charge to be sufficiently large if, for example, V is a Coulomb scattering potential. Specifically, in Coulomb scattering where the charges are ze and $z'e$, the criterion for Born approximation would be simply[32] $zz'e^2/\lambda\!\!\!^- \ll mv^2$, where $\lambda\!\!\!^- = \hbar/mv$ is the de Broglie wavelength. The requirement is then $\beta = v/c \gg zz'e^2/\hbar c = zz'\alpha$ or that the particle energy is $E \gg (zz')^2\text{Ry}$.

With both V and H_q' acting once, the process can be represented in terms of the diagrams in Figure 3.7. In our non-covariant formulation, there would also be pair production-annihilation diagrams (see Figure 3.6), but these give a negligible contribution in the non-relativistic limit (see previous section). The effective perturbation Hamiltonian matrix element for the process is [see Equation (3.44)]

$$H_{f0}' = \sum_I \frac{\langle f|H_q'|I\rangle\langle I|V|0\rangle}{E_0 - E_I} + \sum_{I'} \frac{\langle f|V|I'\rangle\langle I'|H_q'|0\rangle}{E_0 - E_{I'}}. \tag{3.116}$$

In the evaluation of matrix elements, let us, for convenience, set the normalization volume L^3 equal to unity in all subsequent formulae; factors involving L always cancel in final expressions for probabilities, cross sections, etc. For outgoing pho-

[32]This is a requirement that the characteristic scattering potential is small compared with the particle kinetic energy. The probability current associated with the amplitude of the scattered wave would, as a result, be small compared with that of the wave incident on the scattering center. The Born approximation is essentially that of a perturbation or iteration procedure.

tons (wave vector $\boldsymbol{k}$ and polarization $\boldsymbol{\varepsilon}$), the matrix element associated with the photon-emission vertex (between plane-wave states a and b) will be given by [see Equation (3.20)]

$$\langle b|H_q'|a\rangle = (2\pi\hbar^3/\omega)^{1/2}(q/m)(\boldsymbol{\varepsilon}\cdot\boldsymbol{k}_a)\int e^{i(\boldsymbol{k}_a-\boldsymbol{k}_b-\boldsymbol{k})\cdot\boldsymbol{r}}d^3\boldsymbol{r}, \tag{3.117}$$

where $\boldsymbol{k}_a$ and $\boldsymbol{k}_b$ are the wave vectors of the charged particle. The integral is just a three-dimensional, momentum-conservation, Dirac δ-function, which can also be expressed as a Kronecker δ:

$$\int e^{i(\boldsymbol{k}_a-\boldsymbol{k}_b-\boldsymbol{k})\cdot\boldsymbol{r}}d^3r = (2\pi)^3\delta^{(3)}(\boldsymbol{k}_a-\boldsymbol{k}_b-\boldsymbol{k}) = \delta_{\boldsymbol{k}_a,\boldsymbol{k}_b+\boldsymbol{k}}. \tag{3.118}$$

The Dirac or Kronecker δ-functions are eventually used in the intermediate-state summations in Equation (3.116). Leaving out the trivial δ-function factor, the matrix element (3.117) can be written very simply:

$$\langle b|H_q'|a\rangle = (2\pi\hbar^3/\omega)^{1/2}q(\boldsymbol{\varepsilon}\cdot\boldsymbol{v}_a), \tag{3.119}$$

where $\boldsymbol{v}_a$ is the velocity of the charge in state a. The matrix element of V does not involve the photon and is only a function of the charge wave vectors:

$$\langle d|V|c\rangle \equiv V_{dc} = V_{dc}(\boldsymbol{k}_d,\boldsymbol{k}_c). \tag{3.120}$$

Further, we allow the source of the potential V to have "internal structure" such that it can possess an excitation energy χ. For example, it could be an atom or it could be a nucleus when the potential V causing a β-decay.

The energies of the various states are given by

$$\begin{aligned} E_0 &= (\hbar k_0)^2/2m + \chi_0, \\ E_f &= (\hbar k_f)^2/2m + \hbar ck + \chi_f, \\ E_I &= (\hbar k_I)^2/2m + \chi_f, \\ E_{I'} &= (\hbar k_{I'})^2/2m + \hbar ck. \end{aligned} \tag{3.121}$$

Since $E_0 = E_f$, there is also the relation

$$(\hbar k_0)^2/2m + \chi_0 = (\hbar k_f)^2/2m + \hbar ck + \chi_f. \tag{3.122}$$

Moreover, because of momentum conservation at the photon-emission vertices,

$$\boldsymbol{k}_I = \boldsymbol{k}_f + \boldsymbol{k}, \quad \boldsymbol{k}_{I'} = \boldsymbol{k}_0 - \boldsymbol{k}. \tag{3.123}$$

Making use of the identities (3.122) and (3.123), we obtain the energy denominators

$$\begin{aligned} E_0 - E_I &= \hbar ck - \hbar^2\boldsymbol{k}\cdot\boldsymbol{k}_f/m - \hbar^2k^2/2m, \\ E_0 - E_{I'} &= -\hbar ck + \hbar^2\boldsymbol{k}\cdot\boldsymbol{k}_0/m - \hbar^2k^2/2m. \end{aligned} \tag{3.124}$$

We now make the assumption that the photon energy ($\hbar ck$) is of the order of or less than that of the charged particle [say, $E_\alpha = (\hbar k_\alpha)^2/2m$]; then, [see also Equation (3.46)]

$$k/k_\alpha \lesssim \hbar k_\alpha/mc = v_\alpha/c = \beta_\alpha \ll 1. \tag{3.125}$$

In other words, for non-relativistic particles, when the photon and particle energies are comparable, the photon momentum is small. Under these conditions, the photons are always soft—at least as regards their momenta. Also, as we see easily, the three terms on the right-hand side of each of the Equations (3.124) are successively smaller in the ratio β_α.

There are two especially simple and important applications of the above results. The first is to the problem of particle production such as in β-decay. We ask for the probability that a photon (energy within $\hbar d\omega$) accompanies the production of a particle of given velocity $\beta_f c$. Except for summing over photon states $dN = d^3\boldsymbol{k}/(2\pi)^3$, this differential probability is given by, in Born approximation,

$$dw = \frac{\left|H'_{f0}\right|^2}{\left|V_{f0}\right|^2} dN. \tag{3.126}$$

In this process, there is only diagram (1) of Figure 3.7, and the energy denominator is given by $\hbar ck$ in good approximation. Further, since $\boldsymbol{\varepsilon} \cdot \boldsymbol{k} = 0$, by the first equation (3.124), $\boldsymbol{\varepsilon} \cdot \boldsymbol{k}_I = \boldsymbol{\varepsilon} \cdot \boldsymbol{k}_f$, and the photon-emission matrix element is given by

$$\langle f|H'_q|I\rangle = (2\pi\hbar/\omega)^{1/2} q(\boldsymbol{\varepsilon} \cdot \boldsymbol{v}_f). \tag{3.127}$$

Because of the inequality (3.125) and the equation (3.120), the matrix element involving V is given by

$$\langle I|V|0\rangle = V_{I0} \approx V_{f0}. \tag{3.128}$$

Thus,

$$dw = \frac{|\langle f|H'_q|I\rangle|^2}{(\hbar ck)^2} dN, \tag{3.129}$$

with

$$dN = (2\pi)^{-3} k^2 dk\, d\Omega, \tag{3.130}$$

in terms of the solid angle element Ω for the outgoing photon. In terms of ω ($= kc$) and $z = q/e$, we have, very simply,

$$dw = z^2(\alpha/4\pi^2)(d\omega/\omega)(\boldsymbol{\varepsilon} \cdot \boldsymbol{\beta}_f)^2 d\Omega, \tag{3.131}$$

where α is the fine-structure constant.

The expression (3.131) is the result of a quantum-mechanical derivation of a formula identical to the semi-classical expression (2.142). However, now we see that the expression is not restricted to the soft-photon limit when the Born approximation holds. We can also sum over polarizations for the outgoing photon. The two possible (linear) polarization states are perpendicular to $\boldsymbol{k}$. With $\boldsymbol{k}$ and $\boldsymbol{\beta}$ defining a plane, one of these polarization reference directions ($\boldsymbol{\varepsilon}_1$) can be taken perpendicular to this plane. The other polarization unit vector ($\boldsymbol{\varepsilon}_2$) is in the plane and perpendicular to $\boldsymbol{k}$ and gives the whole contribution (the reader can construct a simple diagram to indicate this). Thus,

$$\sum_{\text{pol}}(\boldsymbol{\varepsilon} \cdot \boldsymbol{\beta}_f)^2 = (\boldsymbol{\varepsilon}_2 \cdot \boldsymbol{\beta}_f)^2 = \beta_f^2 \sin^2\theta, \tag{3.132}$$

where θ is the angle between $\boldsymbol{k}$ and $\boldsymbol{\beta}_f$. Summed over polarizations, the angular distributions of the photon emission is then

$$dw = z^2(\alpha/4\pi^2)(d\omega/\omega)\beta_f^2 \sin^2\theta \, d\Omega. \tag{3.133}$$

Finally, integrating over angles for the outgoing photon,

$$dw = z^2(2\alpha/3\pi)\beta_f^2 d\omega/\omega. \tag{3.134}$$

The three formulas (3.131), (3.133), and (3.134) exhibit the characteristic "infrared divergence" factor $d\omega/\omega$, indicating an infinite probability of emitting infinitesimally soft photons. However, we emphasize again that, in the Born approximation, these non-relativistic expressions are not restricted to the soft-photon limit. That is, $\hbar\omega$ does not have to be small compared with the charged-particle kinetic energy. The general nature of these formulas has not been emphasized in the literature.

Let us turn to another important application of the above general formulation. This is to the problem of photon emission (bremsstrahlung) in the scattering of a charged particle. We need not specify the nature of the scattering potential, although the important application would be for Coulomb scattering (the photon emission would then be Coulomb bremsstrahlung). The Born approximation is assumed, however, since the scattering potential is treated as a perturbation that acts only once. Both diagrams in Figure 3.7 give contributions to the amplitude for the process, so that

$$H'_{f0} = H'_1 + H'_2, \tag{3.135}$$

and the energy denominators are given by

$$\begin{aligned} E_0 - E_I &\approx \hbar ck, \\ E_0 - E_{I'} &\approx -\hbar ck. \end{aligned} \tag{3.136}$$

The matrix elements of the electromagnetic perturbation (photon emission) have the form (3.127); that is, $\langle f|H'_q|I\rangle \propto \boldsymbol{\varepsilon}\cdot\boldsymbol{v}_f$ and $\langle I'|H'_q|0\rangle \propto \boldsymbol{\varepsilon}\cdot\boldsymbol{v}_0$. Both matrix elements of the scattering potential can be approximated by the (radiationless) expressions (3.128), because of the inequality (3.125). Because the two energy denominators (3.136) are equal but opposite in sign, we have the important result

$$H'_{f0} \propto \boldsymbol{\varepsilon}\cdot(\boldsymbol{v}_f - \boldsymbol{v}_0) = \boldsymbol{\varepsilon}\cdot\Delta\boldsymbol{v}. \tag{3.137}$$

The formulas for the photon-emission probabilities are then the same as the results (3.131), (3.133), and (3.134) with $\boldsymbol{\beta}_f$ replaced by $\Delta\boldsymbol{\beta}$. They are again identical to the semi-classical expressions (2.142) and (2.143). However, we now see that, basically because of the inequality (3.125), they are not restricted to the soft-photon limit—at least when the Born approximation holds. We return to make use of these formulas when we consider the general bremsstrahlung process in a later chapter.

3.5.2 Emission from Spin Transitions

Phenomena having to do with a particle's intrinsic magnetic moment associated with its spin have only a limited classical analog. This intrinsic moment is without spatial extent in a quantum-mechanical formulation, and the special case of spin-$\frac{1}{2}$ is particularly interesting and unique. The effects associated with photon coupling to this moment are small, especially in the soft-photon limit. It is well to demonstrate this, if only as an illustration of the general theory developed in this chapter. Interest in the problem should be more than academic, however; it is important to demonstrate the magnitude of the effects in comparison with those involving photon interactions with particle "orbital" motion (interaction with the charge). There is, in fact, a particle in nature that has a magnetic moment but no charge. The neutron is electrically neutral but has a magnetic moment of magnitude[33] $\mu_n = -1.913\mu_M$, where $\mu_M = e\hbar/2M_p c$ is the nuclear Bohr magneton. The neutron moment—as well as most of the proton's moment—can be thought to be a result of the cloud of virtual pions that is associated with the nucleon. To regard the neutron's moment as point-like and "rigid" is thus an approximation, but to do so using the experimental magentic moment should take into account the virtual pion field. The approximation can be expected to be valid for photon energies much less than the pion rest energy ($\sim$ 100 MeV); this is in the non-relativistic domain for neutron motion. The other relevant application for spin-$\frac{1}{2}$ particles is to the electron, but photon interactions with that particle's charge are much greater.

For particles of spin-$\frac{1}{2}$, the spin angular momentum is given by $\boldsymbol{J}_{\text{spin}} = \hbar \boldsymbol{s}$ with $\boldsymbol{s} = \frac{1}{2}\boldsymbol{\sigma}$, $\boldsymbol{\sigma}$ being the set of 2 × 2 Pauli spin matrices: $\boldsymbol{\sigma} = (\sigma_x, \sigma_y, \sigma_z)$. The spin wave function has two components representing the amplitudes for the two possible values for the spin component in some specified direction. In terms of its spin magnitude μ_0, the spin magnetic moment is given by

$$\boldsymbol{\mu}_s = \mu_0 \boldsymbol{\sigma}, \tag{3.138}$$

and in the (σ^2, σ_z) representation where the z-direction is chosen to describe one component of $\boldsymbol{\sigma}$, the three Pauli matrices are (see any book on quantum mechanics)

$$\sigma_x = \begin{pmatrix} 0 & 1 \\ 1 & 0 \end{pmatrix}, \quad \sigma_y = \begin{pmatrix} 0 & -i \\ i & 0 \end{pmatrix}, \quad \sigma_z = \begin{pmatrix} 1 & 0 \\ 0 & -1 \end{pmatrix}. \tag{3.139}$$

The matrix σ_z is diagonal and the eigenvalues are ± 1. That is, the two eigenfunctions are

$$\Lambda_+ = \begin{pmatrix} 1 \\ 0 \end{pmatrix}, \quad \Lambda_- = \begin{pmatrix} 0 \\ 1 \end{pmatrix}, \tag{3.140}$$

and we can use the simplified notation

$$\Lambda_\pm = |\pm\rangle . \tag{3.141}$$

The Pauli matrices anticommute; that is, $\sigma_x\sigma_y + \sigma_y\sigma_x = 0$, etc. Also, $\sigma_x\sigma_y = i\sigma_z$, with other relations obtained from a cyclic permutation of x, y, z; in general, we can write

$$\boldsymbol{\sigma} \times \boldsymbol{\sigma} = 2i\boldsymbol{\sigma}. \tag{3.142}$$

Further, $\sigma_x^2 = \sigma_y^2 = \sigma_z^2 = \boldsymbol{\sigma}/3 = 1$. The eigenfunctions are orthogonal: $\langle m'|m\rangle = \delta_{m'm}$. The operators σ_x, σ_y, and σ_z yield

$$\begin{aligned} \sigma_x |\pm\rangle &= |\mp\rangle , \\ \sigma_y |\pm\rangle &= \pm i\, |\mp\rangle , \\ \sigma_z |\pm\rangle &= \pm |\pm\rangle . \end{aligned} \tag{3.143}$$

[33]The minus sign means that the magnetic moment is opposite in direction to the spin.

Instead of the form (3.22) for the spin magnetic moment interaction (with a photon) Hamiltonian, we can rewrite it more simply as

$$S = i(2\pi\hbar\omega)^{1/2}\boldsymbol{\mu}_s \cdot \boldsymbol{\varepsilon}' \tag{3.144}$$

where $\boldsymbol{\varepsilon}'$ is a unit polarization vector in the direction of the photon's magnetic field (that is, $\boldsymbol{\varepsilon}' = \boldsymbol{k} \times \boldsymbol{\varepsilon}/k$, where $\boldsymbol{\varepsilon}$ is the polarization vector for $\boldsymbol{A}$ and the electric field); $\boldsymbol{\varepsilon}'$, like $\boldsymbol{\varepsilon}$, has two fundamental polarization directions perpendicular to $\boldsymbol{k}$. As in Section 3.5.1, we consider the action of perturbations V and H'_μ, where now V is independent of particle spin, so that the total perturbation Hamiltonian is $H' = V + H'_\mu$. We allow both V and H'_μ to act once on a single particle with spin-$\frac{1}{2}$ and magnitude μ_0 of its magnetic moment (see Figure 3.7). The particle is initially unpolarized, so we average over initial spin-polarization states and sum over final spin polarizations; we also sum over polarizations for the outgoing photon. The photon emission probability will be determined by

$$\bar{S} = \frac{1}{2}\sum_{\text{pol}}\sum_{m'}\sum_{m}\left|\langle m'|\boldsymbol{\varepsilon}' \cdot \boldsymbol{\sigma}|m\rangle\right|^2 = \sum_{\text{pol}}(\varepsilon'^2_x + \varepsilon'^2_y + \varepsilon'^2_z) = 2. \tag{3.145}$$

In the evaluation of the photon-emission probability, we again employ the basic equation (3.126) in terms of the effective Hamiltonian (3.116) for the combined process. The diagrams are of the form in Figure 3.7. In the perturbation Hamiltonian (3.116), the energy denominators can be approximated by ($\kappa = mc/\hbar$)

$$\begin{aligned} E_0 - E_I &= \hbar ck(1 - \boldsymbol{k} \cdot \boldsymbol{k}_f/\kappa k), \\ E_0 - E_{I'} &= -\hbar ck(1 - \boldsymbol{k} \cdot \boldsymbol{k}_0/\kappa k), \end{aligned} \tag{3.146}$$

in terms of the initial and final wave vectors ($\boldsymbol{k}_0$ and $\boldsymbol{k}_f$) of the scattered particle (with magnetic moment). The matrix elements of the scattering potential V are

$$\begin{aligned} \langle \boldsymbol{k}_I|V|\boldsymbol{k}_0\rangle &= \langle \boldsymbol{k}_f + \boldsymbol{k}|V|\boldsymbol{k}_0\rangle, \\ \langle \boldsymbol{k}_f|V|\boldsymbol{k}_{I'}\rangle &= \langle \boldsymbol{k}_f|V|\boldsymbol{k}_0 - \boldsymbol{k}\rangle, \end{aligned} \tag{3.147}$$

which are the same in Born approximation. The combination of energy denominators is given by

$$\begin{aligned} \frac{1}{E_0 - E_I} + \frac{1}{E_0 - E_{I'}} &\cong \frac{1}{\hbar ck}\left(1 + \frac{\boldsymbol{k} \cdot \boldsymbol{k}_f}{\kappa k} - 1 - \frac{\boldsymbol{k} \cdot \boldsymbol{k}_0}{\kappa k}\right) \\ &= -\frac{\boldsymbol{k} \cdot \Delta\boldsymbol{k}_\mu}{mc^2k^2}, \end{aligned} \tag{3.148}$$

where $\Delta\boldsymbol{k}_\mu = \boldsymbol{k}_0 - \boldsymbol{k}_f$ is the change in the wave vector of the scattered particle (having a magnetic moment). Employing the above results, including Equation (3.145), we obtain

$$dw_\mu = 2\mu_0^2(2\pi\hbar\omega)\left(\frac{\boldsymbol{k} \cdot \Delta\boldsymbol{k}_\mu}{mc^2k^2}\right)^2 \frac{d^3\boldsymbol{k}}{(2\pi)^3}. \tag{3.149}$$

Setting $d^3\mathbf{k} = k^2 dk\, d\Omega$ and integrating over angles of emission, we have

$$dw_\mu = \frac{2}{3\pi}\frac{\mu_0^2}{m^2c^5}(\Delta k_\mu)^2\hbar\omega\, d\omega. \tag{3.150}$$

For the case of scattering of an electron ($\mu_0 = e\hbar/2mc$), the result can be written

$$dw_\mu = \tfrac{1}{4}(\hbar\omega/mc^2)^2 dw_q, \tag{3.151}$$

where dw_q is the corresponding emission probability associated with photon interactions with the charge [see Equation (3.134) with $\beta_f \to \Delta\beta$]. The emission from spin interactions is small for non-relativistic particles. Moreover, we note that there is no infrared divergence effect as $\omega \to 0$; in fact, $dw_\mu/d\omega \to 0$ in this limit. Finally, it should be mentioned that the results (3.149)–(3.151) are not confined to the soft-photon limit.

3.5.3 Relativistic Particles without Spin

Consideration of soft-photon emission associated with processes involving relativistic particles provides our first application of the methods of covariant QED outlined in Section 3.4. First, for scalar charged particles, we derive an expression for the photon-emission probability corresponding to the non-relativistic formulae (3.134) and (3.150). As in the derivation of those results, two perturbations are allowed to act on the particle (see Figure 3.7); one can correspond to, say, a scattering potential or some perturbation that creates the charge at high energy, while the other is the purely electromagnetic perturbation (photon-emission vertex). In our elementary sketch of covariant QED, we concerned ourselves with the establishment of the form of the perturbation amplitudes and Feynman propagators corresponding to the perturbation Hamiltonian and energy denominator in the non-covariant, non-relativisitic theory. Although the multiplying factors for these terms can be derived formally in a systematic exposition of covariant perturbation theory, we shall merely leave them as undetermined constants. This is not a serious problem; in fact, the undetermined factors can usually be established through a comparison with known results in some special case such as the non-relativistic limit or the classical limit.

The fundamental factors that are needed in a covariant perturbtion calculation are the Feynman propagator (3.86), $P_F(\text{scalar}) = (p^2 + m^2c^2)^{-1}$, and the term associated with the photon-emission vertex. For such a vertex (see Figure 3.2), the matrix element between initial state a and final state b has the form (3.73) in the momentum representation:

$$M_{ba} = a_s q \varepsilon_\mu (p_a + p_b)_\mu, \tag{3.152}$$

where a_s is some numerical factor for this scalar-charge interaction; $p_{a\mu}$ and $p_{b\mu}$ are the four-momenta for the incoming and outgoing charge (q), and ε_μ is the unit polarization four-vector. The amplitude M_{ba} is, like P_F, a Lorentz invariant and is the analog of the non-relativistic formula (3.20) associated with the same type of single-photon vertex. With the combined interactions (3.152) plus the perturbation

V, the matrix element for the process between initial state 0 and final state f will be (see Figure 3.7)

$$
\begin{aligned}
M_{f0} = a_s q \varepsilon_\mu (p_f + p_I)_\mu (p_I^2 + m^2c^2)^{-1} V_{I0} \\
+ V_{fI'}(p_{I'}^2 + m^2c^2)^{-1} a_s q \varepsilon_\mu (p_{I'} + p_0)_\mu.
\end{aligned} \tag{3.153}
$$

At the vertices, there is four-momentum conservation so that

$$
p_{I\mu} = p_{f\mu} + k_\mu, \quad p_{I'\mu} = p_{0\mu} - k_\mu, \tag{3.154}
$$

k_μ referring to the photon. Since $p_f^2 = (p_\mu p_\mu)_f = p_0^2 = (p_\mu p_\mu)_0 = -m^2c^2$ and $k_\mu k_\mu = 0$, the Feynman propagators for the two intermediate states are, very simply,

$$
P_I = \frac{1}{2(p_f \cdot k)}, \quad P_{I'} = -\frac{1}{2(p_0 \cdot k)}, \tag{3.155}
$$

where the notation $(p \cdot k) = p_\mu k_\mu$ is employed for the four-dimensional scalar products. Further, because of the relations (3.154),

$$
\begin{aligned}
(p_f + p_I)_\mu = (2p_f + k)_\mu, \\
(p_{I'} + p_0)_\mu = (2p_0 - k)_\mu.
\end{aligned} \tag{3.156}
$$

Also, the Lorentz gauge condition

$$
\varepsilon_\mu k_\mu = 0 \tag{3.157}
$$

holds. In the soft-photon limit

$$
V_{I0} \approx V_{fI'} \approx V_{f0}, \tag{3.158}
$$

which is the matrix element in the radiationless problem. The matrix element for the combined process involving V and the (soft-) photon perturbation is then the invariant

$$
M_{f0} \approx a_s q V_{f0} \left[\frac{(\varepsilon \cdot p_f)}{(k \cdot p_f)} - \frac{(\varepsilon \cdot p_0)}{(k \cdot p_0)} \right]. \tag{3.159}
$$

The photon-emission probability is [see Equation (3.126)]

$$
dw = \frac{|M_{f0}|^2}{|V_{f0}|^2} dN', \tag{3.160}
$$

where dN' is now the invariant phase-space factor (3.113) or number of photon final states divided by $2E$. That is, with k designating the magnitude of the energy or momentum of the outgoing photon,[34]

$$
dN' \propto k^2 dk\, d\Omega / k = k\, dk\, d\Omega. \tag{3.161}
$$

[34]The appearance of the additional factor of $1/k$ could be understood in another way. Instead of employing the (more natural) invariant phase-space factor, the photon-emission invariant amplitude (3.152) could be employed with an additional factor $k^{-1/2}$. This perturbation amplitude would then be essentially a relativistic generalization of the corresponding non-relativistic interaction Hamiltonian (3.20). Alternately, the factor could be regarded as associated with the relativistic normalization [see Equation (3.108)] of the photon function.

If, in addition, a sum over photon final-state polarizations is performed, the probability (3.160) is given by an expression of the form

$$dw = Kq^2 \sum_{\text{pol}} \left| \frac{(\varepsilon \cdot p_f)}{(k \cdot p_f)} - \frac{(\varepsilon \cdot p_0)}{(k \cdot p_0)} \right|^2 k\, dk\, d\Omega, \tag{3.162}$$

where K is a constant.

Although the photon polarization unit four-vector ε_μ has four components, only two contribute in the polarization sum. There is a covariant (Lorentz) transversality condition (3.158), but within this class of gauges there can be a (convenient) choice of gauge (see Section 2.1.1). The transformed vector potential $A'_\mu = A_\mu + \partial_\mu \chi$ could be used if χ satisfies $\Box^2 \chi = 0$. This is satisfied by, for example, the choice $\chi = a \exp(ik_\nu x_\nu)$, where $k_\nu = (ik, \boldsymbol{k})$ is the photon four-momentum. The transformed polarization four-vector would then be

$$\varepsilon'_\mu = \varepsilon_\mu + iak_\mu, \tag{3.163}$$

and the choice $a = \varepsilon_0/k$ could be made so that $\varepsilon'_0 = 0$. We assume that this gauge is employed so that (dropping the primes) ε_μ has a vanishing time component, the relation (3.157) giving the usual transversality for the space part of ε_μ:

$$\varepsilon_\mu k_\mu = (\boldsymbol{\varepsilon} \cdot \boldsymbol{k}) = 0. \tag{3.164}$$

For general $\boldsymbol{p}_f$ and $\boldsymbol{p}_0$, the expression (3.162) does not yield a simple formula on integrating over $d\Omega$ and summing over polarizations. Only for a frame with $\boldsymbol{p}_0 = 0$ (or $\boldsymbol{p}_f = 0$) does the result simplify. Designating either of these as $\boldsymbol{p}$ (with the other zero), we have

$$\sum_{\text{pol}} |\varepsilon \cdot p|^2 = \sum_{\text{pol}} (\boldsymbol{\varepsilon} \cdot \boldsymbol{p})^2 = \boldsymbol{p}^2 - (\boldsymbol{p} \cdot \boldsymbol{k})^2/k^2, \tag{3.165}$$

and

$$\sum_{\text{pol}} \left| \frac{(\varepsilon \cdot p)}{(k \cdot p)} \right|^2 = \frac{\beta^2}{k^2} \frac{1 - \cos^2\theta}{(1 - \beta\cos\theta)^2}, \tag{3.166}$$

in terms of β $(= v/c)$ for the outgoing charge and the angle θ between $\boldsymbol{p}$ and $\boldsymbol{k}$. This expression (3.166) gives the angular distribution $(dw/d\Omega)$ of the emission. Integrating over $d\Omega = 2\pi \sin\theta\, d\theta$, we get the emission probability in any direction:

$$dw = 2\pi K q^2 \frac{dk}{k} \beta^2 \int_{-1}^{1} \frac{1 - x^2}{(1 - \beta x)^2} dx. \tag{3.167}$$

The integral is elementary [in fact, it was already encountered in Equation (2.146)] and we obtain

$$dw = 4\pi K q^2 \frac{dk}{k} \left(\frac{1}{\beta} \ln \frac{1 + \beta}{1 - \beta} - 2 \right). \tag{3.168}$$

For $\beta \ll 1$, the parenthesis above is

$$\frac{1}{\beta} \ln \frac{1 + \beta}{1 - \beta} - 2 = \frac{2}{3} \beta^2 \left(1 + \frac{3}{5} \beta^2 + \cdots \right), \tag{3.169}$$

which allows a comparison with the non-relativistic formula (3.134). For $q = ze$, we then have $K = (4\pi^2\hbar c)^{-1}$ and

$$dw = \frac{\alpha}{\pi} z^2 \left(\frac{1}{\beta} \ln \frac{1+\beta}{1-\beta} - 2 \right) \frac{dk}{k}, \tag{3.170}$$

which is identical to the result obtained in Equation (2.149). The formula exhibits the usual infrared divergence factor (dk/k). It is valid in the soft-photon limit, that is, for $\hbar ck \ll (\gamma - 1)mc^2$, *except* when $\beta \ll 1$, for which, as we have seen in Section 5.1, it holds for general photon energy.

3.5.4 Relativistic Spin-$\frac{1}{2}$ Particles

Here we wish to derive an expression for the probability that soft-photon emission accompanies some process involving relativistic particles of spin-$\frac{1}{2}$. This case is far more important than that for spin-0 considered in the previous subsection, since it applies to electrons. It will provide a first illustration of the techniques of modern QED and is perhaps the simplest example of methods in the theory. Employing the basic principles, the application is elementary, yielding a general formula that can be compared with that for spin-0 and with the classical expression.

In the covariant theory, the probability of a process will be proportional to the square of a matrix element $\mathcal{M}_{f0}$. This amplitude is a Lorentz invariant and is constructed from the one-particle Dirac wave functions ψ_0 and ψ_f for the initial and final states and a product (M) of factors associated with interaction vertices and Feynman propagators. In the momentum representation, the invariant amplitude is constructed from

$$\mathcal{M}_{f0} = \overline{u}_f M u_0, \tag{3.171}$$

where u_0 and u_f are the momentum-space Fourier amplitudes for the initial and final states, respectively. That is, u is related to ψ by

$$\psi(\boldsymbol{r}, t) = u\, e^{i(p\cdot x)}, \tag{3.172}$$

where, again, $(p \cdot x) = p_\mu x_\mu$. Both ψ and u are four-component (spinor) wave functions satisfying the free-particle Dirac equation in the position and momentum representation, respectively.

Now, however, we have to develop the basic theory a little further. For example, what, precisely, is the conjugate wave function $\overline{u}$? While ψ is the column wave function (3.94) with four components, $\overline{\psi}$ is *not* the row $\psi^\dagger = (\psi_1^*, \psi_2^*, \psi_3^*, \psi_4^*)$, which is the Hermitian conjugate[35] wave function, but is defined by[36]

$$\overline{\psi} = \psi^\dagger \beta = \psi^\dagger \gamma_0; \tag{3.173}$$

[35]The Hermitian conjugate or adjoint of a matrix is formed by interchanging rows and columns and taking the complex conjugate of its elements: $(A^\dagger)_{jk} = (A_{kj})^*$. A matrix is Hermitian if it is equal to its Hermitian conjugate, that is, if $A^\dagger = A$. While β and the α matrices are Hermitian, of the γ matrices only γ_0 $(= \beta)$ is Hermitian; γ_1, γ_2, and γ_3 are anti-Hermitian.

[36]The operation on the right in Equation (3.173) is a matrix multiplication of the single-row matrix $\psi^\dagger$ on β $(= \gamma_0)$ and yields another single-row matrix $(\overline{\psi})$. This can be easily seen from the matrix multiplication rule $(AB)_{kj} = A_{kl}B_{lj}$. Here $A_{kl} = 0$ unless $k = 1$ and $(AB)_{kj} = (AB)_{1j}$, that is, a single-row matrix.

also, of course, $\overline{u} = u^\dagger \gamma_0$. The reason for introducing the functions $\overline{\psi}$ and $\overline{u}$ (rather than $\psi^\dagger$ and $u^\dagger$) to form the invariant (3.171) can be seen through consideration of the continuity equation for probability density and current in the Dirac theory. To obtain this relation, we first derive the equations that $\psi^\dagger$ and $\overline{\psi}$ satisfy. These manipulations will make use of the Hermitian character of the α and β matrices: $\alpha_j^\dagger = \alpha_j$; $\beta^\dagger = \beta$. That β and $\boldsymbol{\alpha}$ are Hermitian is expected because of the Hermitian character of the (four-momentum) operators $i\partial_\mu$ in the Dirac equation (3.93) (multiplied by i). Alternatively, the character can be seen from the specific representation

$$\boldsymbol{\alpha} = \begin{pmatrix} 0 & \boldsymbol{\sigma} \\ \boldsymbol{\sigma} & 0 \end{pmatrix}, \quad \beta = \begin{pmatrix} I & 0 \\ 0 & -I \end{pmatrix}, \tag{3.174}$$

where the σ_j are the three Pauli spin matrices (3.139). The representation (3.174), in terms of the three anti-commuting Pauli matrices and the identity matrix, satisfies the identities (3.95).

Returning now to the problem of the equations for $\psi^\dagger$ and $\overline{\psi}$, we take the Hermitian conjugate of the Dirac equation (3.93), making use of the Hermitian property of α_j and β:

$$(1/c)\partial_t \psi^\dagger + \partial_j \psi^\dagger \alpha_j - i\kappa \psi^\dagger \beta = 0. \tag{3.175}$$

Multiplying from the right by β and making use of the anti-commutation relation $\alpha_j \beta = -\beta \alpha_j$, we have, in terms of the definition (3.173) for $\overline{\psi}$,

$$(1/c)\partial_t \overline{\psi} - \partial_j \overline{\psi} \alpha_j - i\kappa \overline{\psi} \beta = 0. \tag{3.176}$$

Again multiplying on the right by β and using the definitions (3.97) and (3.100), there results an equation for $\overline{\psi}$ in covariant form:

$$\overline{\psi}(\not{p} + imc) = 0, \tag{3.177}$$

where the gradient ∂_μ in $\not{p}$ is meant to operate on the function $\overline{\psi}$ to the left. This equation is the free-particle Dirac equation for the conjugate function $\overline{\psi}$, and the sign difference in parentheses should be noted when comparison is made with the Dirac equation (3.99) for ψ. On the other hand, the conjugate momentum component amplitude $\overline{u}$ is given by

$$\overline{\psi} = \overline{u} e^{-i(p\cdot x)}, \tag{3.178}$$

with a sign difference in the exponent (as if $p_\mu \to -p_\mu$). That is, $\overline{u}_{(-p)}$ satisfies $\overline{u}_{(-p)}(\not{p} + imc)$ and $\overline{u}_p = \overline{u}$ satisfies

$$\overline{u}(\not{p} - imc) = 0, \tag{3.179}$$

the parenthesis having the same sign combination as in

$$(\not{p} - imc)u = 0. \tag{3.180}$$

The sign difference should also be noted in the terms in Equation (3.175) for $\psi^\dagger$ and in Equation (3.176) for $\overline{\psi}$, indicating how $\overline{\psi}$ is a more natural conjugate function than $\psi^\dagger$ in a covariant theory.

The continuity equation for probability density and current can be obtained by multiplying equation (3.93) on the left by $\psi^\dagger$ and adding the result to Equation (3.175) multiplied on the right by ψ. We then have

$$\frac{\partial}{\partial t}(\psi^\dagger\psi) + \frac{\partial}{\partial x_j}(c\psi^\dagger\alpha_j\psi) = 0, \tag{3.181}$$

which is a conservation equation with density $\rho = \psi^\dagger\psi$ and current density $\boldsymbol{j} = c\psi^\dagger\boldsymbol{\alpha}\psi$. However, in terms of the covariant Dirac matrices γ_μ, since $\overline{\psi} = \psi^\dagger\beta$ and so $\psi^\dagger = \overline{\psi}\beta$, while $\gamma_j = i\alpha_j\beta = -i\beta\alpha_j$, the continuity equation (3.181) can be written in covariant form in terms of $\overline{\psi}$, ψ, and the γ_μ:

$$\partial_\mu(\overline{\psi}\gamma_\mu\psi) = 0. \tag{3.182}$$

Thus, we see how, in a covariant theory, the function $\overline{\psi}$ is a more natural conjugate function than $\psi^\dagger$.

Finally, for the basic formulation, we derive a relation between the normalization for $\overline{u}u$ and that for $u^\dagger u$. If we write the Dirac equation (3.93) in terms of the amplitude u defined in Equation (3.172) and then multiply the equation from the left by $\overline{u}$ $(= u^\dagger\beta)$, we obtain

$$Eu^\dagger\beta u = cu^\dagger\beta\boldsymbol{\alpha}\cdot\boldsymbol{p}u + mc^2u^\dagger u. \tag{3.183}$$

Taking the Hermitian conjugate of this equation, and making use of the anti-Hermitian character of the operator $\beta\boldsymbol{\alpha}$, there results an equation identical to the relation (3.183) except that the first term on the right has a minus sign. Adding Equation (3.183) then yields the result

$$\overline{u}u = (mc^2/E)u^\dagger u, \tag{3.184}$$

which will be referred back to later.

The free-particle Dirac equations (3.183) and (3.184) in the momentum representation are necessary to prove the important expression for the soft-photon emission probability associated with a process involving relativistic electrons. We again consider the combined action of a perturbation V and a photon-emission perturbation on a spin-$\frac{1}{2}$ charge (see Figure 3.7). The photon-emission vertex is associated with a coupling of the form $\not{\epsilon}$ [the $(q/c)\gamma_\mu A_\mu$ term in Equation (3.105)]. There are two types of intermediate states, just as in the spin-0 case in Section 5.3, for which the Feynman propagator is of the form (3.106) or (3.107). The matrix element for the process is given by the form (3.171) in terms of an invariant amplitude M. According to the Feynman rules for spin-$\frac{1}{2}$ charges, the amplitude corresponding to the spin-0 case (3.153) is now given by

$$M = \not{\epsilon}(\not{p}_I - imc)^{-1}V(I, 0) + V(f, I')(\not{p}_{I'} - imc)^{-1}\not{\epsilon}. \tag{3.185}$$

As in the relativistic spin-0 formulation, there is four-momentum conservation at the photon-emission vertex and $p_{I\mu}$ and $p_{I'\mu}$ are again given by the relations (3.154). The V-perturbation amplitudes are associated with particle "transitions" $0 \rightarrow I$

and $I' \to f$ as indicated in the simplified notation in Equation (3.185). With the propagators expressed in the form (3.107), the amplitude M is given by

$$M = \frac{\not{\varepsilon}(\not{p}_f + \not{k} + imc)}{2(k \cdot p_f)} V(I, 0) - V(f, I') \frac{(\not{p}_0 - \not{k} + imc)}{2(k \cdot p_0)} \not{\varepsilon}. \tag{3.186}$$

In the soft-photon limit, we neglect the terms $\not{k}$ in the numerators and we also approximate

$$\begin{aligned} V(I, 0) &\approx V(f, 0), \\ V(f, I') &\approx V(f, 0). \end{aligned} \tag{3.187}$$

Employing the identity (3.103), we rewrite $\not{\varepsilon}\not{p}_f$ and $\not{p}_0\not{\varepsilon}$:

$$\begin{aligned} \not{\varepsilon}\not{p}_f &= -\not{p}_f\not{\varepsilon} + 2(\varepsilon \cdot p_f), \\ \not{p}_0\not{\varepsilon} &= -\not{\varepsilon}\not{p}_0 + 2(\varepsilon \cdot p_0). \end{aligned} \tag{3.188}$$

Since $\bar{u}_f \not{p}_f = imc\bar{u}_f$ and $\not{p}_0 u_0 = imcu_0$, we have

$$\mathcal{M}_{f0} = \bar{u}_f M u_0 \approx \bar{u}_f V(f, 0) u_0 \left[\frac{(\varepsilon \cdot p_f)}{(k \cdot p_f)} - \frac{(\varepsilon \cdot p_0)}{(k \cdot p_0)} \right]. \tag{3.189}$$

This modified amplitude for the combined process has the same relation to that for the radiationless problem as in the spin-0 case. Thus, the photon-emission probability is identical to the formulas for that case. In other words, the angular distribution for the soft photons is again given by the form (3.166), and the probability for emission in all directions is given by the result (3.170). The establishment of the multiplying constant requires, as in the spin-0 case, a comparison with the non-relativistic formulas that were given a detailed systematic derivation.

Spin seems to be unimportant for emission in the soft-photon limit. It should be emphasized that in the covariant spin-$\frac{1}{2}$ formulation employed in this subsection, the spin effects are included in the single electromagnetic perturbation $\not{\varepsilon}$ (or γ_μ). That is, as we have emphasized near the end of Section 4.3, this perturbation includes the three perturbation terms $\boldsymbol{A} \cdot \boldsymbol{p}$, $\boldsymbol{A} \cdot \boldsymbol{A}$, and $\boldsymbol{\mu} \cdot \text{curl}\, \boldsymbol{A}$ in the non-covariant theory. The result that spin is unimportant in soft-photon emission involving relativistic (and non-relativistic) charges can perhaps be understood in terms of the result (3.150) for spin transitions as obtained in a non-covariant formulation. Finally, it should be emphasized again that all of these results have been derived in the Born approximation. The results will be employed later in various applications.

There is one feature of the soft-photon emission formula (3.170) for relativistic particles that should perhaps be noted at this point. In the limit of ultrarelativistic energies where $\beta \to 1$, the argument of the logarithm approaches the large number $(2E/mc^2)^2$. The emission probability then approaches

$$dw \to \frac{2z^2\alpha}{\pi} \left(\ln \frac{2E}{mc^2} - 1 \right) \frac{dk}{k}. \tag{3.190}$$

Further, if we integrate over photon energies or wave numbers from k_1 to k_2, the total photon-emission probability is approximately

$$w(k_1 < k < k_2) \to \frac{2z^2\alpha}{\pi} \ln \frac{E}{mc^2} \ln \frac{k_2}{k_1}. \tag{3.191}$$

Although $\alpha = 1/137$, the logarithmic factors can compensate to make the photon emission probability appreciable (that is, not small). This does not occur in the non-relativistic case where instead of the first logarithmic factor in Equation (3.191), there is a factor v^2/c^2 [see Equation (3.134)].

3.6 SPECIAL FEATURES OF ELECTROMAGNETIC PROCESSES

There are several characteristic of electromagnetic processes that can be described in a general way and that represent features that are of great importance in determining the details and qualitative aspects of various processes. Some of these, like radiative corrections and renormalization techniques, are really beyond the scope of this book for a substantive discussion; we only mention these fundamental and subtle topics. However, certain physical processes are very closely related and can be described in a unified way with the help of a property called *crossing symmetry*, which will be discussed later in this section. Also, for some kinds of processes there are special kinematic invariants that are useful in relating cross sections through crossing symmetry, and these will be introduced. Of great importance in categorizing processes is their "order" or characteristic magnitude of the associated cross sections. This can be done quite simply, and it is useful to be aware of the elementary ideas that allow a determination of the characteristic magnitudes of various cross sections.

3.6.1 "Order" of a Process

One of the basic properties of the electromagnetic coupling is that it is a fairly weak interaction—not *very* weak like that (called "weak," in fact) associated with neutrino processes, etc., but weak enough that a perturbation-theory approach is useful. We have discussed this aspect of the coupling in Section 3.5.1 in connection with soft-photon emission. In fact, treatment of that phenomenon yielded results that exhibit the characteristic strength of the coupling very clearly. We have seen [see Equations (3.134), (3.170), (3.190)] that the differential probability for soft-photon emission (energy within $\hbar\, d\omega$) to accompany some process involving a charged particle is of the form

$$dw \sim \begin{cases} \alpha\beta^2(d\omega/\omega) & \text{(NR)}, \\ \alpha \ln\gamma\,(d\omega/\omega) & \text{(ER)}, \end{cases} \tag{3.192}$$

in the non-relativistic (NR) and extreme-relativistic (ER) limits. Aside from the factor $d\omega/\omega$ and the kinematic factors β^2 and $\ln\gamma$, the probability is determined by the dimensionless coupling constant $\alpha = e^2/\hbar c$. The factor α can be considered the square of a dimensionless charge:

$$\overline{q} = e/(\hbar c)^{1/2} = \alpha^{1/2}. \tag{3.193}$$

The probability [Equation (3.192), for example] is the square of an amplitude M corresponding to some perturbation, in this case associated with a photon-emission vertex. For a single-photon vertex, the perturbation (interaction Hamiltonian or

Lagrangian) is proportional to the charge q (or the dimensionless $\overline{q}$), and the probability is

$$W = |M|^2 \propto \overline{q}^2 = \alpha. \tag{3.194}$$

A general relation can be written down for the total amplitude and probability associated with a purely electromagnetic process that can be represented by a collection of diagrams representing the actions of perturbations and the roles of propagators (or energy denominators). Let us now consider processes and diagrams that allow both charged particles and/or photons in intermediate states as well as in initial and final states; up to now we have regarded the photons to be "real" and occurring in intermediate states only if they were the same as in initial or final states. That is, we now consider processes that can correspond to "virtual" photons—photons that are only part of intermediate states. In this more general classification of electromagnetic processes the *order* of the various diagrams can be designated by the corresponding number of vertices they contain. This designation is sufficient in a covariant formulation for spin-$\frac{1}{2}$ charges for which there is only one kind of vertex, namely that representing the γ_μ interaction [see Equation (3.105)]. This is the vertex of Figure 3.2 (or 3.3) with one photon line and two fermion lines, and the strength of the perturbation is proportional to q (or $\overline{q}$). In a non-relativistic formulation, there are two types of vertices (Figures 3.2 and 3.4) associated with the interactions (3.20) and (3.21), respectively (H'_q and H'_{q^2}), as well as the magentic-moment interaction H'_μ given by the form (3.22), but this latter perturbation is usually not important [see Equation (3.25)]. In fact, as we have seen, for spin-0 charges a covariant formulation also introduces two types of vertices [see Equation (3.73)]. However, for these cases we find that the two-photon perturbation H'_{q^2} (and its relativistic counterpart) is proportional to q^2 or to $\overline{q}^2$, so it is higher order by the same factor $\alpha^{1/2}$ in its amplitude. That is, the two-photon vertex yields a factor α^2 in the squared amplitude—like two single-photon vertices.

In classifying the order of a general electromagnetic process, we can say that its probability is of the form

$$W \propto \alpha^n, \tag{3.195}$$

where

$$n = n_1 + 2n_2, \tag{3.196}$$

being the number (n_1) of single-photon vertices plus twice the number (n_2) of two-photon vertices associated with the process. Processes that are complex and require a larger number of interactions, or vertices in a diagrammatic representation, are less probable and have smaller rates. Usually the important parameter associated with some process is a cross section and, if n is its order, the cross section can be written

$$\sigma \sim \alpha^n l^2, \tag{3.197}$$

where l is some characteristic length. We can think of four such lengths: the (electron) Compton wavelength ($\Lambda = \hbar/mc$), the de Broglie wavelength ($\lambda = \hbar/p$), the classical electron radius ($r_0 = e^2/mc^2$), and the Bohr radius ($\alpha_0 = \hbar^2/me^2$). Of these, the only possible candidates for l in Equation (3.197) are Λ and $\lambda\!\!\!^{-}$ if we are considering free-particle processes,[37] for both r_0 and a_0 involve the electronic charge e and all such dependence should be contained within the factor α^n. The squared ratio $(\Lambda/\lambda\!\!\!^{-})^2$ approaches

$$(\Lambda/\lambda\!\!\!^{-})^2 \to \begin{cases} \beta^2 & \text{(NR)} \\ \gamma^2 & \text{(ER)} \end{cases} \tag{3.198}$$

in the non-relativistic and extreme-relativistic limits. The factor β^2 and γ^2 are kinematic in nature and in the intermediate domain where c.m. energies are of the order mc^2 the ratio $\Lambda/\lambda\!\!\!^{-}$ is of order unity. At this "characteristic" energy, the appropriate choice for l is the constant Λ, and so the characteristic cross section is of the order

$$\sigma \sim \alpha^n \Lambda^2. \tag{3.199}$$

Figure 3.8 exhibits Feynman diagrams for a variety of electromagnetic processes, with those on each row having the same order and so the same characteristic cross section. For each process, only one diagram is shown; some processes have a number of other diagrams that contribute to the total amplitude. The top three rows exhibit diagrams with only one kind of vertex, corresponding to the charged spin-$\frac{1}{2}$ case. In the bottom three rows are diagrams for processes involving spin-0 charges for which there are two kinds of vertices. For spin-0 charges, there would also be the diagrams on the top half contributing to the total amplitude for the corresponding processes. Also, in a non-relativistic formulation, there would be both the single- and double-photon vertex in a diagrammatic representation—even for spin-$\frac{1}{2}$ charges. Finally, it should be noted that, for simplicity, the arrows have been left off the diagrams in Figure 3.8. It is clear, for example, that a V-like vertex involves production of a particle-antiparticle pair, etc. The description of the processes below the diagrams will also clarify the nature of the processes represented. In the processes (bremsstrahlung, for example) where there is an "extra" photon produced, the differential cross section will always be proportional to $d\omega/\omega$, as is indicated, for these cases, in parentheses for the corresponding cross section on the right.

We can be a little more sophisticated in expressing the cross section for an electromagnetic process. Formally, at least, the origin of the energy dependence of a cross section can be indicated to give, for an individual process, an expression modifying the characteristic value (3.199). The energy dependence can come from three different factors in the cross section. For a process, the product of the cross section and the incident particle flux (essentially its incident velocity v_0) gives the transition probability per unit time ($\Delta W/\Delta t$). This quantity is, in turn, given by the square of the matrix element (M_{f0}) for the process times the final state phase space per unit energy (Φ_f, see Section 1.3.3). Thus, we write

$$\sigma v_0 \propto \left|M_{f0}\right|^2 \Phi_f. \tag{3.200}$$

[37]For processes involving bound electrons, a_0 might be a candidate. In fact, the characteristic cross section for photoionization is of the order αa_0^2.

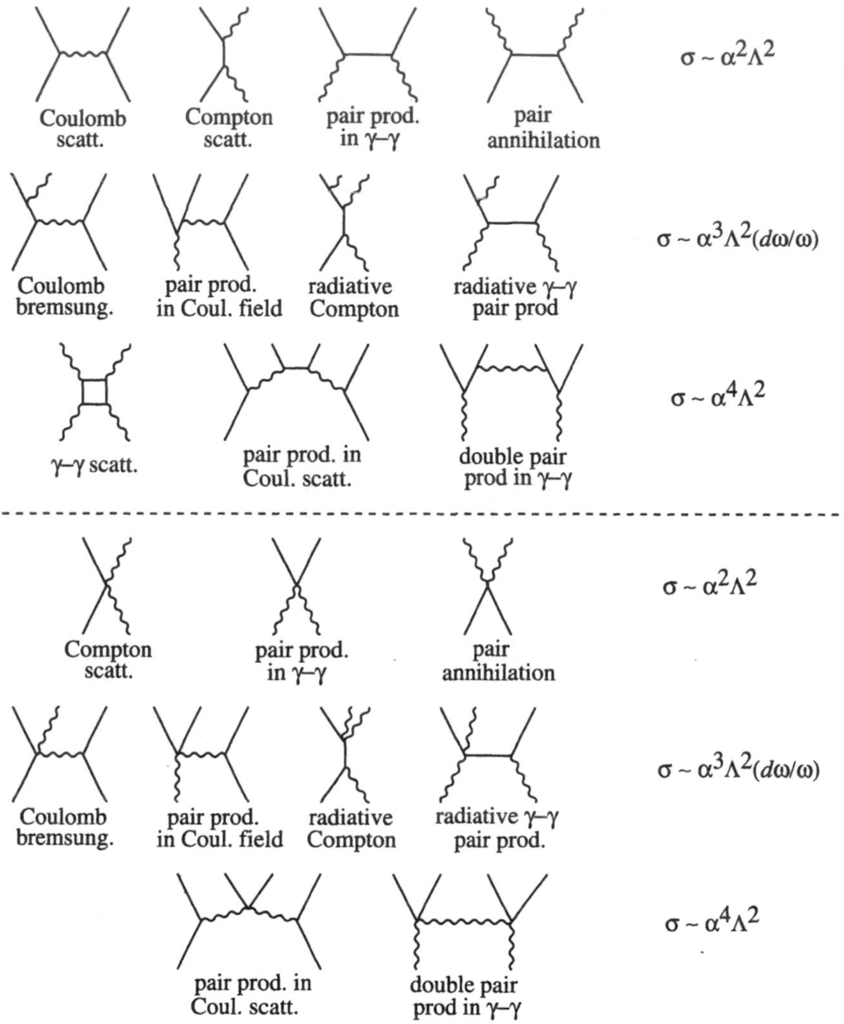

Figure 3.8 A collection of diagrams associated with various processes. There are other diagrams for the particular processes. The top part is for the covariant spin-$\frac{1}{2}$ formulation. The bottom diagrams would be for the covariant spin-0 formulation or a non-relativistic formulation and include the two-photon vertex.

The energy dependence in the cross section for a process is contained in v_0, M_{f0}, and Φ_f. If we express these in dimensionless form by dividing by their values at the characteristic energy $E_c \sim mc^2$, the cross section can be written

$$\sigma \sim \alpha^n \Lambda^2 \overline{\eta}, \tag{3.201}$$

where

$$\overline{\eta} = \overline{\mu}_{f0}^2 \overline{\Phi}_f / \overline{\beta}_0, \tag{3.202}$$

with ($\overline{\beta}_0 = v_0/c$)

$$\overline{\mu}^2_{f0} = \left|M_{f0}(E)\right|^2 / \left|M_{f0}(E_c)\right|^2, \tag{3.203}$$

$$\overline{\Phi}_f = \Phi_f(E)/\Phi_f(E_c). \tag{3.204}$$

The phase-space factor Φ_f always increases with increasing energy; for a single-particle phase space, $\Phi_f \propto v_f$ and E_f^2 in the NR and ER limits, respectively. The matrix element M_{f0} generally decreases with increasing energy at high energies. The energy dependence of $\overline{\eta}$ thus depends on the nature of the process, that is, its amplitude, number of final-state particles, etc.

3.6.2 Radiative Corrections and Renormalization

In Section 3.1, brief mention was made of the problems encountered in QED when attempts were made to compute processes higher order in perturbation theory. After the initial development of QED in 1927, in the early 1930s it was noticed that in the calculation of perturbation amplitudes to higher order, divergences sometimes appeared in the theory when summations over intermediate states were performed. These problems occurred, in particular, when the higher-order amplitude involved perturbations that could be described in terms of diagrams in which a virtual photon is "emitted and reabsorbed" by the same charge (in loose terminology). These "photons" are thus not observed but contribute to the corresponding correction (M_1) to the lowest-order amplitude (M_0). Then, for a process in which the final state is specified, the total amplitude is

$$M = M_0 + M_1 + \cdots. \tag{3.205}$$

Here, for each M_k, the final state is the same, so that the probability for the process is

$$W = |M_0 + M_1 + \cdots|^2 = |M_0|^2 + M_0 M_1^* + M_1 M_0^* + \cdots. \tag{3.206}$$

The type of amplitudes that would be part of M_1 can be indicated in terms of the diagrams[38] in Figure 3.9. These are the types[39] modifying the lowest-order amplitude for which M_0 is part of the total diagram for the complete process. The characteristic magnitude for each amplitude is determined by $\overline{e}^n$, where $\overline{e} = \alpha^{1/2}$ is the dimensionless (electron) charge and n is the number of vertices. Moreover, there is a special terminology for these types of diagram parts, as is indicated. Because of its role in giving rise to the electron self-energy due to electromagnetic effects, parts with virtual photons being emitted and reabsorbed before another interaction (vertex) are called "self-energy parts." Then there is the "vertex part" of the same order in $\overline{e}$. The "photon self-energy part" has a virtual electron-positron pair bubble; this type of virtual state is also connected with a phenomenon known as *vacuum polarization*. At the bottom of Figure 3.9 are what we shall call "radiative parts." These perturbation amplitudes (M_r) are associated with a different process, namely,

---

[38] Here we consider only one kind of basic vertex as in the covariant formulation for spin-$\frac{1}{2}$ charges.

[39] For a complete process with associated diagrams, there can be other kinds of higher-order diagrams. For example, in Coulomb scattering (see Figure 3.8), there can be the higher-order amplitude associated with the exchange of two photons.

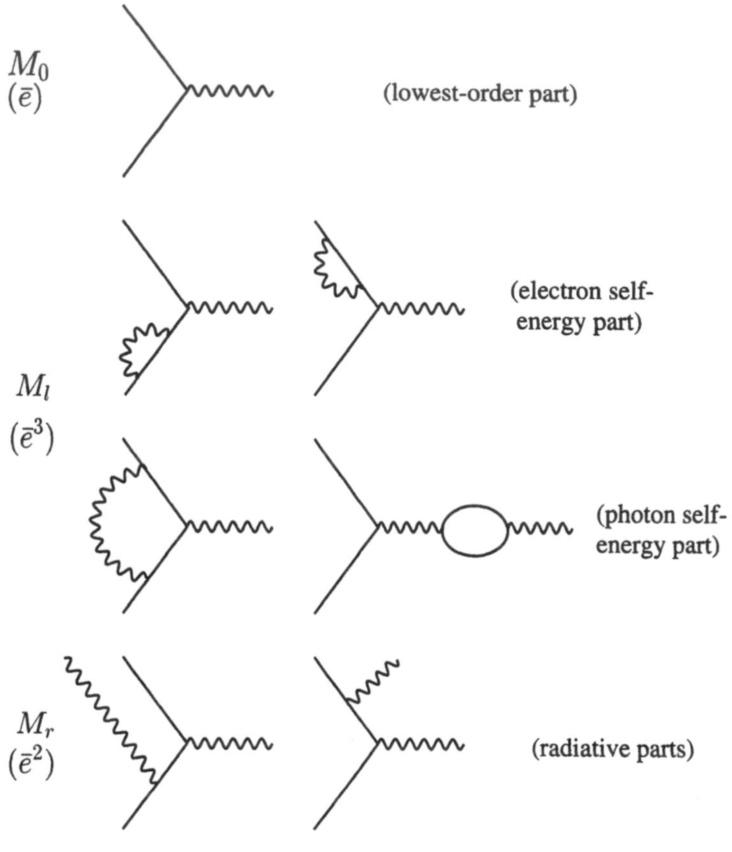

Figure 3.9 Self-energy diagrams.

that in which there is an additional photon in the final state. Thus, they do not contribute to the total amplitude (3.205), and the associated probability is

$$W_r = |M_r|^2 \,. \tag{3.207}$$

We see that, while $|M_0|^2$ is of order α (times something from the rest of the diagram), the lowest-order correction term is of order α^2; W_r is also of order α^2.

When the corrected amplitude M_1 is computed by the standard methods, it is found, for example, in the integration over the four-dimensional volume d^4k' associated with the virtual photon state, that there are divergences. Divergences occur, in different instances, at both the low and high end of the k' integration and came to be called, respectively, the "infrared catastrophe" and the "ultraviolet catastrophe." The former is really no catastrophe at all, as first shown by Bloch and Nordsieck in 1937. They explained that for any process where we consider some particular final state, our ability to observe the details of the process depends on the experimental energy resolution and detector sensitivities. In particular, if an additional soft photon is emitted in the process, our measuring apparatus may not be capable of detecting it. Suppose that the apparatus has a threshold ε_t so that photons of energy below ε_t

cannot be detected. Then, experimentally we are looking at the process without an extra photon *plus* the radiative process with an extra photon of any energy *less than* ε_t. The associated probability would be

$$E_{\text{expt}} = W + W_r = W_0 + W_1 + \cdots + W_r, \tag{3.208}$$

with $W_0 = |M_0|^2$ and $W_1 = M_0 M_1^* + M_1 M_0^*$ from Equation (3.206). If we write

$$W_{\text{expt}} = W_0(1 + \xi), \tag{3.209}$$

we find that the relative correction ξ is of order $\bar{e}^2 = \alpha$ and composed of several parts. In the two terms

$$\xi = W_1/W_0 + W_r/W_0, \tag{3.210}$$

there is a cancellation of the infrared divergence. This can be shown to order α when explicit processes are considered, such as bremsstrahlung or Compton scattering, and also for a general process.[40] That is, if we take ε_0 as the lower limit in the (extra) photon spectrum in W_r and also in the lower limit of the virtual photon spectrum in the (infrared) divergent part of W_1, we find that both integrals have the same coefficient except for a sign difference. In W_r there is a term of the form $\ln(\varepsilon_t/\varepsilon_0)$ (see Section 3.5) and in W_1 there is one of the form $\ln(\varepsilon_0/\varepsilon_{\text{char}})$, where $\varepsilon_{\text{char}}$ is some characteristic energy in the problem (such as mc^2 or some charged particle energy). The terms have the same coefficient and the sum is then of the form $\ln(\varepsilon_t/\varepsilon_{\text{char}})$ and contains no divergence for $\varepsilon_0 \to 0$.

The occurrence of infrared divergences is a consequence of the particular mathematical formulation of the subject. When the results are expressed in a form more suited to experience or measurement, the divergence disappears. Not that the soft photons emitted are unreal; there *are* an infinite number of infinitesimally soft photons emitted during, say, the scattering of a charged particle. However, the very soft photons have no physical effect on the process. In fact, in the case of very soft photons when the wavelength is very large, there is little distinction between real and virtual photons. The real photons are eventually reabsorbed someplace and in that sense are virtual in their temporary existence. Thus, it is understandable that there is a cancellation of the infrared divergences, their occurrence being simply a mathematical artifact. In fact, in most problems involving soft photons, it is actually preferable to treat phenomena by classical electrodynamics, which is a limiting domain of QED. In the classical theory, the photon concept does not appear and the mathemtical formulation does not result in an infrared "catastrophe."

The ultraviolet divergence is a more serious problem in QED and was handled in calculations of phenomena for the first time in 1947—twenty years after the subject was originally developed. The first such calculation was by Bethe in an evaluation of the Lamb shift, and the techniques of the method have been developed fully and incorporated into modern relativistic QED. The resolution of this type of divergence problem with the theory is not clean like that for the infrared divergence, and it is generally regarded as a rather unsatisfactory procedure even though it seems to allow

[40] See J. M. Jauch and F. Rohrlich, *The Theory of Photons and Electrons*, 2nd ed., Berlin: Springer-Verlag, 1976. This book gives an excellent discussion of the divergences in QED and, in particular, of the infrared divergence.

accurate higher-order calculations of any observable electromagnetic phenomenon. In the end, perhaps the ultraviolet divergence will, in some future formulation of QED, like the infrared problem, be shown to be a mathematical artifact. This has yet to be achieved, and the present theory still suffers from an uncomfortable subtraction of infinities.

The method for handling the ultraviolet divergence makes use of *renormalization* techniques. This idea, already mentioned in Section 2.6 of the previous chapter on classical electrodynamics, prescribes that results be expressed in terms of the observed or phenomenological mass (m_0) and charge (e_0) of the particles involved in the process. The values m_0 and e_0 contain electromagnetic parts (δm and δe) due to emission and absorption of virtual photons and charged pairs, and these contributions should, in principle, be capable of computation by the theory. The problem is that the theory gives infinity for δm and δe when it is applied. However, when radiative corrections to processes are computed and then formally expressed in terms of m_0 and e_0, the ultraviolet divergences can be handled by introducing covariant cutoff or convergence factors, so that the whole computational procedure is made systematic. In a practical sense, the whole technique seems to work very well. However, the parameters m_0 and e_0 are measured quantities, and the theory really gives infinity for both when it is applied.

We go no further into this topic. Generally, the coefficients of the higher-order corrections in α^n have numerical values of order unity when the theory is handled as prescribed above. The nature of the infrared and ultraviolet divergences seem to be quite different. The former difficulty is completely understood and resolved, while the latter is handled only with some embarrassment. For a more extensive discussion, the reader is referred to the standard textbooks on the subject.

3.6.3 Kinematic Invariants

In covariant perturbation theory, the amplitude $\mathcal{M}_{f0}$ for a process is itself a Lorentz invariant, being a function of invariants involving the energies and momenta of the various particles that are present in the initial and final states. These are the so-called kinematic invariants constructed from the four-momenta of the incoming and outgoing particles. In particular, for the case where there are two incoming and two outgoing particles, certain of these invariants are particularly convenient and have some physical significance. For the process

$$a + b \rightarrow c + d, \tag{3.211}$$

kinematic invariants can be constructed from the individual four-momenta $(p_\mu)_\alpha = (iE/c, \mathbf{p})_\alpha$, and there are four such quantities. Designating an individual four-momentum simply as p_α, we can express the conservation of four-momentum in the reaction (3.211) as

$$\sum_{\alpha=1}^{4} q_\alpha = 0, \tag{3.212}$$

if $q_1 = p_a$, $q_2 = p_b$, $q_3 = -p_c$, and $q_4 = -p_d$.

In addition to the squares of the individual q_α [$(q_\mu q_\mu)_\alpha = q_\alpha^2 = -m_\alpha^2 c^4$], which are just constants, six invariants $q_\alpha \cdot q_\beta$ can be formed from the six combinations of α and β. These quantities are kinematic invariants, but only two are independent because of the four conservation relations (3.212) for each component ($\mu = 0, 1, 2, 3$) of four-momentum. However, instead of the $q_\alpha \cdot q_\beta$, it is more convenient to introduce the following three invariants:

$$\begin{aligned} s &= (q_1 + q_2)^2 = (q_3 + q_4)^2, \\ t &= (q_1 + q_3)^2 = (q_2 + q_4)^2, \\ u &= (q_1 + q_4)^2 = (q_2 + q_3)^2. \end{aligned} \tag{3.213}$$

Here the squares represent four-dimensional scalar products and, since, as is easily seen,

$$s + t + u = -\sum_\alpha m_\alpha^2 c^4, \tag{3.214}$$

only two of the three invariants are independent.

The quantities s, t, and u are what we shall be referring to when we speak of kinematic invariants. Although only two are independent, it is convenient to introduce all three. This is because in addition to the process (3.211) for specific particles a, b, c, and d, there are also reactions involving the corresponding anti-particles. If, for example, b and c are transferred to the other sides of the arrow and made to represent their anti-particles (designated by a bar), the reaction is $a + \bar{c} \to \bar{b} + d$. In fact, designating the particles by numbers instead of letters, we can have the following three "channels" for reactions with two incoming and two outgoing particles[41]:

$$\begin{aligned} &1 + 2 \to 3 + 4 \quad (s), \\ &1 + \bar{3} \to \bar{2} + 4 \quad (t), \\ &1 + \bar{4} \to 3 + \bar{2} \quad (u). \end{aligned} \tag{3.215}$$

The channels are sometimes referred to as the "s", "t", and "u" channels, as indicated above. The reason for this is that s, t, and u have a very physical significance for the corresponding channels. They are, for the channels, just minus the total c.m. energies (divided by c^2).

In the case of elastic scattering, the expressions for s, t, and u are particularly simple. Then the type of outgoing particle is the same as the incoming ones: $m_1 = m_3$ and $m_2 = m_4$. In terms of c.m. quantities, the q's are

$$\begin{aligned} q_1 &= (iE_1/c, \boldsymbol{p}_s), & q_2 &= (iE_2/c, -\boldsymbol{p}_s), \\ q_3 &= -(iE_3/c, \boldsymbol{p}_s'), & q_4 &= -(iE_4/c, -\boldsymbol{p}_s'), \end{aligned} \tag{3.216}$$

where $\boldsymbol{p}_s$ and $\boldsymbol{p}_s'$ (or their negative) refer to the initial and final momenta, respectively. Since the scattering is elastic, $|\boldsymbol{p}_s| = |\boldsymbol{p}_s'| \equiv p_s$ and so $E_1 = E_3$ and $E_2 = E_4$. The values for the three invariants can then be written

$$\begin{aligned} s &= -(E_1 + E_2)^2/c^2, \\ t &= 2p_s^2(1 - \cos\theta_s), \\ u &= -(E_1 - E_2)^2/c^2 + 2p_s^2(1 + \cos\theta_s), \end{aligned} \tag{3.217}$$

where θ_s is the c.m. scattering angle between $\boldsymbol{p}_s$ and $\boldsymbol{p}_s'$.

[41]The reactions can also procede in either direction, these additional processes being simply the time reverse of the others.

3.6.4 Crossing Symmetry

The kinematic invariants are particularly useful in connection with a fundamental property of covariant perturbation theory. Because the amplitude $\mathcal{M}_{f0}$ introduced in Equation (3.109) is an invariant, it must be a function of invariants. These invariants must be associated with the observable characteristics of the process, that is, with parameters of the initial and final states. For a reaction of the type (3.211), these would be the invariants s, t, and u. More properly, for the case where the particles have spin, if we are not interested in the spin states of the incoming and outgoing particles, we should average over initial spin states and sum over final states. Then the only remaining kinematic parameters for these states are the particle four-momenta, and the invariant squared amplitude can be written

$$\sum_{\text{all spins}} \left|\mathcal{M}_{f0}\right|^2 = f(s, t, u). \tag{3.218}$$

One of the significant and useful features of covariant QED is the basic symmetry in its formulation in terms of particle and antiparticle states. The unified description of processes involving, in particular, electrons and positrons has been examined only in a superficial manner in this chapter. A brief discussion was given in Section 4.2, indicating how the theory encompasses, in a convenient and simplified way, the inclusion of positron and pair-production effects in terms of a backward-in-time description of the antiparticle states. As we have seen, it is useful to speak of "entry" and "exit" states in evaluating the total amplitude for the process. The unified treatment is also appropriate in the description of initial and final states for processes, that is, in the diagrams, for the meaning of "external lines." In this formulation, the cases corresponding to an initial-state particle and a final-state antiparticle are equivalent in that a single expression for an amplitude can refer to either case. The symmetric covariant theory yields a useful result that allows a convenient way of obtaining the squared amplitude (3.218), spin-averaged, for any cross channel (3.215) in terms of that for one of the other channels. That is, as a result of the unified formulation of the theory, a single function f yields the squared amplitude for each of the three channels (3.215). The invariants s, t, and u will be different for each channel, but the functional form (f) will be the same. This is what is known by *crossing symmetry*.

Figure 3.10 indicates the replacements for s, t, and u in channels II and III as obtained from those for channel I. When an antiparticle is involved, the q_α has a minus sign as a result of the backward-in-time description. Also exhibited in Figure 3.10 are three examples of specific physical processes in the three corresponding channels. In addition to these, there are the time-reversed processes that have the same squared amplitude. We see, for example, that the processes of Compton scattering, pair annihilation into two photons, and pair production in photon-photon collisions are closely related. Note that, since the photon is identical to its antiparticle ($\gamma = \overline{\gamma}$), we do not distinguish the two. In transferring the photon to the opposite side of the reaction equation (3.211) the processes of absorption and emission are interchanged. We should also keep in mind that for antiparticles, even though they are indicated as going backward in time in a diagram, the actual momentum of the particle is in the opposite direction; in other words, $p_{\text{anti-particle}} = -p_{\text{diagram}}$. In

I: $a + b \rightarrow c + d$

$1 + 2 \rightarrow 3 + 4$

$\gamma + e \rightarrow \gamma + e$

$e + e \rightarrow e + e$

$e + \mu \rightarrow e + \mu$

$s_I = (q_1 + q_2)^2$

$t_I = (q_1 + q_3)^2$

$u_I = (q_1 + q_4)^2$

II: $a + \bar{c} \rightarrow \bar{b} + d$

$1 + \bar{3} \rightarrow \bar{2} + 4$

$\gamma + \gamma \rightarrow \bar{e} + e$

$e + \bar{e} \rightarrow \bar{e} + e$

$e + \bar{e} \rightarrow \bar{\mu} + \mu$

$s_{II} = (q_1 - q_3)^2$

$t_{II} = (q_1 - q_2)^2$

$u_{II} = (q_1 - q_2)^2$

III: $a + \bar{d} \rightarrow c + \bar{b}$

$1 + \bar{4} \rightarrow 3 + \bar{2}$

$\gamma + \bar{e} \rightarrow \bar{e} + \gamma$

$e + \bar{e} \rightarrow \bar{e} + e$

$e + \bar{\mu} \rightarrow e + \bar{\mu}$

$s_{III} = (q_1 - q_4)^2$

$t_{III} = (q_1 + q_3)^2$

$u_{III} = (q_1 - q_2)^2$

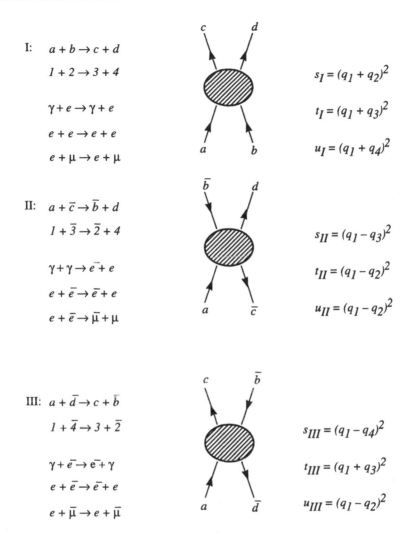

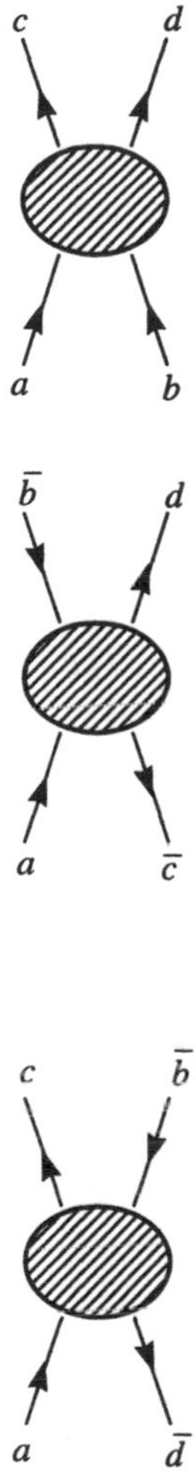

Figure 3.10 Channels associated with related processes.

evaluating the kinematics of the process in terms of diagrams and Feynman rules, the four-momenta are handled as in the diagram.

There are other examples of crossing symmetry besides those associated with the processes in Figure 3.10. For example, bremsstrahlung and pair production are related by crossing symmetry (see Figure 3.8). In fact, this was noted already in 1934 by Bethe and Heitler. Actually, the existence of this type of symmetry is very general and not restricted to electromagnetic processes. Finally, it should be mentioned that the physical domains of s, t, and u for the three channels of a process would be different. Mathematically, it is said that the scattering amplitude, being an analytic function, is "analytically continued" from one domain to another for the three channels.

Bibliographical Notes

A brief historical summary of quantum electrodynamics, along with a collection of fundamental papers may be found in

1. Schwinger, J., *Quantum Electrodynamics*, New York: Dover Publ., Inc., 1958.

To this collection of papers on QED the following could be added:

2. Fermi, E., *Rev. Mod. Phys.* **4**, 87 (1932); see also Reference 8 of Chapter 1.

There are a number of excellent textbooks on modern covariant QED, especially

3. Jauch, J. M. and Rohrlich, F., *The Theory of Photons and Electrons*, 2nd ed., Berlin: Springer-Verlag, 1976.
4. Berestetskii, V. B., Lifshitz, E. M., and Pitaevskii, L. P., *Relativistic Quantum Theory, Parts 1 and 2*, Reading, MA: Addison-Wesley Publ. Co., 1979.
5. Bjorken, J. D. and Drell, S., *Relativistic Quantum Mechanics*, New York: McGraw-Hill, 1964.
6. Feynman, R. P., *Quantum Electrodynamics*, New York: W. A. Benjamin, Inc., 1961.
7. Feynman, R. P., *Theory of Fundamental Processes*, New York: W. A. Benjamin, Inc., 1962.
8. Dyson, F. J., *Advanced Quantum Mechanics* (mimeographed lecture notes), Laboratory of Nuclear Studies, Cornell University, 1951.

Not so modern, but excellent, is

9. Heitler, W., *The Quantum Theory of Radiation*, Oxford, UK: Oxford University Press, 1954.

For the material in Sections 3.5 and 3.6, References 3 and 4 are especially good.

Chapter Four

Elastic Scattering of Charged Particles

Although radiative processes are of prime importance, the scattering of charged particles *without* photon emission is also of interest. Indeed, the scattering cross section is closely related to the corresponding radiative process (Coulomb bremsstrahlung). Here we treat Coulomb scattering classically and quantum mechanically in the Born approximation. The classical and Born limits have their own domains of validity. However, apparently by accident, the classical, Born, and exact quantum-mechanical cross sections for non-relativistic Coulomb scattering without exchange are *identical*. That is, the classical and Born formulas happen to hold outside their domains of applicability. Since, in addition, both formulations contribute to our insight to the overall problem, each will be given in this chapter.

4.1 CLASSICAL COULOMB SCATTERING

4.1.1 Small-Angle Scattering

The simplest type of scattering occurs when a charge (ze is incident on another charge (Ze) at a large impact parameter (b). This is indicated in Figure 4.1, and, to be more general, the scattering problem can be formulated for an arbitrary central force for which the Coulomb interaction is a special case. For simplicity, we assume, at first, that the scattering center is infinitely heavy and thus it acts as an external potential. The case where the recoil of the force center must be included is an elementary generalization and is treated by considering the binary collision in the c.m. frame (see Section 1.3). In this manner, the motion is described in terms of the relative position $\boldsymbol{r}_{12}$ of the particles in the two-body system. This is a well-known technique, which will be employed later; the two-particle problem is transformed into a single-particle one in terms of a reduced mass

$$\mu = m_1 m_2/(m_1 + m_2). \tag{4.1}$$

The technique is employed in classical mechanics and quantum mechanics.

The scattering center is placed at the origin, and since the force is in the radial direction, reference axes can be chosen, as in Figure 4.1, so that the classical orbit is in the x–y-plane. In terms of the y-component of the scattering force, the change in the corresponding component of the particle momentum is

$$\Delta p_y = \int_{-\infty}^{\infty} F_y(t)\, dt, \tag{4.2}$$

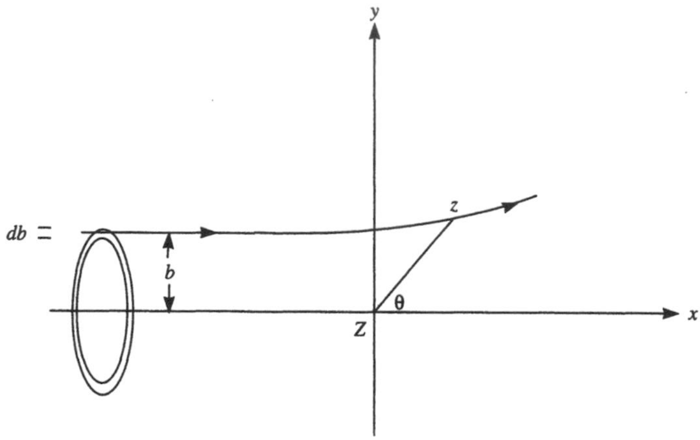

Figure 4.1 The classical orbit in scattering by a fixed potential.

with

$$F_y = F(r)\sin\theta = (y/r)F(r), \tag{4.3}$$

where $F(r)$ is the magnitude of the radial force. The deflection is then

$$\Delta p_y = \int_{-\infty}^{\infty} (y/r)F(r)\,dt, \tag{4.4}$$

with $r^2 = x^2 + y^2$ and $y = y(t)$ and $x = x(t)$; in this form, the expression (4.4) is still exact. For the case of small-angle scattering, Δp_y can be evaluated by setting

$$y(t) \approx b, \qquad x(t) \approx vt, \tag{4.5}$$

where v is the velocity of the charge incident from the left. This corresponds to a perturbation or iteration approach, whereby, in the evaluation of the effect of the perturbing force, we employ the zeroth-order (unperturbed) orbit characteristics. It is the classical analog of the Born approximation, in that it is an iterative approach. The analog has its limitations, however, since the quantum-mechanical Born approximation is less restrictive in the sense that its application is not confined to the case of small-angle scattering. Shortly, we discuss the validity of both the classical limit and the Born approximation.

For the case of Coulomb scattering for which $F(r) = zZe^2/r^2$, the deflection (4.4) is, in the approximation (4.5),

$$\Delta p_y \approx \frac{zZe^2 b}{v}\int_{-\infty}^{\infty} \frac{dx}{(b^2 + x^2)^{3/2}} = \frac{2zZe^2}{bv}. \tag{4.6}$$

Note that the main contribution to the integral comes from $x \sim b$; that is, even for the long-range Coulomb force, the deflection is produced mainly by the acceleration while the particle is near the scattering center around the distance of closest approach ($\approx b$). In terms of the (small) scattering angle θ_{sc}, $\Delta p_y \approx p\theta_{sc}$, where $p = mv$ is the initial momentum. We then have

$$b \approx \frac{2zZe^2}{pv}\frac{1}{\theta_{sc}}, \tag{4.7}$$

being the basic relation between impact parameter and scattering angle. The differential scattering cross section is the area $d\sigma = 2\pi b\,db$ of the ring projected on the y–z-plane (see Figure 4.1). Per unit solid angle element $d\Omega = 2\pi \sin\theta_{sc} d\theta_{sc} \approx 2\pi\theta_{sc} d\theta_{sc}$ describing directions for the outgoing scattered charge, we have, employing the relation (4.7) to evaluate the absolute value of $db/d\theta_{sc}$,

$$d\sigma/d\Omega \approx (b/\theta_{sc})db/d\theta_{sc} \approx (2zZe^2/pv)^2(\theta_{sc})^{-4}. \tag{4.8}$$

The result is the small-θ_{sc} limit of the well-known Rutherford formula (see Section 1.2), exhibiting the divergent form $d\sigma \propto d\theta_{sc}/\theta_{sc}^3$ and yielding an infinite total cross section. The infinity is associated with the "infinite range" (for the scattering problem) of the Coulomb force.

As an *academic* problem, let us now consider the case where the incident charge z is relativistic; that is, its velocity may be close to c. The "academic" nature of the treatment below is emphasized, because it is simply not valid to treat Coulomb scattering classically except in the low-energy limit. In fact, as shown in Section 1.4, the energy domain of the classical limit for Coulomb scattering is $E \ll (zZ)^2(m_z/m_e)\mathrm{Ry}$, which is generally well below relativistic energies. The relativistic result is derived by an (incorrect) classical method because it happens to give a correct answer. A correct quantum-mechanical derivation in the Born approximation will be given in the following sections. For small-angle scattering, the classical deflection of a relativistic charge is computed easily by employing results given in Section 8.1 in Chapter 2. If K is the lab frame and K' is the rest frame of the charge z, we have

$$\Delta p_y = \Delta p'_y = \int F'_y(t')dt'. \tag{4.9}$$

Here $F'_y(t) = zeE'_y(t)$, where $E'_y(t)$ is the electric field experienced by the charge in its rest frame and is given by the expression E_2 in Equation (2.155) with primes added. That is, we have

$$\begin{aligned}\Delta p_y = \Delta p'_y &= zZe^2\gamma b\int_{-\infty}^{\infty}(b^2+\gamma^2v^2t^{2'})^{-3/2}dt' \\ &= 2zZe^2/bv,\end{aligned} \tag{4.10}$$

which is identical to the result (4.6) derived in a non-relativistic treatment. The scattering angle is $\theta_{sc} = v_y/v = \Delta p_y/p$, with $p = \gamma mv$, and the impact parameter $b(\theta_{sc})$ is given by the same form (4.7) with, again, p now equal to γmv. The result for the differential cross section $d\sigma/d\Omega$ is the same as the non-relativistic formula (4.8) (with $p = mv$) except there is an extra factor $1/\gamma^2$.

4.1.2 General Case

For the case where θ_{sc} is not necessarily small, the (exact) treatment of classical Coulomb scattering is only a little more complicated than that in the small-θ_{sc} approximation. Since the problem is covered in just about every textbook in classical mechanics, the treatment here is brief, although the procedure outlined here is a little different from that usually given.

The starting point for the derivation of the "orbit equation" for a particle in a central force field is the radial component of the equation of motion[1]:

$$m(\ddot{r} - r\dot{\theta}^2) = f(r); \tag{4.11}$$

here $f(r)$ is the radial force, equal to zZe^2/r^2 in the Coulomb case (and $-GmM/r^2$ in the gravitational problem). The θ-equation of motion yields the well-known constant of the motion

$$J = mr^2\dot{\theta}, \tag{4.12}$$

which is the particle angular momentum. The result (4.12) is then used to transform time derivatives into derivatives with respect to θ by means of

$$\frac{d}{dt} = \frac{J}{mr^2}\frac{d}{d\theta}. \tag{4.13}$$

The equation (4.11) for $r(t)$ is then transformed into an orbit equation for $r(\theta)$:

$$d^2r/d\theta^2 - (2/r)(dr/d\theta) - r = (m/J^2)r^4 f(r). \tag{4.14}$$

This equation is further simplified through the substitution

$$u = 1/r, \tag{4.15}$$

to yield the simpler form

$$\frac{d^2u}{d\theta^2} + u = -\frac{m}{J^2}\frac{f(u)}{u^2}. \tag{4.16}$$

In the case of a Coulomb field $f(u)/u^2 = zZe^2$, and Equation (4.16) has the particularly simple form

$$d^2u/d\theta^2 + u = -\varkappa. \tag{4.17}$$

In terms of the impact parameter and incident velocity, $J = mvb$, and we can write

$$\varkappa = r_v/b^2, \tag{4.18}$$

where

$$r_v \equiv zZe^2/mv^2 \tag{4.19}$$

is a convenient length.

---

[1] See any book on classical mechanics. The terms in the equation of motion (4.11) have an elementary interpretation. Because a spherical polar coordinate system (with unit vectors $\boldsymbol{i}_r$, $\boldsymbol{i}_\theta$, and $\boldsymbol{i}_\phi$) is a rotating one rather than fixed in space, with $\boldsymbol{i}_r$ always in the radial direction from the origin, the radial equation (4.11) has a "centrifugal" acceleration term ($-r\dot{\theta}^2$) in addition to $\ddot{r}$.

The solution of Equation (4.17) can be written

$$u(\theta) = A\cos\theta + B\sin\theta - \varkappa, \tag{4.20}$$

with the constants A and B determined from the boundary conditions. With the particle incident from infinite distance from the left (see Figure 4.1), we have $u \to 0$ for $\theta \to \pi$, which immediately gives $A = -\varkappa$. Then we can consider the limit where, again on the orbit incident from the left, $\theta \to \pi - \vartheta$ with $\vartheta \ll 1$, for which $\vartheta \to b/r = bu$; that is, $u \to \vartheta/b$. Since in this limit $\cos(\pi - \vartheta) \to -1 + O(\vartheta^2)$ and $\sin(\pi - \vartheta) \to \vartheta + O(\vartheta^3)$; we then obtain $B = 1/b$. The boundary condition $u \to 0$ for $\theta \to \theta_{sc}$ at the other end of the orbit then yields, with the help of the relation (4.18),

$$b = r_v \frac{1 + \cos\theta_{sc}}{\sin\theta_{sc}} = r_v \cot\frac{\theta_{sc}}{2}. \tag{4.21}$$

This is the important relation between impact parameter and scattering angle that is used to derive the cross section

$$\frac{d\sigma}{d\Omega} = \frac{2\pi b\, db}{2\pi \sin\theta_{sc} d\theta_{sc}} = \frac{1}{4} r_v^2 \csc^4\frac{\theta_{sc}}{2}, \tag{4.22}$$

which is the famous Rutherford formula. In terms of the convenient variable

$$s \equiv \sin(\theta_{sc}/2), \tag{4.23}$$

the result can be written very simply as $d\sigma/d\Omega = r_v^2/4s^4$, and the solid angle element is $d\Omega = 8\pi s\, ds$. In the case $\theta_{sc} \ll 1$, the general expression (4.22) agrees with Equation (4.8) derived in that limit.

The length parameter (4.19) can be given a simple physical meaning by relating the distance of closest approach[2] to θ_{sc}. At r_{min}, $\dot{r} = 0$; applying energy and angular momentum conservation as well as Equation (4.21), we easily find

$$r_{min} = r_v\big(1 + \csc(\theta_{sc}/2)\big). \tag{4.24}$$

For a head-on collision for which $\theta_{sc} = \pi$, $r_{min} \to 2r_v$.

4.1.3 Two-Body Problem—Relative Motion

The description of motion in the two-body problem when the masses m_1 and m_2 have arbitrary values is a minor complication on the single-particle problem (when, say, $m_2 \to \infty$). The procedure is well known and involves the introduction of the

---

[2]The relation between r_{min} and θ_{sc} is a useful one in a historic application. In the scattering of α-particles by a heavy nucleus such that the classical limit is applicable, the size of the nucleus can be determined as the r_{min} for which deviation from Rutherford scattering occurs (for corresponding large enough θ_{sc}).

center-of-mass coordinate

$$\boldsymbol{R} = (m_1\boldsymbol{r}_1 + m_2\boldsymbol{r}_2)/(m_1 + m_2), \tag{4.25}$$

and the relative coordinate[3]

$$\boldsymbol{r} = \boldsymbol{r}_1 - \boldsymbol{r}_2. \tag{4.26}$$

With $\boldsymbol{F}_{1(2)} \equiv \boldsymbol{F}$ as the force on particle 1 due to particle 2, since $\boldsymbol{F}_{2(1)} = -\boldsymbol{F}_{1(2)} = -\boldsymbol{F}$, the two equations of motion

$$m_1\ddot{\boldsymbol{r}}_1 = \boldsymbol{F}_{1(2)}, \quad m_2\ddot{\boldsymbol{r}}_2 = \boldsymbol{F}_{2(1)}, \tag{4.27}$$

can be subtracted to yield the "equivalent" one-body equation

$$\ddot{\boldsymbol{r}} = \boldsymbol{F}/\mu, \tag{4.28}$$

where

$$1/\mu \equiv 1/m_1 + 1/m_2 \tag{4.29}$$

is the "reduced mass." This is a convenient procedure, since generally $\boldsymbol{F}$ is an explicit function of $\boldsymbol{r}$ as in the one-body problem; for example, in the Coulomb problem $\boldsymbol{F} = (z_1z_2e^2/r^3)\boldsymbol{r}$. The equation (4.28) for relative motion is then identical to that for one body ($m_2 \to \infty$, say) except that m_1 is replaced by μ. That is, the same equations (4.11) and (4.12) would result[4] with μ appearing instead of m and $\boldsymbol{r}$ referring to the relative position. The solution for $\boldsymbol{r}$ then gives a complete description of the motion of both particles, since the position of each with respect to the center of mass is ($M = m_1 + m_2$)

$$\begin{aligned} \boldsymbol{r}_1' &= \boldsymbol{r}_1 - \boldsymbol{R} = (\mu/m_1)\boldsymbol{r} = (m_2/M)\boldsymbol{r}, \\ \boldsymbol{r}_2' &= \boldsymbol{r}_2 - \boldsymbol{R} = -(\mu/m_2)\boldsymbol{r} = -(m_1/M)\boldsymbol{r}. \end{aligned} \tag{4.30}$$

Specifically, for Coulomb scattering, the cross section would be given by [see Equations (4.19) and (4.22)]

$$\frac{d\sigma}{d\Omega} = \frac{2\pi b\, db}{2\pi \sin\theta_{\rm sc}' d\theta_{\rm sc}'} = \left(\frac{z_1z_2e^2}{2\mu v^2}\right)^2 \csc^4\frac{\theta_{\rm sc}'}{2}; \tag{4.31}$$

here $\theta_{\rm sc}'$ is the c.m. scattering angle (see Figure 4.2) and v is the initial (and final) relative velocity. Note, as indicated in Figure 4.2, that the impact parameter b is that for relative motion; that is, it is the distance between the projected asymptotes of the orbits of the particles in the c.m. system. In the frame where one particle is initially at rest—what is often called the "lab" system—this is also the impact parameter of the incident particle with respect to the target particle initially at rest.[5] The cross section is conveniently represented in terms of the c.m. scattering angle $\theta_{\rm sc}'$.

---

[3]Here an arbitrary choice of sign has been made, such that $\boldsymbol{r}$ points from particle 2 to particle 1.

[4]The constant of the motion J ($= \mu r^2\dot{\theta}$) would now be the total angular momentum in the c.m. system.

[5]Then v in Equation (4.31) is simply the velocity of the incident particle.

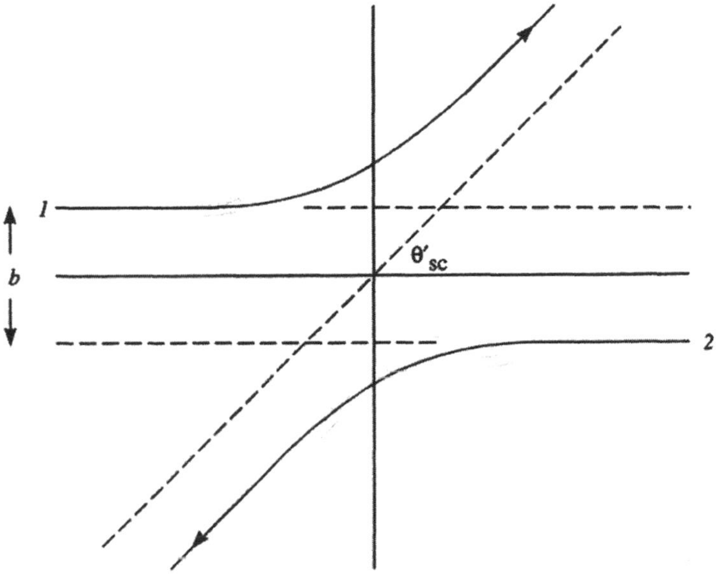

Figure 4.2 Two-body scattering in the center of mass frame.

4.1.4 Validity of the Classical Limit

Now we come to the important question of when classical mechanics applies in the treatment of scattering. Essentially, the validity criterion would be

$$ƛ \ll l, \tag{4.32}$$

where $ƛ = \hbar/p$ is the de Broglie wavelength of the scattered particle and l is some characteristic length associated with the scattering process. The Coulomb potential has no inherent length, but when we set

$$zZe^2/l \sim mv^2, \tag{4.33}$$

the inequality (4.32) yields

$$v/c \ll zZe^2/\hbar c = zZ\alpha. \tag{4.34}$$

In terms of the particle energy and the Rydberg energy $\text{Ry} = \frac{1}{2}\alpha^2 m_e c^2$, this inequality can be written

$$E \ll z^2 Z^2 (m/m_e)\text{Ry}, \tag{4.35}$$

where m/m_e is the ratio of the particle mass to that of an electron. For electron scattering by a nucleus, the requirement (4.35) limits the domain to energies less than the one-electron or K-shell binding energy. For scattering of α-particles by a heavy nucleus, say $Z \sim 80$, the criterion (4.35) implies that the classical domain covers the whole non-relativistic region ($E \lesssim 3$ GeV). Actually, nuclear forces would affect the scattering at a lower energy in α-particle scattering. For proton-proton scattering, the restriction (4.35) would limit the classical domain to less than about 30 keV; this is below the energy at which nuclear forces are effective.

For Coulomb scattering, the classical domain corresponds to low energy. Note also the dependence on Z in the inequalities (4.34) and (4.35); the classical treatment is a better approximation for scattering fields of greater strength. This is a general characteristic of the classical limit.

Let us now consider the validity of the classical limit in somewhat greater depth. One might wonder, for example, whether the classical approximation would hold for large impact parameters even at energies higher than the domain (4.35). It does not; the classical approximation holds for all scattering angles or impact parameters only when the energy is low enough to satisfy the requirement (4.35). In the more careful analysis given below, the validity of the classical limit is considered for a general scattering potential, and the characteristics of the potential that are required for validity are brought out more clearly.

In addition to the requirement (4.32), there is a restriction associated with the fundamental indeterminacy of momentum and position for the path of the scattered particle. If δp_y is this quantity for the y-component of momentum (see Figure 4.1), then for an orbit determined in y to within $\delta y \sim b$,

$$\delta p_y \sim \hbar/b. \tag{4.36}$$

For the validity of the classical description of scattering, we must have

$$\Delta p_y/\delta p_y \gg 1, \tag{4.37}$$

where Δp_y is the deflection given by the classical expression (4.4) in terms of the classical force. Since most of the perturbation comes from $y \sim x \sim b$, the inequality (4.37) requires

$$\frac{\Delta p_y}{\delta p_y} \sim \frac{b^2}{\hbar v}\int_{-\infty}^{\infty}\frac{F(r)}{r}dx \sim \frac{b^2}{\hbar v}F(b) \gg 1. \tag{4.38}$$

The restriction (4.38) shows explicitly how the classical approximation is more easily satisfied for stronger scattering forces. In addition, for the Coulomb case, we see that the restriction again yields the necessity of the inequality (4.34), and this is also required for small-angle scattering or a large impact parameter. That is, the classical limit applies (or does not apply) over the complete range of θ_{sc} as long as the energy restriction (4.34) is satisfied (or is not).

4.2 NON-RELATIVISTIC BORN APPROXIMATION AND EXACT TREATMENT

4.2.1 Perturbation-Theory Formulation

The expression for the scattering cross section in the Born approximation can be obtained most directly by making use of results already derived in Section 3 of Chapter 3 on perturbation theory. In this formulation, we consider a flux of particles (or a single particle) incident on a scattering center and calculate the rate at which the incident particles undergo transitions such that they are scattered in various directions. If the initial (incident) and final (scattered) states of these particles are described or specified by wave vectors $\boldsymbol{k}_0$ and $\boldsymbol{k}_f$, respectively, the normalized single-particle spatial wave functions in these states would be of the form[6]

$$u_0 = L^{-3/2} e^{i\boldsymbol{k}_0 \cdot \boldsymbol{r}}, \quad u_f = L^{-3/2} e^{i\boldsymbol{k}_f \cdot \boldsymbol{r}}. \tag{4.39}$$

Here L^3 is the normalization volume, so that $\int |u|^2 d^3\boldsymbol{r} = 1$; as we have done in the previous chapter, we set the volume L^3 equal to unity since it never appears in final expressions for the cross section. The wave functions (4.39) are of the plane-wave type, corresponding to motion in a force-free region. Near the scattering center, the exact particle wave functions would deviate from this form, being distorted because of the presence of the potential. It is in the nature of a perturbation approach to neglect the effects of the potential. This is also equivalent to an iterative method for solving the wave equation for the scattering problem by employing the zeroth-order (unperturbed) wave function in the term involving the scattering potential. The procedure works when the potential is "effectively" weak, but it is not assumed that the potential is weak over the entire spatial region of the scattering. We return to the question of the validity of the Born approximation in Section 2.5.

The scattering cross section is computed from the transition probability per unit time by means of

$$\Delta W_{f0}/\Delta t = J_{\text{inc}} d\sigma_{f0}, \tag{4.40}$$

where [see Equation (3.34)]

$$\Delta W_{f0}/\Delta t = (2\pi/\hbar) \left|H'_{f0}\right|^2 \delta(E_f - E_0). \tag{4.41}$$

In the Golden Rule formula (4.41), the interaction Hamiltonian H' is just the scattering potential V_{sc}, which is the pertubation causing the transition $0 \to f$ ($\boldsymbol{k}_0 \to \boldsymbol{k}_f$) in the state of the incident particle. The incident flux is $J_{\text{inc}} = |u_0|^2 v_0 = v_0$, since the free-particle states are normalized to unit volume. Employing the plane-wave or zeroth-order wave functions (4.39) to compute the matrix element H'_{f0}, we have

$$H'_{f0} = \int e^{i\boldsymbol{q}\cdot\boldsymbol{r}} V_{\text{sc}}(\boldsymbol{r})\, d^3\boldsymbol{r} \equiv V_{\text{sc}}(\boldsymbol{q}), \tag{4.42}$$

where

$$\boldsymbol{q} = \boldsymbol{k}_0 - \boldsymbol{k}_f. \tag{4.43}$$

The particle momentum change is $\Delta \boldsymbol{p} = -\hbar\boldsymbol{q}$, so that $\hbar\boldsymbol{q}$ is the momentum transferred to the scattering center.

[6]Spin is ignored here. That is, interactions involving the particle spin are neglected in the scattering process; the spin coordinate does not undergo a change. Formally, the total wave function could be introduced by having factors χ_0 and χ_f multiplying the spatial wave functions (4.39). Because of the orthogonality of the spin eigenfunctions, summation over spin coordinates in matrix elements involved in the perturbation calculation would simply yield the result that the spin coordinate would not change in the transition (scattering).

In evaluating the total transition rate from the expression (4.41), we sum (integrate) over the number of final states

$$dN_f = p_f^2\, dp_f\, d\Omega_f/(2\pi\hbar)^3. \tag{4.44}$$

Since we are interested in the differential scattering cross section, we do not integrate over $d\Omega_f$; we do integrate over dp_f, however, using the δ-function:

$$\int p_f^2\, dp_f\, \delta(E_f - E_0) = m \int p_f\, \delta(E_f - E_0)\, dE_f = m^2 v_0. \tag{4.45}$$

It is now implied, then, that in the matrix element (4.42) $|\boldsymbol{k}_f| = |\boldsymbol{k}_0|$. Combining these results,[7] the differential elastic scattering cross section in the Born approximation is given by

$$d\sigma_{\boldsymbol{q}} = (m/2\pi\hbar^2)^2\, |V_{\rm sc}(\boldsymbol{q})|^2\, d\Omega \tag{4.46}$$

(the subscript f has been left off $d\Omega_f$). The cross section is determined by the squared magnitude of the Fourier transform of the scattering potential with the momentum transfer as argument.

For the case where the scattering center is a central potential $[V_{\rm sc}(\boldsymbol{r}) = V_{\rm sc}(r)]$, the Fourier transform can be written more explicitly in terms of a radial integral. If we set up the axes of the coordinate integration (4.42) with, say, the z-axis as the polar axis in a spherical polar coordinate system aligned with q, then

$$\boldsymbol{q}\cdot\boldsymbol{r} = qz = qr\cos\theta. \tag{4.47}$$

The volume element is $d^3\boldsymbol{r} = 2\pi r^2 \sin\theta\, d\theta\, dr$, and the angular integration yields

$$\int_0^{\pi} e^{iqr\cos\theta} \sin\theta\, d\theta = (2/qr)\sin qr, \tag{4.48}$$

and the Fourier transform $V_{\rm sc}(\boldsymbol{q})$ is a function of the magnitude of $\boldsymbol{q}$:

$$V_{\rm sc}(q) = (4\pi/q) \int_0^{\infty} V_{\rm sc}(r)\, r \sin qr\, dr. \tag{4.49}$$

The Born cross section (4.46) with the Fourier transform given by the explicit formula (4.49) allows an evaluation for any radial scattering potential. In particular, for the"screened" Coulomb potential[8]

$$V_{\rm sc}(r) = (zZe^2/r)\, e^{-\varkappa r}, \tag{4.50}$$

the Fourier transform is

$$V_{\rm sc}(q) = 4\pi zZe^2/(\varkappa^2 + q^2). \tag{4.51}$$

In terms of the scattering angle, the momentum transfer is

$$q = (2mv_0/\hbar)\sin(\theta_{\rm sc}/2), \tag{4.52}$$

---

[7]We can also see how the factors involving the volume L^3 cancel. Since $(J_{\rm inc})^{-1} \propto L^3$ and dN_f should include a factor L^3, the product $L^3L^3 = L^6$ will cancel with a factor L^{-6} from the squared matrix element $|H'_{f0}|^2$ employing two wave functions (4.39), each with a factor $L^{-3/2}$.

[8]The exponential screening factor is inserted because without it the integral (4.49) would not be well-defined as it stands. This is a purely mathematical artifact, since, after evaluating the integral, $\varkappa$ can be allowed to approach zero, yielding a meaningful result.

and, in the limit ($\varkappa \to 0$) of a pure Coulomb field, we have, by Equations (4.46), (4.51), and (4.52),

$$d\sigma/d\Omega = (zZe^2/4E_0)^2 \csc^4(\theta_{sc}/2). \tag{4.53}$$

This is *identical* to the expression (4.22) obtained in a purely classical derivation. Note, especially, the cancellation of the factors of $\hbar$, which occurs only in the special case of the Coulomb potential. The equivalent results of the classical and Born derivations for the Coulomb problem are thought to be an accident, since the methods have different (and non-overlapping) domains of applicability.

4.2.2 Sketch of Exact Theory

Coulomb scattering can be treated by quantum mechanics without recourse to the Born approximation. The result of this exact calculation is an expression for the cross section that is the same as the Born (and classical) formula. That is, a single formula holds throughout the whole non-relativistic domain; the classical and Born formulations, by accident, happen to yield this same general expression.

In order to treat the scattering problem for the Coulomb potential, it is necessary to solve the Schrödinger equation

$$\nabla^2 u + (2m/\hbar^2)(E - zZe^2/r)u = 0. \tag{4.54}$$

That is, we are interested in the continuum-state solution of the equation that also describes the bound states of the hydrogen atom. Unfortunately, continuum Coulomb wave functions are not simple expressions but are proportional to a special mathematical form called a *confluent hypergeometric function*. Although these functions have no elementary explicit representation, they are fairly simple in appearance. For example, if the total wave function solving equation (4.54) is written

$$u(r, \theta, \phi) = R_{nl}(r)Y_{lm}(\theta, \phi), \tag{4.55}$$

the radial function R has the form of an oscillating sine wave $A(r)\sin[k(r)r + \Delta(r)]$ with modulating amplitude $A(r)$ and wave vector $k(r)$ varying with radial coordinate r. It is a special peculiarity of the Coulomb problem that the phase $\Delta(r)$ does not approach a definite value as $r \to \infty$. Instead, it has a logarithmic dependence on r; this is an effect of the long-range nature of the Coulomb field. Actually, for the Coulomb-scattering problem, it is not convenient to express the wave function in the form (4.55) in terms of the various angular momentum l-states. Here we only outline the complete solution to the problem and refer the reader to further details given in the excellent book by Mott and Massey[9] (MM).

In terms of the wave vector $k = (2mE/\hbar^2)^{1/2}$ of the incident particle and $\beta \equiv 2mzZe^2/\hbar^2$, the wave equation (4.54) has the form

$$\nabla^2 u + (k^2 - \beta/r)u = 0. \tag{4.56}$$

If the particle is incident on the scattering center in the z-direction, it is suggestive to first mathematically insert this aspect of the problem and write

$$u \propto e^{ikz} F \tag{4.57}$$

---

[9] N. F. Mott and H. S. W. Massey, *Theory of Atomic Collisions*, 3rd ed., Oxford, UK: Oxford University Press, 1965.

to obtain an equation for F:

$$\nabla^2 F + 2ik(\partial F/\partial z) - \beta F/r = 0. \tag{4.58}$$

Now in considering the type of solution we are seeking, we note that the total solution must include a part representing an incident wave (essential form e^{ikz}) and an outgoing scattered wave (of form e^{ikr}/r). The asymptotic forms will not be *exactly* like these terms, because of the Coulomb distortion of the phase, but these will be the dominant forms. Further, as is well known, the so-called parabolic coordinates

$$\begin{aligned} \xi &= r - z = r(1 - \cos\theta), \\ \eta &= r + z = r(1 + \cos\theta), \\ \phi &= \phi, \end{aligned} \tag{4.59}$$

are sometimes useful in the Coulomb problem, rather than the usual r, θ, and ϕ. Since there is axial symmetry, we do not expect a ϕ-dependence, and to obtain a form $u \propto e^{ikr}$, we see that this can result if F is a function of ξ but not of η. Thus, we are led to try a form

$$F = F(\xi) = F(r - z), \tag{4.60}$$

as a solution to Equation (4.58). Transforming the derivatives, we find that this is indeed the case, since the resulting equation

$$\xi \frac{d^2 F}{d\xi^2} + (1 - ik\xi)\frac{dF}{d\xi} - \frac{1}{2}\beta F = 0 \tag{4.61}$$

contains only ξ as an independent variable. Equation (4.61) is, in fact, of a type called a hypergeometric equation, of general form (here z is the complex coordinate $x + iy$)

$$z\frac{d^2 F}{dz^2} + (b - z)\frac{dF}{dz} - aF = 0, \tag{4.62}$$

with a solution $F(a; b; z)$. That is, if we set $ik\xi = z$, Equation (4.61) is precisely of the form (4.62) with $b = 1$ and $a = -i\beta/2k \equiv -in$, where

$$n = zZe^2/\hbar v. \tag{4.63}$$

The solution of Equation (4.62) is then

$$F = F(-in; 1; ik\xi). \tag{4.64}$$

Special aspects of the solution are investigated by standard methods; for example, if a power series

$$F(\xi) = \xi^{\rho}(1 + a_1\xi + a_2\xi^2 + \cdots) \tag{4.65}$$

is written, the "indicial equation" (coefficient of $\xi^{\rho-1}$ in this case) yields the result

$$\rho^2 = 0, \tag{4.66}$$

so that F is finite at the origin. However, we are primarily interested in the asymptotic forms (MM; see also Footnote 9). It is found that the solution can be written as $F = W_1 + W_2$, where W_1 and W_2 are the incident- and scattered-wave parts and have different asymptotic forms. Specifically, it is found (MM) that for an incident wave of unit amplitude (flux $v_0 = v$) the total solution is

$$u(r, \theta) = e^{-\pi n/2}\Gamma(1 + in)e^{ikz}F(-in; 1; ik\xi). \tag{4.67}$$

This expression has the asymptotic (large r) form

$$u \sim I + Sf(\theta), \tag{4.68}$$

where I and Sf are the incident and scattered parts. Through terms of order $1/r$, the forms are (MM)

$$\begin{aligned} I &= [1 - n^2/ik(r - z)]\exp[ikz + in\ln k(r - z)], \\ S &= r^{-1}\exp[ikr - in\ln kr], \\ f(\theta) &= (zZe^2/2mv^2)\csc^2(\theta/2)\exp[-in\ln(1 - \cos\theta) + i\pi + 2i\eta_0], \end{aligned} \tag{4.69}$$

where

$$\exp 2i\eta_0 = \Gamma(1 + in)/\Gamma(1 - in) \tag{4.70}$$

[or $\eta_0 = \arg\Gamma(1 + in)$]. Note that I has unit asymptotic amplitude, which is the reason for the normalization choice with the factors $\exp(-\pi n/2)\Gamma(1 + in)$ in the general solution (4.67).

The differential scattering cross section is obtained from the scattered wave amplitude given in Equation (4.69). From the definition of the cross section in terms of the incident and outgoing fluxes and element of area $dA_{\text{out}} = r^2 d\Omega$ crossed by the outgoing particle,

$$J_{\text{inc}}d\sigma = v\,|u_I|^2\,d\sigma = v\,d\sigma = J_{\text{sc}}dA_{\text{out}} = v\,|u_S|^2\,r^2 d\Omega, \tag{4.71}$$

we have

$$\begin{aligned} d\sigma/d\Omega &= r^2\,|u_S|^2 = |f(\theta)|^2 \\ &= (zZe^2/2mv^2)^2\csc^4(\theta/2). \end{aligned} \tag{4.72}$$

This is exactly the same as the Born and classical formulas.

Some other features of the solution are of significance. First, note that the phases of both the incident and scattered waves do not approach a constant value at infinity, but have a logarithmic dependence on position. Another quantity of interest is the squared amplitude of the wave function at the origin when the (incident) amplitude is normalized to unity at infinity. Since, by the result (4.66), $F(0) \to 1$, the expression (4.68) yields

$$|u(0)|^2 = e^{-\pi n}\,|\Gamma(1 + in)|^2. \tag{4.73}$$

Employing the well-known indentities

$$\Gamma(1 + z) = z\Gamma(z), \tag{4.74}$$

$$\Gamma(z)\Gamma(1 - z) = \pi\csc\pi z, \tag{4.75}$$

we find

$$|u(0)|^2 = e^{-\pi n} in\Gamma(in)\Gamma(1-in) = e^{-\pi n} in\pi \csc i\pi n$$
$$= \frac{\pi n e^{-\pi n}}{\sinh \pi n} = \frac{2\pi n}{e^{2\pi n}-1}. \tag{4.76}$$

For an attractive Coulomb field ($n < 0$),

$$|u(0)|^2 \underset{n<0}{\longrightarrow} 2\pi|n|, \tag{4.77}$$

while for a repulsive field ($n > 0$)

$$|u(0)|^2 \underset{n>0}{\longrightarrow} 2\pi n e^{-2\pi n}. \tag{4.78}$$

On the other hand, when the field is weak ($n << 1$)

$$|u(0)|^2 \underset{n<<1}{\longrightarrow} 1, \tag{4.79}$$

which is the Born-limit result. The exponential in the result (4.78), indicating a very small amplitude at the origin, is sometimes referred to as the Gamow penetration factor. Quantum mechanics gives a finite value for the wave function at the origin, although this region is inaccessible in a classical orbit.

4.2.3 Two-Body Problem

Just as in the classical two-body problem (Section 1.3), when the interparticle potential is a function of the relative coordinate (4.26), the corresponding quantum-mechanical problem is cast into an equivalent one-body description with a reduced mass (4.29) as parameter. It is an elementary exercise to show this, and the formulation is applicable to the case of bound states as well as to scattering problems, and in the latter it is not restricted to the Born limit. For now we do not consider the effects of exchange (see Section 2.4); it is assumed that the particles, with masses m_1 and m_2, are distinguishable. The spin coordinates are ignored, so that the (total) two-particle wave function is

$$u_t(1,2) = u_t(\boldsymbol{r}_1, \boldsymbol{r}_2). \tag{4.80}$$

The stationary-state Schrödinger equation is

$$-\frac{\hbar^2}{2}\left(\frac{1}{m_1}\nabla_1^2 + \frac{1}{m_2}\nabla_2^2\right) u_t(\boldsymbol{r}_1,\boldsymbol{r}_2) + V(\boldsymbol{r}_1,\boldsymbol{r}_2)u_t(\boldsymbol{r}_1,\boldsymbol{r}_2) = E_t u_t(\boldsymbol{r}_1,\boldsymbol{r}_2), \tag{4.81}$$

where E_t is the eigenvalue of the total energy and V is the interparticle potential. The c.m. ($\boldsymbol{R}$) and relative ($\boldsymbol{r}$) coordinates (4.25) and (4.26) are then introduced, and the two Laplacian operators are transformed in terms of $\boldsymbol{R} = (X, Y, Z)$ and $\boldsymbol{r} = (x, y, z)$. The basic operator transformation is (by Equations (4.25) and (4.26)

$$\frac{\partial}{\partial x_1} = \frac{m_1}{M}\frac{\partial}{\partial X} + \frac{\partial}{\partial x},$$
$$\frac{\partial}{\partial x_2} = \frac{m_2}{M}\frac{\partial}{\partial X} - \frac{\partial}{\partial x}, \tag{4.82}$$

with similar transformations in the y- and x-coordinates. We then obtain

$$\frac{1}{m_1}\nabla_1^2 + \frac{1}{m_2}\nabla_2^2 = \frac{1}{M}\nabla_R^2 + \frac{1}{\mu}\nabla_r^2, \tag{4.83}$$

where ∇_R^2 and ∇_r^2 are the Laplacians with respect to the $\boldsymbol{R}$- and $\boldsymbol{r}$-coordinates, respectively. If the total wave function (4.80) is written in the form

$$u_t(\boldsymbol{r}_1, \boldsymbol{r}_2) = U(\boldsymbol{R})u(\boldsymbol{r}), \tag{4.84}$$

since $V = V(\boldsymbol{r})$, the Schrödinger equation (4.81), after division by u_t, can be written

$$\frac{\hbar^2}{2M}\frac{1}{U}\nabla_R^2 U = -E_t + V(\boldsymbol{r}) - \frac{\hbar^2}{2\mu}\frac{1}{u}\nabla_r^2 u. \tag{4.85}$$

Here, since the left-hand side is a function of $\boldsymbol{R}$ and the right-hand side is a function of $\boldsymbol{r}$, each side must be equal to a "separation" constant, which we designate as $-E_M$. Defining

$$E_r = E_t - E_M, \tag{4.86}$$

we see that E_M is just the kinetic energy of the total mass M, and E_r is the energy of motion in the c.m. frame. Dropping the subscript r from E_r and ∇_r^2, we have the wave equation describing relative motion:

$$\left(-\frac{\hbar^2}{2\mu}\nabla^2 + V(\boldsymbol{r})\right)u(\boldsymbol{r}) = Eu(\boldsymbol{r}). \tag{4.87}$$

This is the same as the one-particle wave equation with mass m replaced by the reduced mass μ. That is, when the potential is a function of interparticle (or relative) position, if the single-particle problem is solved, the solution to the two-body problem is obtained. The total wave function, giving a complete description of motion, is then

$$u_t(\boldsymbol{r}_1, \boldsymbol{r}_2) = e^{i\boldsymbol{K}\cdot\boldsymbol{R}}u(\boldsymbol{r}), \tag{4.88}$$

where $\hbar\boldsymbol{K} = M\boldsymbol{V}$ is the momentum of the center of mass; that is, $U(\boldsymbol{R})$ is a plane-wave state and $u(\boldsymbol{r})$ is the solution of Equation (4.87).

The description in terms of $\boldsymbol{R}$ and $\boldsymbol{r}$ with M and μ as parameters is also convenient in time-dependent problems. The scattering problem is of this type, in which V is a perturbation and $u(1, 2)$ is a two-particle wave function describing plane-wave states. The total zeroth-order, time-dependent wave function is the of the form

$$\psi(\boldsymbol{r}_1, \boldsymbol{r}_2, t) = \exp(iE_M t/\hbar)U(\boldsymbol{R})\exp(-iEt/\hbar)u(\boldsymbol{r}), \tag{4.89}$$

with both U and u as spatial plane waves $\exp(i\boldsymbol{K}\cdot\boldsymbol{R})$ and $\exp(i\boldsymbol{k}\cdot\boldsymbol{r})$, respectively. More generally, the two-particle system—including its interparticle potential—can be subjected to an additional perturbation H', while described by a zeroth-order wave function (4.89) in which $u(\boldsymbol{r})$ is the solution of Equation (4.87). An example of such a problem would be the photoionization of positronium (bound e^+–e^- system) in which the perturbation H' results from an incident electromagnetic wave. The states u would then be (reduced-mass) hydrogenic functions for the bound state and the continuum.

4.2.4 Scattering of Identical Particles

When the particles involved in a binary collision are identical, it is necessary to bring in the effects of exchange in describing the process. The particle indistinguishability is fundamental, and the associated application of the superposition principle yields results that are quantum-mechanical in nature. In fact, for Coulomb scattering, the interference term involving the direct and exchange amplitudes has an explicit form whose effect is such as to vanish in the limit $\hbar \to 0$. This interference term has a dependence on the spin of the particles involved, even though spin interactions are negligible in the scattering. The reason for this is that the precise form of the interference term is dependent on the exchange symmetry required for the total particle wave function and amplitude for the process. The mathematical statement of the general superposition principle, which is at the basis of the analysis of the scattering problem (and many other processes), can be expressed as

$$W_{\text{process}} = \sum_s \Big|\sum_p A_p^{(s)}\Big|^2. \tag{4.90}$$

That is, we are interested in the probability (i.e., a cross section) for some process that can take place in various ways or "paths" (p). We do not try to determine the path that is taken but only specify the initial and final states (0 and f). The process $0 \to f$ could include various combinations of observed or specified initial and final substates that we want to include, and we sum[10] over these (s). For example, in the scattering we may not be interested in particle spins (although we may be capable of determining the various spins in 0 and f), so we sum over final spins and average over initial spins. The quantum-mechanical interference effects come into the (squared) sum over p, and the form of this sum depends on the individual combinations s—hence the notation $A_p^{(s)}$.

The simplest example of a scattering process with exchange effects is perhaps the problem of the collision of two α-particles or helium nuclei. The α-particle has spin-0, so it is a boson, and the wave function for two α-particles must be symmetric under exchange. The wave function has no spin part and only describes the spatial coordinates ($\boldsymbol{r}_1$ and $\boldsymbol{r}_2$) of the two particles. Considering the process in the (convenient) c.m. frame, we can obtain the cross section from the wave function describing relative motion.[11] Without correcting for exchange, the cross section would be given by $d\sigma/d\Omega = |f(\theta_{\text{sc}})|^2$, where $f(\theta_{\text{sc}})$ is the amplitude of the scattered part of the total wave function (4.69). For simplicity, although we are treating the problem in the c.m. frame, we do not add primes to θ_{sc} and other variables as was done in Section 1.3. The function $f(\theta_{\text{sc}})$ would be obtained from the expression for the corresponding one-particle problem (fixed heavy force center) with the mass replaced by the reduced mass ($m \to \mu = m/2$).

In considering the scattering in the c.m. frame (Figure 4.3), we must recognize that the process consists of two incident particles being scattered through an angle

---

[10] Note that we sum over these *after* squaring the sum over p.

[11] See Equations (4.71) and (4.72). Here the wave functions u_I and u_S represent the relative-motion function $u(\boldsymbol{r})$ [see Equation (4.88)], which can be considered, in the c.m. frame, as describing the motion of both particle 1 and particle 2. The flux of relative motion for the incident particles would then be $|u_I|^2(v_1 + v_2) = |u_I|^2 v$, where v is the magnitude of the relative velocity.

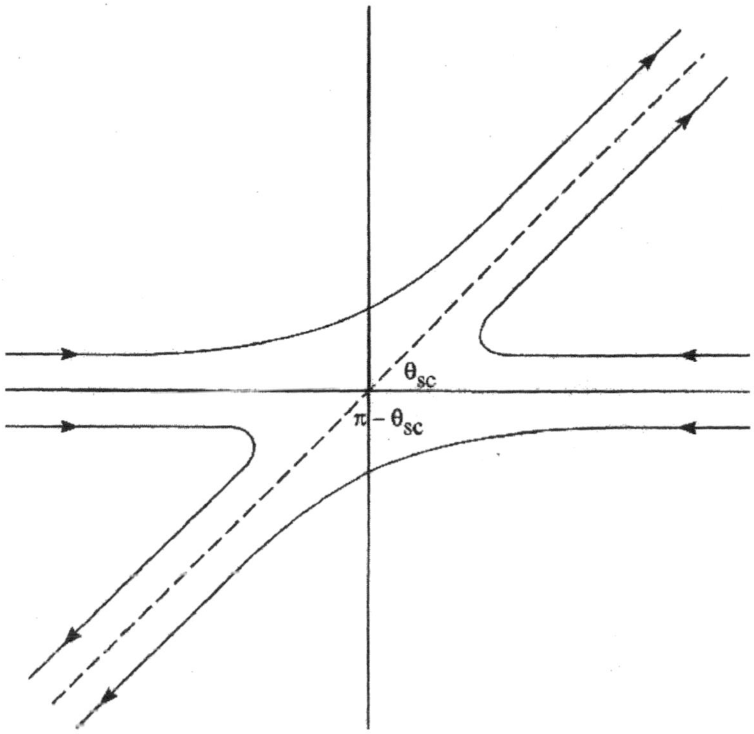

Figure 4.3 Two-body scattering with classical orbits corresponding to direct and exchange amplitudes.

θ_{sc}. We do not inquire about the motions in the region of the scattering, and the scattering through an angle $\pi - \theta_{sc}$, indicated by the other pair of classical orbits in Figure 4.3, describes the same process. Since this involves the "other" pair of particles emerging from the scattered center, the factor $f(\pi - \theta_{sc})$ represents an *exchange amplitude*. The $\alpha - \alpha$ total scattering cross section would then be given by

$$\begin{aligned}(d\sigma/d\Omega)_{\alpha-\alpha} &= |f(\theta_{sc}) + f(\pi - \theta_{sc})|^2 \\ &= |f(\theta_{sc})|^2 + |f(\pi - \theta_{sc})|^2 \\ &\quad + f(\theta_{sc})f^*(\pi - \theta_{sc}) + f^*(\theta_{sc})f(\pi - \theta_{sc}).\end{aligned} \tag{4.91}$$

In this expression, the last two terms involve the direct and exchange amplitudes, and can be written

$$f(\theta_{sc})f^*(\pi - \theta_{sc}) + f^*(\theta_{sc})f(\pi - \theta_{sc}) = 2\mathrm{Re}[f(\theta_{sc})f^*(\pi - \theta_{sc})]. \tag{4.92}$$

The interference term (4.92) would not appear in a classical formulation without the superposition principle; in a classical treatment, which assumes well-defined orbits and inherent distinguishability, only the squared amplitudes $|f(\theta_{sc})|^2$ and $|f(\pi - \theta_{sc})|^2$ would contribute to $(d\sigma/d\Omega)$. Further, it should be noted that in computing effects involving the scattering of identical particles, only the range $0 \leqslant \theta_{sc} \leqslant \frac{1}{2}\pi$ is to be considered. Because the contribution from the exchange amplitude has been included in this range, to involve the interval $\frac{1}{2}\pi$ to π would describe the same process and (erroneously) double the computed process. Sometimes, in fact, it is convenient to consider the scattering process in a "lab" frame where one of the particles is initially at rest. It is an elementary exercise [see also Equation (1.90)] to show that the scattering angle in this frame is related to the c.m. angle by

$$(\theta_{sc})_{lab} = \tfrac{1}{2}\theta_{sc}, \tag{4.93}$$

so that the lab-frame scattering angles would be restricted to the range 0 to $\frac{1}{4}\pi$.

In the Born limit, the scattering amplitude is very simple and can be obtained from Equation (4.54) with $z = Z$, $E_0 \to \frac{1}{2}\mu v^2 = \frac{1}{4}mv^2$, with v the relative velocity before (or after) the collision, and θ_{sc} now the c.m. scattering angle:

$$f_{Born}(\theta_{sc}) = (z^2e^2/2\mu v^2)\csc^2(\theta_{sc}/2). \tag{4.94}$$

The spin-0 Born cross section with exchange is then, by Equation (4.91),

$$\left(\frac{d\sigma}{d\Omega}\right)_{\text{spin-0}} \xrightarrow[\text{(Born)}]{} \left(\frac{z^2e^2}{2\mu v^2}\right)^2 \left(\csc^4\frac{\theta_{sc}}{2} + \sec^4\frac{\theta_{sc}}{2} + 2\csc^2\frac{\theta_{sc}}{2}\sec^2\frac{\theta_{sc}}{2}\right); \tag{4.95}$$

here the first term in parentheses is the square of the direct amplitude, the second is the square of the exchange amplitude, and the third is twice the product of the direct and exchange amplitudes.[12]

In the general case, including both the Born and classical limits, the cross section would again be given by Equation (4.91) with, now, $f(\theta_{sc})$ given by the exact formula (4.69); that is,

$$f(\theta_{sc}) = f_{Born}(\theta_{sc})\exp[-in\ln(1-\cos\theta_{sc}) + i\pi + 2i\eta_0]. \tag{4.96}$$

The result for the cross section is the same as the Born formula (4.95) except that the term $2\csc^2(\theta_{sc}/2)\sec^2(\theta_{sc}/2)$ is multiplied by an additional factor

$$\Phi = \cos[n\ln\tan^2(\theta_{sc}/2)], \tag{4.97}$$

where [see Equation (4.64)] $n = z^2e^2/\hbar v$. Thus, in the classical limit, $n \gg 1$ [see Equation (4.34)], and the function Φ is highly oscillatory, averaging to zero over even small ranges of θ_{sc}. In this limit, the cross section would, effectively, be given by the result (4.95) with only the terms $\csc^4(\theta_{sc}/2) + \sec^4(\theta_{sc}/2)$ in the parentheses. Although it is a classical formula, it does include the square of the exchange amplitude (but not the interference term) since, if the particles are the same, they cannot be distinguished even in a classical experiment (and detector). In other words, if we acknowledge this even in the classical limit and only consider $0 \leqslant \theta_{sc} \leqslant \frac{1}{2}\pi$, we must include the square of the exchange amplitude in the cross section. Alternatively—again, only in the classical limit—we could consider θ_{sc} also in the range $\frac{1}{2}\pi$ to π, but now only include the square of the direct amplitude in the cross section.

[12]Here, for α–α scattering, only the Coulomb force has been included, and the effects of the nuclear force have been neglected.

In the limit $n \ll 1$, the function Φ goes to unity and the general formula reduces to its form in the Born limit. The validity of the Born approximation will be discussed in Section 2.5, and it will be seen that small-n is indeed the domain of this approximation for Coulomb scattering.

The problem of the scattering of two identical spin-$\frac{1}{2}$ particles is only a little more complicated than that for spin 0, but it is interesting as a further application of the superposition principle (4.90). The two-particle wave function must now describe both the position and spin coordinates of each particle and, since the particles are fermions, it must be anti-symmetric under exchange of particle labels. When there is no coupling between the space and spin coordinates, the total wave function can be written as a product

$$u = u(\boldsymbol{r}_1, \boldsymbol{r}_2)\chi(s_1, s_2), \tag{4.98}$$

where χ is the wave function describing the two spin coordinates s_1 and s_2. If there are negligible spin-spin interactions, χ can further be written as the product $\chi_1(s_1)\chi_2(s_2)$. The spin states can be specified in terms of the value of the spin "component" or projection (s_z) along some (z) axis designating a direction. We have considered a calculation involving a spin-$\frac{1}{2}$ particle in Section 5.2 of Chapter 3 and introduced these two states, referred to there as the orthonormal functions χ_+ and χ_- in terms of a column representation of the spin wave function. A convenient notation for the spin-$\frac{1}{2}$ eigenfunctions is to write, say, $\chi_1(s_{1z} = +\frac{1}{2})\chi_2(s_{2z} = +\frac{1}{2}) = (+)(+)$; that is, the first parenthesis refers to the state of particle 1 and the second to particle 2. For two spin-$\frac{1}{2}$ particles, there are then four possible states: $(+)(+)$, $(-)(-)$, $(+)(-)$, and $(-)(+)$. Of these, the first two are symmetric under exchange, but the last two do not have a symmetry. The spin functions can be rearranged in groups by, instead, forming linear combinations of these two states. Then four two-particle spin functions can be formed:

$$\left.\begin{array}{l} (+)(+) \\ (-)(-) \\ 2^{-1/2}[(+)(-) + (-)(+)] \end{array}\right\} \quad \text{(symmetric)} \tag{4.99}$$
$$2^{-1/2}[(+)(-) - (-)(+)] \quad \text{(anti-symmetric).}$$

The grouping (4.99) is simply the "(S, S_z) representation" of the spin states for the two-particle system; that is, it is a designation in terms of a specification of the magnitude (S) of the total spin and its component (S_z) along some axis. This is often a convenient way of designating a spin function rather than in terms of a specification of the s_z amplitudes for each particle. The symmetric states (4.99) are, in fact, those for $S = 1$, and there are three $(2S + 1)$ substates corresponding to the triplet $S_z = \pm 1$ and 0; the single anti-symmetric state is that for $S = 0$ which is singlet (only $S_z = 0$).

We now consider the scattering of two spin-$\frac{1}{2}$ particles and assume that there is no polarization of the spins; then each of the spin states (4.99) are equally probable. This means that a fraction $w_1 = \frac{1}{4}$ of the time the particles will be coming together in the singlet state, and a fraction $w_3 = \frac{3}{4} = 1 - w_s$ of the time they will be in the triplet state. Since the total wave function (4.98) must be anti-symmetric, the singlet anti-symmetric spin state must go with a symmetric (s) spatial wave function and scattering amplitude (A_s); the triplet symmetric spin function must be associated with an anti-symmetric (a) scattering amplitude (A_a). That is, the total transition probability (cross section) must be computed from

$$W = w_1 |A_s|^2 + w_3 |A_a|^2 . \tag{4.100}$$

If $A = f(\theta_{sc})$ is the direct amplitude and $\overline{A} = f(\pi - \theta_{sc})$ is the exchange amplitude, $A_s = A + \overline{A}$ and $A_a = A - \overline{A}$, and

$$\begin{aligned} W &= w_1 \left|A + \overline{A}\right|^2 + w_3 \left|A - \overline{A}\right|^2 \\ &= |A|^2 + \left|\overline{A}\right|^2 + 2(w_1 - w_3)\mathrm{Re}\{A\overline{A}\}. \end{aligned} \tag{4.101}$$

In the Born approximation, we have [see Equation (4.95) for spin 0] a scattering cross section

$$\left(\frac{d\sigma}{d\Omega}\right)_{\text{spin-}\frac{1}{2}} \xrightarrow[\text{(Born)}]{} \left(\frac{z^2e^2}{2\mu v^2}\right)^2 \left(\csc^4 \frac{\theta_{sc}}{2} + \sec^4 \frac{\theta_{sc}}{2} - \csc^2 \frac{\theta_{sc}}{2} \sec^2 \frac{\theta_{sc}}{2}\right), \tag{4.102}$$

again in terms of c.m. variables (and v is the asymptotic relative velocity).

Actually, it is possible to write down a general expression for the scattering cross section, valid for the scattering of unpolarized particles of arbitrary spin s and for arbitrary energy (i.e., not just in the Born approximation). All that is necessary is a general expression for the fraction of spin substates that are associated with the symmetric and anti-symmetric scattering states. The result[13] is that the difference $w_1 - w_3$ is to be replaced by $(-1)^{2s}/(2s + 1)$. Also, instead of the Born amplitude, the more general Coulomb amplitude (4.96) is employed. We then obtain

$$\frac{d\sigma}{d\Omega} = \left(\frac{z^2e^2}{2\mu v^2}\right)^2 \left(\csc^4 \frac{\theta_{sc}}{2} + \sec^4 \frac{\theta_{sc}}{2} + \frac{2(-1)^{2s}}{2s+1} \Phi \csc^2 \frac{\theta_{sc}}{2} \sec^2 \frac{\theta_{sc}}{2}\right), \tag{4.103}$$

where Φ is the function (4.97). The result (4.103) holds for any kind of particle and over the whole classical-to-Born energy domain. When $s = 0$ and $s = \frac{1}{2}$, the general formula (4.103) is seen to reduce to the expressions already derived for these special cases.

4.2.5 Validity of the Born Approximation

Basically, the Born approximation implies that the scattering amplitude is small, so that a small fraction of the incident probability current is converted into an outgoing spherical wave associated with the scattered particle. A weak scattering potential V guarantees this, and the rough criterion $V \ll E$, where E is the particle energy,

[13]Cf. L. D. Landau and E. M. Lifshitz, *Quantum Mechanics, Non-Relativistic Theory*, 2nd ed., Reading, MA: Addison-Wesley, 1965; L. I. Schiff, *Quantum Mechanics*, 3rd ed., New York: McGraw-Hill, 1968. Essentially, what is needed is the fraction (c) of two-particle spin states having the required spin-exchange symmetry to go with an even spatial coordinate total amplitude (squared). For bosons, $c = (s + 1)/(2s + 1)$; for fermions, $c = s/(2s + 1)$.

may be considered. If we set V equal to $zZe^2/\lambda\!\!\!^{-}$, with $\lambda\!\!\!^{-}$ the particle de Broglie wavelength, the criterion yields

$$v/c \gg zZ\alpha, \tag{4.104}$$

which is the same as the criterion for the classical limit except that the inequality goes the other way. That the limit (4.104) corresponds to the Born approximation has already been seen in the results for the general treatment of Coulomb scattering. The limit corresponds to the case $n \ll 1$ for which the factor Φ defined in Equation (4.97) approaches unity.

A more general and precise formulation can be given for the validity of the Born approximation. Again, when we consider the scattering potential as a perturbation affecting the incident particle and its associated wave, the validity criterion can be written

$$\Delta\phi = (1/\hbar)\int \Delta p\, ds \ll 1. \tag{4.105}$$

Here $\Delta\phi$ is the change in the phase of the incident wave as a result of propagation through the region of the potential, ds is the path element, and

$$\Delta p = (2m)^{1/2}[(E - V)^{1/2} - E^{1/2}] \tag{4.106}$$

is the difference in the classical momentum with and without the scattering potential. The criterion (4.105) is really very general for Born-type approximate treatments of wave propagation; it is employed, for example, in optics in wave propagation through regions of variable index of refraction. If the potential has a characteristic strength Δ and range a, for $\overline{V} \ll E$, the term in brakets in Equation (4.106) can be expanded, and

$$\Delta p \approx (m/2E)^{1/2}\overline{V}. \tag{4.107}$$

In terms of the particle incident velocity, the criterion (4.105) yields, on integration over a path length of order a,

$$\overline{V}a/\hbar v \ll 1. \tag{4.108}$$

For the Coulomb case $\overline{V}a \sim zZe^2$, and the inequality (4.108) is the same as that in Equation (4.104).

For Coulomb scattering, although the classical and Born domains are complimentary in the sense that one limit corresponds to low energy and the other to high energy, the domains do not overlap. That is, there is an intermediate energy domain where neither approximation holds, and an exact quantum-mechanical treatment must be given. We have seen, however, that except in the case of scattering of identical particles where exchange effects come in, the classical and Born formulas are exact and happen to hold outside their domains of applicability. This is so only for the pure radiationless-scattering problem, however. In Coulomb bremsstrahlung, which is a combined scattering- and photon-emission problem, the classical and Born formulas for the cross section are *not* identical. In fact, in some problems involving (radiationless) scattering, even though the cross section is the same in the classical and Born domains, it is necessary to recognize the inherent limit of applicability. An example of this type would be the phenomenon of the energy loss through Coulomb interactions of, say, a proton traversing a plasma. The loss rate is determined by the product of the individual energy loss (ΔE) in an e-p collision and the scattering cross section $d\sigma$, then integrated over momentum transfers. The formulas for the loss rate are different in the classical and Born limits even though the cross-section formulas are the same. The problem requires a different treatment in the two limits for the case of small momentum transfers where it is necessary to consider the excitation of collective modes (plasma waves).

4.3 SCATTERING OF RELATIVISTIC PARTICLES OF ZERO SPIN

For the scattering of relativistic charged particles, it is necessary to include the effects associated with the interaction of the particles' magnetic moment with the scattering field. In the non-relativistic case treated in the last section, particle spin comes in only in the scattering of identical particles in fixing the exchange symmetry of the spatial-coordinate scattering amplitude. However, even in the scattering by a fixed force center (external potential), spin- or magnetic-moment effects come in at relativistic energies. As a result, the general cross section for the scattering of a spin-$\frac{1}{2}$ particle (for example, an electron) by a central Coulomb potential differs from that for a spin-0 particle (say, a pion[14]). There is, on the other hand, no difference in the limit of small scattering angle, and we discuss later why this is so. Although the spin-$\frac{1}{2}$ problem is far more important than the spin-0 case, the latter is treated first in some detail because it is much simpler and provides a good illustration of the general techniques needed in computing processes involving relativistic particles.

4.3.1 Coulomb Scattering

The scattering of a charge ze by a fixed central charge Ze is a simple problem, even in the relativistic case, when the charge is spinless with no intrinsic magnetic moment. To calculate the cross section, the same non-covariant formulation can be employed that was developed for the non-relativistic Born-approximation case in Section 2.1. For, as has been emphasized earlier, the Fermi Golden Rule formula (4.41) is very general in that it is based only (physically) on the validity of the time-dependent Schrödinger equation $H\psi = i\hbar\,\partial\psi/\partial t$ and on the associated perturbation theory. The relationship (4.40) to the cross section is also completely general, and the only difference from the non-relativistic problem involves the proper employment of kinematic relations involving relativistic energy and momentum. That is, the one-particle wave functions (4.39) are still applicable, and the cross section is determined by the Fourier transform of the scattering potential $V = zZe^2/r$. Now, however, the momentum-transfer argument is given by the relativistic expression

$$q = (2\gamma mv/h)\sin(\theta_{\rm sc}/2), \tag{4.109}$$

which is simply γ times the non-relativistic result (4.52). The phase space factor in the relativistic case is

$$p^2 dp\, d\Omega = (1/c^2) pE\, dE\, d\Omega, \tag{4.110}$$

[14]The pion might also have non-electromagnetic interactions with the Coulomb force center—for example, by the strong interaction.

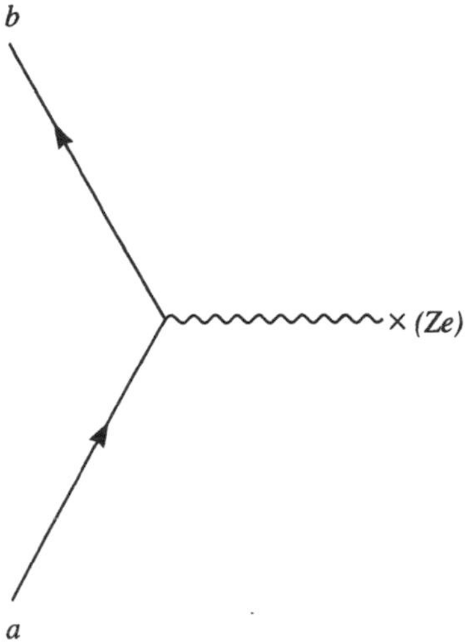

Figure 4.4 Feynman diagram for Coulomb scattering by a fixed central charge.

and the energy-conservation δ-function, on integration over dE ($= dE_f$) simply equates the initial and final energies (and magnitude of momenta and velocity). The resulting cross section is given by, since $p = \gamma m v = Ev/c^2$,

$$d\sigma/d\Omega = (zZe^2/2\gamma m v^2)^2 \csc^4(\theta_{sc}/2), \tag{4.111}$$

which is just γ^{-2} times the non-relativistic formula. The general expression (4.111) is valid over the entire energy domain $E \gg (zZ)^2(m/m_e)\text{Ry}$ for which the Born approximation applies [see Equation (4.104)]. In fact, because of the (accidental) validity of the non-relativistic Born result at lower energies into the intermediate and low-energy classical domain, the formula (4.111) can be considered to apply at essentially *any* energy if $zZ\alpha = zZ/137 \ll 1$, where the restriction on $zZ\alpha$ is required only at relativistic energies.

For purposes of illustration, let us now outline the derivation of the relativistic formula (4.111) by the methods of covariant QED discussed in the previous chapter. In a QED formulation, the scattering is considered to result from the exchange of a "virtual" photon between the central charge Ze and the incident charge ze as is indicated in Figure 4.4. The form of the interaction for a scalar spin-0 particle is such that the vertex corresponds to an associated invariant matrix element of the form [see Equations (3.73), (3.76), and (3.80)]

$$M_{ba} = a_s z e A_\mu (p_a + p_b)_\mu, \tag{4.112}$$

where a_s is a constant, A_μ is the Fourier component of the four-vector potential describing the field of the static central charge, and $p_{a\mu}$ and $p_{b\mu}$ are the four-momenta of the incident and scattered charges, respectively. In terms of the spatial amplitude of the field,

$$A_\mu = A_\mu(k_\nu) = (2\pi)^{-4} \int d^4x \, e^{ik_\nu x_\nu} A_\mu(x_\nu), \tag{4.113}$$

and, in the frame of the charge center, $A_\mu(x_\nu)$ would have only the zero component $A_0 = i\Phi$ with ($\varkappa \to 0$ eventually)

$$\Phi = (Ze/r)e^{-\varkappa r}. \tag{4.114}$$

The Fourier amplitude of the virtual photon field is then

$$A_\mu(k_\nu) = A_\mu(q_\nu) \propto (Z/\boldsymbol{q}^2)\delta(q_0)\delta_{\mu 0}, \tag{4.115}$$

where the notation q_ν (instead of k_ν) is employed for comparison with results already derived. That is, $\hbar\boldsymbol{q}$ is the momentum transferred, via the virtual photon, to the fixed scattering center. The Dirac δ-function $\delta(q_0)$ has the physical meaning that there is no energy exchanged; although this factor can be handled mathematically in the computation, we simply ignore it. The Kronecker δ-function $\delta_{\mu 0}$ in A_μ has the effect, on substituting in the expression (4.112) for the matrix element, of yielding a factor $E_0 + E_f$ in terms of the initial and final energies of the scattered particle. Thus, the total invariant matrix element for the process (see Section 3.4 of Chapter 3) is of the form

$$\mathcal{M}_{f0} \propto (E_0 + E_f)/\boldsymbol{q}^2. \tag{4.116}$$

In the covariant formulation of perturbation theory [see Equation (3.109)], the squared invariant matrix element is divided by factors of E_0 and E_f for the incoming and outgoing particles. The scattering cross section is then computed from

$$v_0\, d\sigma \propto \int (|\mathcal{M}_{f0}|^2/E_0 E_f)\, p_f^2\, dp_f\, d\Omega_f\, \delta(E_0 - E_f). \tag{4.117}$$

Since $p_f^2\, dp_f = p_f E_f\, dE_f = m^2\gamma_f^2 v_f\, dE_f$, we obtain the basic result

$$d\sigma \propto (\gamma^2/\boldsymbol{q}^4)\, d\Omega. \tag{4.118}$$

We see that, with $|\boldsymbol{q}|$ again given by Equation (4.109), the expression (4.111) is obtained. Although we have simplified the derivation by ignoring numerical factors, these can be recovered for the final formula for the cross section by comparing the result for $\gamma \to 1$ with the expession already derived in the non-relativistic limit. In fact, the multiplying constant could be established even by comparison with the limit of small θ_{sc}—a special case of the non-relativistic formula.

4.3.2 Scattering of Two Distinguishable Charges

In gradually increasing the complexity of the general charged-particle scattering problem, we now consider the case where both the masses are finite, so that there is both momentum and energy exchanged. The particles have masses m and M and charges ze and Ze, respectively, and could be, say, a pion and a kaon. Although these particles can interact by other means (through some combinations of strong couplings, for example), we only compute the effects of the electromagnetic coupling. That is, the treatment is meant to illustrate certain aspects and methods of QED. The more important class of problems involves spin-$\frac{1}{2}$ charges, but the spin-0 case is simpler and can help us understand the case of spin-$\frac{1}{2}$, which we come to shortly.

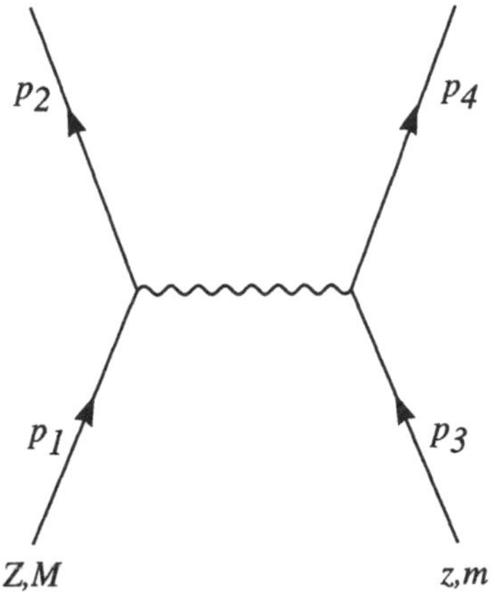

Figure 4.5 Diagram for two-body Coulomb scattering of distinguishable charges.

The most important feature of this problem that was not present in previous processes that we have considered is the necessity of including the "photon propagator" in the total amplitude for the process. Up to now we have dealt only with real or radiation-field photons and have not had to handle problems involving the exchange of a virtual photon. Thus, treatment of this problem represents a significant step forward in a comprehension of general QED. The case of scattering of two distinguishable relativistic spin-0 charges is probably the simplest example involving (virtual) photon exchange or transfer, since there is only one Feynman diagram (Figure 4.5). That is, there is, in covariant QED, only one amplitude or matrix element for the process. The amplitude is an invariant and involves three factors associated with (i) the vertex involving the photon interaction with the charge Z, (ii) the photon "propagation" between Z and z, and (iii) the vertex involving the photon interaction with z. In covariant QED, a single diagram and amplitude accounts for[15] the cases where the photon is exchanged "in either direction" (from Z to z or from z to Z). However, in writing down the amplitude by the Feynman rules, we have to make a choice between one or the other, simply to make the expression for the total amplitude self-consistent. For definiteness then, we choose the Z-to-z case; if k (or k_μ) designates the virtual photon four-momentum, energy and momentum conservation at the vertices require that

$$\begin{aligned} p_1 &= p_2 + k, \\ p_4 &= p_3 + k, \end{aligned} \tag{4.119}$$

for each component of four-momentum. The set of relations (4.119) also imply the energy-momentum conservation relation

$$p_1 + p_3 = p_2 + p_4 \tag{4.120}$$

for the whole process.

[15] See, especially, Section 4.1 of Chapter 3.

With the Z-to-z convention chosen for the photon propagation, the amplitude for the process can be written

$$\mathcal{M} = \sum_I M_z P M_Z, \tag{4.121}$$

where the sum is over the intermediate states associated with the photon propagating between Z and z. The amplitudes M_z and M_Z are of the form [see Equation (3.152)]

$$\begin{aligned} M_z &= a_s z e (p_3 + p_4)_\mu \varepsilon_\mu, \\ M_Z &= a_s Z e (p_1 + p_2)_\nu \varepsilon_\nu, \end{aligned} \tag{4.122}$$

where ε_μ and ε_ν refer to the photon polarization. We have not yet given an expression for the photon propagator, although we have obtained the result for scalar particles with mass [see Equations (3.86)]. The expression for a photon, which is also a boson but one with spin-1 and zero mass, is very similar. The photon is a vector particle, being described in terms of its wave vector $\boldsymbol{k}$ and polarization ε_μ, and the propagator must involve the polarization as well as its kinematics. Modifying the corresponding expression $P_F(\text{scalar}) = (p^2 + m^2c^2)^{-1}$ for scalar particles, we take

$$P(\text{photon}) = \delta_{\mu\nu}/k^2. \tag{4.123}$$

Here the Kronecker $\delta_{\mu\nu}$ simply notes that the polarization state of the photon is the same at the "endpoints of its propagation." The invariant amplitude associated with the diagram in Figure 4.5 is then

$$\mathcal{M} = \sum_{\text{pol}} a_s z e (p_3 + p_4)_\mu \varepsilon_\mu (\delta_{\mu\nu}/k^2) a_s Z e (p_1 + p_2)_\nu \varepsilon_\nu. \tag{4.124}$$

Summing over the four[16] components of the unit polarization components, the amplitude becomes

$$\mathcal{M} = a_s^2 z Z e^2 \frac{(p_3 + p_4) \cdot (p_1 + p_2)}{(p_1 - p_2)^2}, \tag{4.125}$$

in terms of Lorentz-invariant scalars involving only the kinematic energy-momentum four-vectors of the incoming and outgoing particles.

The invariant amplitude (4.125) is then employed to compute the cross section from the invariant transition rate (see Section 4.4 of Chapter 3)

$$\frac{\Delta W_{f0}}{\Delta t} = \int \frac{|\mathcal{M}|^2}{\prod_{(0)}(2E)} \prod_{(f)} \frac{d^3 p}{(2E)(2\pi)^3} \delta^{(4)}\Big(\sum_{(0)} p_\mu - \sum_{(f)} p_{mu}\Big). \tag{4.126}$$

This is a standard general formula in covariant perturbation theory.[17] It is convenient to evaluate the cross section in the c.m. frame, for which

$$\Delta W_{f0}/\Delta t = (v_1 + v_3)\, d\sigma. \tag{4.127}$$

[16]A photon in a (virtual) intermediate state has four polarization components corresponding to the unit vector $\varepsilon_\mu = (1, 1, 1, 1)$. That is, it is, in general, not transverse with the possibility of eliminating an additional component by a gauge transformation as in the case of real or radiation-field photons.

[17]For covenience here, we set $c = \hbar = 1$ and take a unit normalization volume ($L^3 = 1$). In the end of the calculation, it will be clear where c and $\hbar$ should be inserted to express the result in terms of more physical (c.g.s.) units.

In the c.m. frame, the particles are incident and outgoing with the same magnitude of momentum:

$$\boldsymbol{p}_1 = -\boldsymbol{p}_3, \qquad \boldsymbol{p}_2 = -\boldsymbol{p}_4. \tag{4.128}$$

Moreover, because of energy conservation,[18] the initial and final momenta are of equal magnitude; that is,

$$|\boldsymbol{p}_1| = |\boldsymbol{p}_3| = |\boldsymbol{p}_2| = |\boldsymbol{p}_4| \equiv p. \tag{4.129}$$

The velocity sum in Equation (4.127) is then given by

$$v_1 + v_3 = p(1/E_1 + 1/E_3) = p(E_m + E_M)/E_m E_M, \tag{4.130}$$

where we have written $E_1 = E_M$; $E_3 = E_m$. The invariant two-body phase space factor[19] in Equation (4.126) is, in terms of c.m. quantities,

$$\Phi_{(2)} = \iint (d^3\boldsymbol{p}_2/E_2)(d^3\boldsymbol{p}_4/E_4)\delta^{(3)}(\boldsymbol{p}_2 + \boldsymbol{p}_4)\delta(E_2 + E_4 - E_T), \tag{4.131}$$

with $E_T = E_m + E_M$ equal to the total energy. Integrating over, say, $d^3\boldsymbol{p}_4$ using the momentum-conservation $\delta^{(3)}$, and setting $d^3\boldsymbol{p}_2 = p^2 dp\, d\Omega$, we have

$$\Phi_{(2)} = d\Omega \int p^2 dp \frac{\delta[(p^2 + M^2)^{1/2} + (p^2 + m^2)^{1/2} - E_T]}{(p^2 + M^2)^{1/2}(p^2 + m^2)^{1/2}}. \tag{4.132}$$

The integral is easily evaluated by making the change of variable from p to $f(p) = (p^2 + M^2)^{1/2} + (p^2 + m^2)^{1/2}$. We then have

$$\Phi_{(2)} = d\Omega \int p(f) f^{-1} \delta(f - E_T) df = p(E_m + E_M)^{-1} d\Omega \tag{4.133}$$

where p is the value (4.129) and E_m and E_M are the energies for this p.

In the c.m. frame, the invariant amplitude (4.125) is given by

$$\mathcal{M} = -a_s^2 z Z e^2 \frac{E_m E_M + p^2 \cos^2(\theta/2)}{p^2 \sin^2(\theta/2)}, \tag{4.134}$$

and by Equations (4.126), (4.127), (4.130), (4.133), and (4.134),

$$\frac{d\sigma}{d\Omega} \propto \left| \frac{\mathcal{M}}{E_m + E_M} \right|^2 = a_s^4 (zZe^2)^2 \left[\frac{E_m E_M + p^2 \cos^2(\theta/2)}{(E_m + E_M) p^2 \sin^2(\theta/2)} \right]^2. \tag{4.135}$$

The multiplying constant can then be determined by taking the limit $M \to \infty$ and comparing with the result (4.111). At the same time, we can express the cross section in c.g.s. units by inserting a factor $1/c^2$ to multiply $E_m E_M$ in the denominator of the parentheses on the right of Equation (4.135). Since, as $M \to \infty$, $p/E_m \to v/c^2$, and $p \to \gamma m v$, the result (4.111) is recovered if we ignore a_s^4 and multiply by $\frac{1}{4}$. That is, we obtain

$$\frac{d\sigma}{d\Omega} = \left[\frac{zZe^2}{2} \frac{E_m E_M/c^2 + p^2 \cos^2(\theta/2)}{(E_m + E_M) p^2 \sin^2(\theta/2)} \right]^2. \tag{4.136}$$

---

[18]Of course, momentum and energy conservation are imposed in the transition-rate formula (4.126) in the $\delta^{(4)}$ factor.

[19]This expression, with momentum-space elements $d^3\boldsymbol{p}$ divided by E to yield invariant factors, is not to be confused with the conventional phase-space functions introduced in Chapter 1.

This general expression for distinguishable spin-0 particles reduces to the result (4.111) (for $M \to \infty$) and also to certain non-relativistic formulas derived earlier. In fact, the multiplying constant could have been established through comparison with the non-relativistic expressions.

4.3.3 Two Identical Charges

When the charges are identical, it is necessary to include exchange effects. That is, there is a direct and an exchange amplitude, and these are represented by the two Feynman diagrams (A and A_e) exhibited in Figure 4.6. Since the particles are bosons with zero spin, the two amplitudes add with the same sign. In terms of a photon "propagation" from left to right in diagram A (inherently, the opposite direction is also included in covariant perturbation theory), the four-momenta are related by

$$p_1 = p_2 + k, \qquad p_4 = p_3 + k, \tag{4.137}$$

as a result of conservation at the vertices. These relations hold for diagram A and, in the exchange diagram A_e a similar pair is obtained by the interchange $2 \leftrightarrow 4$. For particles of charge e, the total amplitude for the process would be[20]

$$\begin{aligned} \mathcal{M}/a_s^2 e^2 &= \sum_{\text{pol}} (p_3 + p_4)_\nu \varepsilon_\nu (\delta_{\mu\nu}/k^2)(p_1 + p_2)_\mu \varepsilon_\mu \\ &\quad + \sum_{\text{pol}} (p_3 + p_4)_\nu \varepsilon_\nu (\delta_{\mu\nu}/k^2)(p_1 + p_2)_\mu \varepsilon_\mu \\ &= \frac{(p_3 + p_4)\cdot(p_1 + p_2)}{(p_1 - p_2)^2} + \frac{(p_3 + p_2)\cdot(p_1 + p_4)}{(p_1 - p_4)^2}. \end{aligned} \tag{4.138}$$

This invariant amplitude is most conveniently expressed in terms of the c.m. magnitude of the momenta (p) and energy $E = (p^2 + m^2)^{1/2}$ of (each of) the particles. We then obtain

$$\frac{\mathcal{M}}{a_s^2 e^2} = -\frac{E^2 + p^2 \cos^2(\theta/2)}{p^2 \sin^2(\theta/2)} - \frac{E^2 + p^2 \sin^2(\theta/2)}{p^2 \cos^2(\theta/2)}, \tag{4.139}$$

where θ is the c.m. scattering angle.

Note that, as in the non-relativistic case, the exchange amplitude for the relativistic spin-0 problem is related to the direct amplitude $A(\theta)$ by $A_e(\theta) = A(\pi - \theta)$. Further, we see that, in terms of the invariants (3.217) introduced at the end of Chapter 3, the amplitude (4.139) can be written

$$\mathcal{M}/a_s^2 e^2 = (s - u)/t + (s - t)/u. \tag{4.140}$$

The scattering cross section is obtained from the relations (4.126), (4.127), (4.130), and (4.133), yielding

$$2v\, d\sigma \propto \frac{|\mathcal{M}|^2}{E^2} \frac{p\, d\Omega}{E}, \tag{4.141}$$

[20]Again, as in formula (4.124), the left-to-right product of factors in the amplitude corresponds to a top-to-bottom reading of the Feynman diagrams. Because of the convention (4.137) adopted for the photon propagation, the vertices on the right in the diagrams are taken to be "above" those on the left.

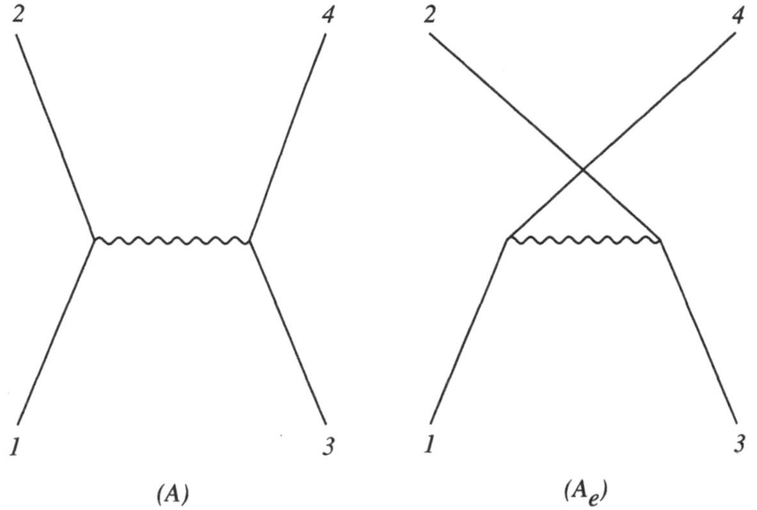

Figure 4.6 Diagrams for two-body scattering of identical charges. There is a direct and an exchange amplitude.

or, since $p = vE$,

$$d\sigma/d\Omega \propto |\mathcal{M}|^2/E^2. \tag{4.142}$$

Employing the result (4.139) for m and establishing the multiplying factor by comparison with the non-relativistic limit (4.95), the cross section can be expressed in terms of the dimensionless $\beta = v/c$ and $\gamma^{-2} = 1 - \beta^2$ for the particles in the c.m. frame[21]:

$$\frac{d\sigma}{d\Omega} = \frac{e^4}{16\gamma^2 m^2 c^4}\left[\frac{1+\beta^2\cos^2(\theta/2)}{\beta^2\sin^2(\theta/2)} + \frac{1+\beta^2\sin^2(\theta/2)}{\beta^2\cos^2(\theta/2)}\right]^2. \tag{4.143}$$

If there were only the electromagentic coupling, this formula would apply, for example, in π^+–π^+ and π^-–π^- scattering.

4.3.4 Scattering of Charged Antiparticles

In scattering of two charged spin-0 bosons of equal but opposite charge, there are no exchange effects, but there is an additional mechanism or combination of interactions that allow the process to procede. The scattering can take place by means of a "temporary annihilation" into a single photon in an intermediate state followed by a "pair production" into a final state with the same kind of particles. This is represented by the Feynman diagram on the right in Figure 4.7. In both diagrams in this figure, the type of vertex is basically the same, having an interaction of the form

$$M \propto [(qp_\mu)_a + (qp_\mu)_b]\varepsilon_\mu, \tag{4.144}$$

[21]This v ($= \beta c$) is not to be confused with the relative velocity ($2v$ when $m_1 = m_2$) in Equation (4.95).

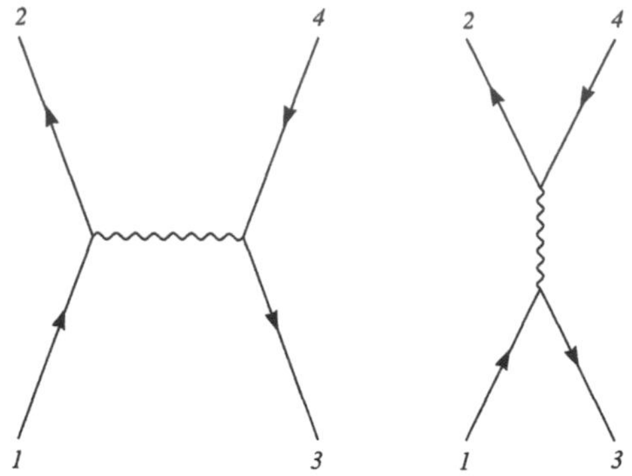

Figure 4.7 Diagrams associated with the scattering of particles and antiparticles (e^+-e^- scattering).

where a and b refer to the two particle paths involved in the vertex (with charges q_a and q_b).

There are two ways of evaluating the amplitude corresponding to the diagrams in Figure 4.7, and both are equivalent. One method—the more conventional procedure in covariant perturbation theory using Feynman diagrams—treats the anti-particle as a particle moving backward in time, as is indicated by the arrows associated with particles 3 and 4 in the diagrams. The other approach gives up the backward-in-time idea but notes that the antiparticle has a charge opposite in sign to that of the other particle involved in the scattering. In the latter method, the four-momentum representing the anti-particle in the formulation is the same as its actual value (in the former method, it is minus the value), and we lay out this formulation first.

Let us take particle 1,2 to be positively charged and particle 3,4 as negative, and consider the negative particle as the anti-particle.[22] The four-momentum of particle a, say, is denoted by $P_{\mu a}$ (we reserve the lowercase $p_{\mu a}$ for the more unified backward-in-time formulation); this will be its actual kinematic value. Summing over the polarization states of the virtual photon exchanged, the total amplitude from the two diagrams in Figure 4.7 will be

$$\frac{\mathcal{M}}{a_s^2 e^2} = -\frac{(P_3 + P_4) \cdot (P_1 + P_2)}{k^2} + \frac{(P_2 - P_4) \cdot (P_1 - P_3)}{k^2}. \tag{4.145}$$

This formula results from an application of the basic expression (4.144) for the amplitude associated with an interaction vertex. For the diagram on the left in Figure 4.7, we take left-to-right as the reference direction for the virtual photon propagation. Then in the first term in the amplitude (4.145) the photon four-momentum will be given by $k = P_1 - P_2$ (or $P_4 - P_3$), while in the second term we would substitute $k = P_1 + P_3$ (or $P_2 + P_4$). The minus signs in the terms in the amplitude (4.145) should be particularly noted; they arise because the factors P_3 and P_4 in the interaction (4.141) are multiplied by the charge $q = -e$. In the c.m. frame $\boldsymbol{P}_1 = -\boldsymbol{P}_3$ and

[22]Actually, it would be more conventional to take the positively charged particle as the antiparticle. For this problem, the designation or choice of one or the other is arbitrary.

$\boldsymbol{P}_2 = -\boldsymbol{P}_4$; all four particles have energy E and magnitude of (three-) momentum p. If θ is the c.m. scattering angle, we obtain[23]

$$\frac{\mathcal{M}}{a_s^2 e^2} = \frac{E^2 + p^2\cos^2(\theta/2)}{p^2\sin^2(\theta/2)} - \frac{p^2\cos\theta}{E^2}, \tag{4.146}$$

or, in terms of the dimensionless velocity $\beta = p/E$,

$$\frac{\mathcal{M}}{a_s^2 e^2} = \frac{1 + \beta^2\cos^2(\theta/2)}{\beta^2\sin^2(\theta/2)} - \beta^2\cos\theta. \tag{4.147}$$

Now let us derive the result (4.147) using the backward-in-time idea for the antiparticle in the diagrams in Figure 4.7. In this method, we do not assign a negative value to the charge of the antiparticle (since it is regarded as the same as the particle), but in writing down the expression for the amplitude for the process, its four-momentum is negative. We now take the lowercase $p_{\mu a}$ to designate four-momenta, and the amplitude for the process is again given by an application of the coupling term (4.144) for each vertex and a Feynman propagator $\delta_{\mu\nu}/k^2$ for the exchanged photon. In the diagram on the left in Figure 4.7, $k = p_1 - p_2$ (or, now, $p_3 - p_4$), while, for the one on the right, we now have $k = p_1 - p_3$ (or $p_2 - p_4$). Summing over polarizations for the virtual photon, we obtain an amplitude

$$\frac{\mathcal{M}}{a_s^2 e^2} = \frac{(p_3 + p_4)\cdot(p_1 + p_2)}{(p_1 - p_2)^2} + \frac{(p_2 + p_4)\cdot(p_1 + p_3)}{(p_1 - p_3)^2}. \tag{4.148}$$

Since, for p_3 and p_4 the *actual* four-momenta will be minus their values, in the c.m. frame,

$$\begin{aligned} \boldsymbol{p}_1 &= \boldsymbol{p}_3, \qquad E_1 = -E_3 = E, \\ \boldsymbol{p}_2 &= \boldsymbol{p}_4, \qquad E_2 = -E_4 = E, \end{aligned} \tag{4.149}$$

with all four particles having the same magnitude $p = |\boldsymbol{p}|$. Substituting these values into the amplitude (4.148), we again obtain the results (4.146) and (4.147). Perhaps seeing the two formulations (which are really equivalent) employed to derive the amplitude, the reader can appreciate the convenience of the backward-in-time idea for antiparticles and have confidence in its employment.

As in the problem in the last subsection, the invariant amplitude can be expressed in terms of the invariants s, t, and u [see Equation (4.140)]; we now have

$$\mathcal{M}/a_s^2 e^2 = (u - s)/t + (u - t)/s. \tag{4.150}$$

The cross section is related to the amplitude by the same type of expression as Equation (4.141), and we can again fix the multiplying constant by referring to the non-relativistic limit (4.95). The result is then, with $\gamma^{-2} = 1 - \beta^2$,

$$\frac{d\sigma}{d\Omega} = \frac{e^4}{16\gamma^2 m^2 c^4}\left(\frac{1 + \beta^2\cos^2(\theta/2)}{\beta^2\sin^2(\theta/2)} - \beta^2\cos\theta\right)^2. \tag{4.151}$$

[23] In the expression given in R. P. Feynman, *Theory of Fundamental Processes*, New York: W. A. Benjamin, Inc., 1962, there appears to be a misprint in that the second term has a $\cos^2\theta$ instead of $\cos\theta$. The treatment of relativistic scattering in this section has followed that by Feynman.

As we see, in the non-relativistic limit the term $\beta^2 \cos\theta$ from the diagram on the right in Figure 4.7 is negligible. If there were only the Coulomb interaction, formula (4.151) would apply in π^+–π^- scattering.

The comparison of the π^+–π^- cross section with that for $\pi^\pm$–$\pi^\pm$ merits further comment. The amplitude (4.148) in the backward-in-time formulation for the antiparticle is identical to the expression (4.138) except that there is a label exchange $3 \leftrightarrow 4$. Also, the form (4.150) is the same as expression (4.140) except that the invariants s, t, u are replaced by u, t, s. This is an example of *crossing symmetry*, discussed in Section 6.4 of Chapter 3. That is, the two processes are essentially the same, being just different channels of a single basic process. For processes related in this manner, it is really necessary to calculate only a single fundamental amplitude, or one for a particular channel. The amplitudes for the other channels are related by crossing-symmetry tranformations.

4.4 SCATTERING OF RELATIVISTIC SPIN-$\frac{1}{2}$ PARTICLES

For charged particles with spin, in particular spin-$\frac{1}{2}$, it is necessary to include the effects of interactions involving the particles' magnetic moments. The theory required for the calculation of processes involving electromagnetic interactions of spin-$\frac{1}{2}$ charges is what is generally known as modern QED, and we have given a sketch of its foundations in the previous chapter. Unfortunately, spin-$\frac{1}{2}$ QED is more complicated than the spin-0 case, and to consider scattering problems analogous to those treated in the last section it is necessary to develop the theory more fully. These developments will also be useful in calculations of other processes described in later chapters.

4.4.1 Spin Sums, Projection Operators, and Trace Theorems

Basically, the need for treating spin sums can be seen if we have to calculate the squared amplitude $|\mathcal{M}|^2$ for some process that takes place by means of an interaction total amplitude M, which is a combination of operators involving Dirac matrices. That is, if, for example, u_1 and u_2 are Dirac spinors associated with the initial and final states of a spin-$\frac{1}{2}$ charge, the invariant amplitude for the process will be given by

$$\mathcal{M} = \overline{u}_2 M u_1, \tag{4.152}$$

where, in terms of the Hermitian conjugate[24] $u^\dagger$, $\overline{u} = u^\dagger \gamma_0$. Then

$$\mathcal{M}^\dagger = u_1^\dagger M^\dagger \overline{u}_2^\dagger, \tag{4.153}$$

and, since $\gamma_0^\dagger = \gamma_0$ and $\gamma_0^2 = I$ (identity matrix), $u^\dagger = \overline{u}\gamma_0$ and $\overline{u}^\dagger = \gamma_0 u$. Thus, the squared amplitude can be written

$$|\mathcal{M}|^2 = \mathcal{M}^\dagger \mathcal{M} - \overline{u}_1 M' u_2 \overline{u}_2 M' u_1, \tag{4.154}$$

[24] See Chapter 3, especially Section 5.4.

where

$$M' = \gamma_0 M^\dagger \gamma_0. \tag{4.155}$$

Usually we are not interested in the spin states of the outgoing particle (2), and the incoming particle (1) is unpolarized, so we would average over its spin states. If we designate the particle states by the Dirac spinor $u^{(r)}$, where r runs from 1 to 4, corresponding to the four possible eigenstates associated with the two energy states and the two spin states, and, if the incoming and outgoing states represent particles rather than anti-particles, the quantity to be computed is

$$X = \frac{1}{2}\sum_{r=1}^{2}\sum_{s=1}^{2} \overline{u}_1^{(r)} M' u_2^{(s)} \overline{u}_2^{(s)} M u_1^{(r)}. \tag{4.156}$$

This final-state sum and initial-state average can be written in a more convenient form with the help of the so-called projection operators. Let us now see how these quantities arise, first in evaluating the sum over s in $u_2^{(s)}\overline{u}_2^{(s)}$. Note that this factor is a matrix, since each $u_2^{(s)}$ has four components, while the factors in reverse order ($\overline{u}u$) is a scalar determined by normalization convention. Concerning the normalization, we adopt the convention[25] $|u|^2 = u^\dagger u = 2|E|/mc^2$, so that

$$\overline{u}u = (mc^2/E)u^\dagger u = |E|/E = \varepsilon = \pm 1. \tag{4.157}$$

That is, $\varepsilon = +1$ for the positive-energy (particle) states and $\varepsilon = -1$ for the negative-energy (anti-particle) states. For a Dirac spinor $u^{(r)}$, say, we can adopt the notation

$$\overline{u}^{(r)} u^{(r)} = \varepsilon_r = \begin{cases} +1 & (r = 1, 2) \\ -1 & (r = 3, 4). \end{cases} \tag{4.158}$$

with $r = 1$ and 2 referring to the two particle spin states and 3 and 4 designating the two spin states for the antiparticle.

The eigenstates are particularly simple in the rest frame of the particle (or antiparticle) for which $p_\mu = (i\varepsilon_r mc, 0)$ and the Dirac equation reduces to

$$\gamma_0 u^{(r)} = \varepsilon_r u^{(r)} \tag{4.159}$$

where [see Chapter 3, Equation (3.174)]

$$\gamma_0 = \beta = \begin{pmatrix} 1 & \cdot & \cdot & \cdot \\ \cdot & 1 & \cdot & \cdot \\ \cdot & \cdot & -1 & \cdot \\ \cdot & \cdot & \cdot & -1 \end{pmatrix}, \tag{4.160}$$

[25] See Equations (3.108) and (3.184) of Chapter 3. We are now eliminating the factor of 2 in the relation (3.108); if we did not, the result (4.157) would have this factor on the right. This minor inconsistency in notation choice is made to simplify the equations in the present section and to make them correspond to those in most common usage and in the principal references for the material in this section. We, of course, have not been concerning ourselves with numerical factors in our simplified formulation of relativistic QED, since for the evaluation of the various processes, the factors can be fixed by comparison with the non-relativistic limit.

the dots representing zeroes in the matrix. The solutions to Equation (4.159) are the four independent orthogonal functions

$$u^{(1)} = \begin{pmatrix} 1 \\ \cdot \\ \cdot \\ \cdot \end{pmatrix}, \qquad u^{(2)} = \begin{pmatrix} \cdot \\ 1 \\ \cdot \\ \cdot \end{pmatrix}, \qquad u^{(3)} = \begin{pmatrix} \cdot \\ \cdot \\ 1 \\ \cdot \end{pmatrix}, \qquad u^{(4)} = \begin{pmatrix} \cdot \\ \cdot \\ \cdot \\ 1 \end{pmatrix}. \tag{4.161}$$

The corresponding functions $\overline{u} = u^\dagger \gamma_0$ are then

$$\begin{aligned} \overline{u}^{(1)} &= \begin{pmatrix} 1 & \cdot & \cdot & \cdot \end{pmatrix}, \\ \overline{u}^{(2)} &= \begin{pmatrix} \cdot & 1 & \cdot & \cdot \end{pmatrix}, \\ \overline{u}^{(3)} &= \begin{pmatrix} \cdot & \cdot & -1 & \cdot \end{pmatrix}, \\ \overline{u}^{(4)} &= \begin{pmatrix} \cdot & \cdot & \cdot & -1 \end{pmatrix}. \end{aligned} \tag{4.162}$$

Returning to the problem of the sum over positive-energy states s in Equation (4.156), we see that, in the particle rest frame,

$$\sum_{s=1}^{2} u^{(s)}\overline{u}^{(s)} = \begin{pmatrix} 1 & \cdot & \cdot & \cdot \\ \cdot & 1 & \cdot & \cdot \\ \cdot & \cdot & \cdot & \cdot \\ \cdot & \cdot & \cdot & \cdot \end{pmatrix}. \tag{4.163}$$

this matrix is also equal to $\frac{1}{2}(\gamma_0 + I)$, and a covariant generalization valid in any reference frame would be the matrix operator

$$\Lambda_+ = (\not{p} + imc)/2imc. \tag{4.164}$$

Similarly, if, instead, a summation is made over the negative energy states,

$$\sum_{s=3}^{4} u^{(s)}\overline{u}^{(s)} = \tfrac{1}{2}(\gamma_0 - I). \tag{4.165}$$

A convenient operator to introduce is minus the covariant operator, which reduces to the result (4.165) in the antiparticle rest frame. The positive- and negative-energy projection operators can thus be written

$$\Lambda_\pm = (\not{p} \pm imc)/(\pm 2imc) = (\not{p} + i\varepsilon_r mc)/2i\varepsilon_r mc. \tag{4.166}$$

If $u_\pm$ are the positive- and negative-energy Dirac states, since

$$(\pm\not{p} - imc)u_\pm = 0, \tag{4.167}$$

the two operators (4.166) have the convenient properties

$$\begin{aligned} \Lambda_\pm u_\pm &= u_\pm \\ \Lambda_\pm u_\mp &= 0 \\ \Lambda_+ + \Lambda_- &= I. \end{aligned} \tag{4.168}$$

Finally, for products of the type (4.163), there is the important identity

$$\sum_{s=1}^{4} u^{(s)}\overline{u}^{(s)}\varepsilon_s = I, \tag{4.169}$$

which is essentially a "completeness relation" for the eigenfunctions. By virtue of the identities (4.168) and (4.169), there is an additional significant relation: if O is some operator,

$$\begin{aligned}\sum_{r=1}^{2} \overline{u}^{(+)} O u^{(r)} &= \sum_{r=1}^{2} \overline{u}^{(r)} O u^{(r)}\varepsilon_r \\ &= \sum_{r=1}^{4} \overline{u}^{(r)} O \Lambda_+ u^{(r)}\varepsilon_r = \sum_{r=1}^{4} \overline{u}^{(r)} O_+ u^{(r)}\varepsilon_r = \mathrm{Tr}\, O_+,\end{aligned} \tag{4.170}$$

where $O_+ = O\Lambda_+$ and Tr is the trace[26] or sum of the diagonal elements.

It should be noted, in particular, how, through the employment of the projection operators, the sum over r from 1 to 2 is extended to 1 to 4. Because of the above identities, the invariant (4.156) can be written

$$X = \frac{1}{2}\mathrm{Tr}\left(M' \frac{\not{p}_2 + imc}{2imc} M \frac{\not{p}_1 + imc}{2imc}\right). \tag{4.171}$$

In the evaluation of trace (4.171), in which the operator in parentheses is a collection of terms that are products of γ-matrices, a number of useful theorems can be employed. These theorems are based on the identities (3.101) and (3.103) of Chapter 3:

$$\begin{aligned}\gamma_\mu \gamma_\nu + \gamma_\nu \gamma_\mu &= 2\delta_{\mu\nu}, \\ \not{A}\not{B} + \not{B}\not{A} &= 2\,A \cdot B.\end{aligned} \tag{4.172}$$

Other fundamental properties of the trace are also used, such as the well-known cyclic property

$$\mathrm{Tr}\,(AB) = \mathrm{Tr}\,(BA), \tag{4.173}$$

and its generalization

$$\mathrm{Tr}\,(a_1 a_2 \cdots a_n) = \mathrm{Tr}\,(a_n a_1 a_2 \cdots a_{n-1}). \tag{4.174}$$

It then follows that[27]

$$\mathrm{Tr}\,(\gamma_\mu \gamma_\nu) = 4\delta_{\mu\nu}. \tag{4.175}$$

If $\gamma_\mu \gamma_\nu \cdots \gamma_\lambda$ contains an odd number of factors,

$$\mathrm{Tr}\,(\gamma_\mu \gamma_\nu \cdots \gamma_\lambda) = 0. \tag{4.176}$$

---

[26] Sometimes the notation Sp (for "spur") is used instead of Tr.

[27] See, for further details, J. D. Bjorken and S. D. Drell, *Relativistic Quantum Mechanics*, New York: McGraw-Hill, Inc., 1964; see also F. Mandl, *Introduction to Quantum Field Theory*, New York: Interscience Publ., Inc., 1959. Mandl's book is a good introduction for modern QED, and the notation in this section is similar to his.

Also,

$$\mathrm{Tr}\,(\not{A}\not{B}) = 4(A \cdot B), \tag{4.177}$$

$$\mathrm{Tr}\,(\not{A}\not{B}\not{C}\not{D}) = 4[(A \cdot B)(C \cdot D) - (A \cdot C)(B \cdot D) + (A \cdot D)(B \cdot C)]. \tag{4.178}$$

Finally, in the manipulations of Dirac matrices, it is convenient to define

$$\gamma_5 = \gamma_0\gamma_1\gamma_2\gamma_3. \tag{4.179}$$

This matrix satisfies $\gamma_5^2 = 1$ and anticommutes with any of the γ-matrices:

$$\gamma_5\gamma_\mu + \gamma_\mu\gamma_5 = 0, \tag{4.180}$$

as is easily seen for μ equal to 0, 1, 2, 3.

4.4.2 Coulomb Scattering

In the scattering of an unpolarized electron by a static Coulomb field, the perturbation (M) is of the form $e\not{A}$ [see Equation (3.105) of Chapter 3] and $A_\mu = (i\Phi, 0)$, where Φ is the scalar potential. We need its Fourier tranform (4.115), so that

$$M \propto \gamma_0 Ze^2/q^2, \tag{4.181}$$

in terms of the momentum transfer $q = |\boldsymbol{q}|$. Then

$$M' = \gamma_0 M^\dagger \gamma_0 = M, \tag{4.182}$$

and, after an initial average and final sum over spins,

$$|\mathcal{M}|^2 \propto \left(\frac{Ze^2}{q^2}\right)^2 \mathrm{Tr}\left(\gamma_0 \frac{\not{p}_2 + imc}{2imc}\gamma_0 \frac{\not{p}_1 + imc}{2imc}\right), \tag{4.183}$$

where the fundamental result (4.171) has been employed. The trace theorems (4.175)–(4.178) then yield the simple result

$$\begin{aligned} \mathrm{Tr}\,[\gamma_0(\not{p}_2 + imc)\gamma_0(\not{p}_1 + imc)] &= \mathrm{Tr}\,(\gamma_0\not{p}_2\gamma_0\not{p}_1 - m^2c^2 I) \\ &= -4[E_2E_1/c^2 - (\boldsymbol{p}_1 \cdot \boldsymbol{p}_2) - m^2c^2]. \end{aligned} \tag{4.184}$$

The cross section is computed as in Equation (4.117) for the spin-0 case:

$$v_1\, d\sigma \propto \int (|\mathcal{M}|^2/E_1E_2)p_2^2\, dp_2\, d\Omega\, \delta(E_2 - E_1). \tag{4.185}$$

Substituting the results (4.183) and (4.184), integrating over $dE_2 \propto p_2 dp_2/E_2$ with the δ-function, we have, in particular, for the trace factor,

$$2(E/c)^2 - p^2(1 - \cos\theta_{\mathrm{sc}}) = 2(E/c)^2[1 - \beta^2 \sin^2(\theta_{\mathrm{sc}}/2)], \tag{4.186}$$

where θ_{sc} is the scattering angle and $\beta = v/c$. The momentum transfer is again given by the expression (4.109) in the spin-0 case. Comparison with the non-relativistic limit (4.54) yields the multiplying factor in the final form, which is then

$$d\sigma/d\Omega = (Ze^2/2mc^2\gamma\beta^2)^2 \csc^4(\theta_{\mathrm{sc}}/2)[1 - \beta^2 \sin^2(\theta_{\mathrm{sc}}/2)]. \tag{4.187}$$

The result[28] is the same as the spin-0 formula (4.111) except for an additional factor $1 - \beta^2 \sin^2(\theta_{sc}/2)$, which arises in the evaluation of the trace (4.186). This correction factor can thus be attributed to the interaction effects of the electron's magnetic moment.

Two properties of the magnetic-moment correction factor should be noted: it approaches unity (i) in the non-relativistic limit and (ii) in the limit of small scattering angles. Let us now see if these properties can be understood, at least in the non-relativistic limit, in terms of simple physical arguments. The relative correction $\eta = -\beta^2 \sin^2(\theta_{sc}/2)$ can be regarded as the result of a correction to the perturbation causing the scattering. If the magnetic perturbation is weak, we can write, in terms of the electric and magnetic perturbations,

$$\eta \sim H'_{mag}/H'_{el}. \qquad (4.188)$$

The electric perturbation is just the electrostatic interaction $H'_{el} = e\Phi = Ze^2/r$, and $H'_{mag} \sim -\boldsymbol{\mu} \cdot \boldsymbol{B}$, where $\boldsymbol{\mu}$ is the magnetic moment and $\boldsymbol{B}$ is an effective magnetic field seen by the electron. We take $B \sim \Delta\beta E \sim \Delta\beta\Phi/r$, where $\Delta\beta$ is the change in the electron velocity. Setting the characteristic r equal to $\hbar/\Delta p$, we have $\eta \sim -\mu\Delta\beta\Delta p/e\hbar$. Since $\mu \sim e\hbar/mc$, and $\Delta\beta = \Delta p/mc \sim \beta\sin(\theta_{sc}/2)$, we obtain

$$\eta \sim -\beta^2 \sin^2(\theta_{sc}/2). \qquad (4.189)$$

Although the above analysis is non-relativistic in nature, it does yield the correct angular dependence for the correction in the relativistic formula.

4.4.3 Møller and Bhabha Scattering

We have treated the spin-0 analogs of Møller and Bhabha scattering in Sections 3.3 and 3.4, respectively. Møller scattering refers to the scattering of two relativistic electrons (or positrons): $e^\pm + e^\pm \to e^\pm + e^\pm$; Bhabha scattering is the scattering of an electron by a positron: $e^+ + e^- \to e^+ + e^-$. We have already derived the Møller and Bhabha cross section in the non-relativistic limit. In the NR limit, the magnetic moment interactions are negligible, as are, for e^+–e^- scattering, the effects of intermediate states where the particles have "annihilated to a photon state." Exchange is still important in the NR Born limit, and the cross section (4.102) would have to be a limiting form of the Møller formula for NR energies. The NR limit of the Bhabha cross section would be given by the result (4.31), or by the first term in the formula (4.102), or by the NR limit of the cross section (4.136) for distinguishable spinless particles.

The Feynman diagrams for Møller scattering are the same as in Figure 4.6 for the spin-0 case. However, since electrons are fermions, the exchange amplitude appears with a minus sign. Note that, in contrast to the non-relativistic formulation, in which the particle spin description was not included in the states $u(\boldsymbol{r})$ or $u(\boldsymbol{p})$, the relativistic spinors give a complete specification of the particle state. The vertices

[28]First obtained by N. F. Mott.

in Figure 4.6, when describing a process involving spin-$\frac{1}{2}$ particles, correspond to the interaction $\not\varepsilon = \gamma_\mu \varepsilon_\mu$. When a summation is performed over polarization states for the virtual photon, the amplitude for the process is given by the form[29]

$$\mathcal{M} = \frac{\overline{u}_4 \gamma_\mu u_3 \overline{u}_2 \gamma_\mu u_1}{(p_1 - p_2)^2} - \frac{\overline{u}_2 \gamma_\mu u_3 \overline{u}_4 \gamma_\mu u_1}{(p_1 - p_4)^2}. \tag{4.190}$$

The second term is the exchange amplitude; it appears with a minus sign and is the same as the direct amplitude with the replacement $2 \leftrightarrow 4$. That is,

$$\mathcal{M} = A - A_e, \tag{4.191}$$

where

$$A_e = A(2 \leftrightarrow 4). \tag{4.192}$$

The NR limit of Møller scattering can be obtained very easily from the covariant form (4.190). This is the limit of low energy ($E \ll mc^2$) and, in lowest order, the rest-frame expressions (4.161) can be used for the Dirac spinors. Since in the scattering $1, 3 \to 2, 4$ all of the states are the positive-energy (electron) functions, they may be taken as the two spin states $u_a = u^{(1)}$ and $u_b = u^{(2)}$:

$$u_a = \begin{pmatrix} 1 \\ 0 \\ 0 \\ 0 \end{pmatrix}, \qquad u_b = \begin{pmatrix} 0 \\ 1 \\ 0 \\ 0 \end{pmatrix}. \tag{4.193}$$

For the representation (3.174) of Chapter 3 for the Dirac matrices, the γ_μ are, specifically,

$$\gamma_0 = \begin{pmatrix} 1 & \cdot & \cdot & \cdot \\ \cdot & 1 & \cdot & \cdot \\ \cdot & \cdot & -1 & \cdot \\ \cdot & \cdot & \cdot & -1 \end{pmatrix}, \quad \gamma_1 = \begin{pmatrix} \cdot & \cdot & \cdot & -i \\ \cdot & \cdot & -i & \cdot \\ \cdot & i & \cdot & \cdot \\ i & \cdot & \cdot & \cdot \end{pmatrix},$$
$$\gamma_2 = \begin{pmatrix} \cdot & \cdot & \cdot & -1 \\ \cdot & \cdot & 1 & \cdot \\ \cdot & 1 & \cdot & \cdot \\ -1 & \cdot & \cdot & \cdot \end{pmatrix}, \quad \gamma_3 = \begin{pmatrix} \cdot & \cdot & -i & \cdot \\ \cdot & \cdot & \cdot & i \\ i & \cdot & \cdot & \cdot \\ \cdot & -i & \cdot & \cdot \end{pmatrix}. \tag{4.194}$$

With the states (4.194) for the particles 1, 2, 3, 4, the only non-vanishing combinations of matrices are

$$\overline{u}_a \gamma_0 u_a = \overline{u}_b \gamma_0 u_b = 1. \tag{4.195}$$

The amplitudes involving γ_1, γ_2, and γ_3 are all zero, and $\overline{u}_a \gamma_0 u_b = \overline{u}_b \gamma_0 u_a = 0$. We introduce the denominators

$$[(p_1 - p_2)^2]^{-1} = d, \qquad [(p_1 - p_4)^2]^{-1} = d_e. \tag{4.196}$$

[29]Compare the corresponding spin-0 amplitude (4.138).

Of the total amplitude $m = (1, 3 \to 2, 4)$, the non-vanishing spin combinations are then

$$\begin{aligned}(aa \to aa) &= d - d_e,\\ (ab \to ab) &= d,\\ (ab \to ba) &= -d_e,\\ (bb \to bb) &= d - d_e,\\ (ba \to ba) &= d,\\ (ba \to ab) &= -d_e.\end{aligned} \tag{4.197}$$

The sum of the squared amplitudes is

$$\sum |\mathcal{M}(1, 3 \to 2, 4)|^2 = 4(d^2 + d_e^2 - dd_e). \tag{4.198}$$

The cross section is most conveniently expressed in terms of c.m. frame quantities, and the relationship of the (summed) squared amplitudes (4.198) to the cross section is given by the same formula as in the spin-0 case. In the NR limit, however, $E \to mc^2$ and $d\sigma/d\Omega$ is simply proportional to the sum (4.198). In terms of the c.m. momentum (of each particle) p and the c.m. scattering angle,

$$\begin{aligned}d &= (4p^2)^{-1} \csc^2(\theta_{sc}/2),\\ d_e &= (4p^2)^{-1} \sec^2(\theta_{sc}/2),\end{aligned} \tag{4.199}$$

and the NR Born formula (4.102) is again obtained (after fixing the multiplying constant).

This method of obtaining the NR cross section may illustrate some convenient aspects of the covariant formulation. Note, for example, as remarked earlier, how the required anti-symmetry is contained in the single amplitude (4.190), since spin is contained in the Dirac eigenstates.

In the general case for arbitrary relativistic energy, it is necessary to form the product $\mathcal{M}\mathcal{M}^\dagger$ and sum and average over final and initial spins, respectively. We have [see Equations (4.154) and (4.155)]

$$\mathcal{M}^\dagger = \frac{\bar{u}_3\gamma_0\gamma_\mu\gamma_0 u_4 \bar{u}_1\gamma_0\gamma_\mu\gamma_0 u_2}{(p_1 - p_2)^2} - (2 \leftrightarrow 4), \tag{4.200}$$

and, employing the projection operators introduced in Section 4.1, we get

$$\sum_{\text{spins}} \mathcal{M}\mathcal{M}^\dagger = \frac{\operatorname{Tr}\left(\Lambda_+(4)\gamma_\mu\Lambda_+(3)\gamma_0\gamma_\nu\gamma_0\right)\operatorname{Tr}\left(\Lambda_+(2)\gamma_\mu\Lambda_+(1)\gamma_0\gamma_\nu\gamma_0\right)}{(p_1 - p_2)^4}$$
$$- \frac{\operatorname{Tr}\left(\Lambda_+(4)\gamma_\mu\Lambda_+(3)\gamma_0\gamma_\nu\gamma_0\Lambda_+(2)\gamma_\mu\Lambda_+(1)\gamma_0\gamma_\nu\gamma_0\right)}{(p_1 - p_2)^2(p_1 - p_4)^2} + (2 \leftrightarrow 4). \tag{4.201}$$

Inside the trace factors, since $\gamma_\nu\gamma_0 = 1$ for $\nu = 0$ and $-\gamma_0\gamma_\nu$ for $\nu \neq 0$, and since $\gamma_0\gamma_\nu\gamma_0$ appears twice in a product, in Equation (4.201) it can be set equal to γ_ν. The evaluation of the traces in Equation (4.201) is accomplished with the help of the trace theorems of Section 4.1, and the details will not be reproduced here. The phase-space and kinematic factors are the same as in the spin-0 case [see Equation (4.142)]. There are various ways of expressing the result for the Møller cross section; in terms of c.m. quantities, we have ($\Lambda = \hbar/mc$)

$$\left(\frac{d\sigma}{d\Omega}\right)_M = \frac{\alpha^2\Lambda^2}{4}\left(\frac{1+\beta^2}{\gamma\beta^2}\right)^2 f(\theta;\beta), \tag{4.202}$$

where β and γ are the (dimensionless) c.m. velocity and energy of each of the four electrons involved in the scattering, and (the subscript "sc" has been left off the scattering angle)

$$f(\theta;\beta) = 4\csc^4\theta - 3\csc^2\theta + [\beta^2/(1+\beta^2)]^2(1+4\csc^2\theta). \tag{4.203}$$

The above forms allow an easy evaluation of the NR ($\gamma \to 1$; $\beta \ll 1$) and extreme-relativistic (ER) ($\gamma \gg 1$; $\beta \to 1$) limits, for which

$$f(\theta;0) = 4\csc^4\theta - 3\csc^2\theta = \csc^4\theta(1+3\cos^2\theta), \tag{4.204}$$

$$f(\theta;1) = 4\csc^4\theta - 2\csc^2\theta + \tfrac{1}{4}. \tag{4.205}$$

Note that the cross section has been written as a function of θ rather than $\theta/2$, and the reader may not, at first, recognize the NR limit. Actually, the NR formula[30] (4.102) itself is more compact when expressed as a function of θ; from the half-angle identities,

$$\csc^4(\theta/2) + \sec^4(\theta/2) - \csc^2(\theta/2)\sec^2(\theta/2) = 4\csc^4\theta(1+3\cos^2\theta). \tag{4.206}$$

For arbitrary energy, in the limit $\theta \ll 1$, the cross section also has a simple form:

$$\left(\frac{d\sigma}{d\Omega}\right)_M \underset{\theta\ll 1}{\longrightarrow} \alpha^2\Lambda^2\left(\frac{1+\beta^2}{\gamma\beta^2}\right)^2\frac{1}{\theta^4}, \tag{4.207}$$

which is identical to the corresponding limit of the spin-0 formula (4.143). Exchange and magnetic-moment effects are negligible at small scattering angles; that is, they do not contribute to the dominant term in the cross section $d\sigma/d\Omega$. As will be shown in applications in later chapters, however, the small-θ dominant term can lead to a logarithmic factor in a formula,[31] and the spin and exchange corrections are of order unity (compared with, say, the dominant term $\ln N$).

As in the spin-0 analog, the cross section for relativistic e^+–e^- (Bhabha) scattering can be obtained from that from Møller scattering through an application of crossing symmetry. That is, from the Møller formula expressed in terms of the kinematic invariants s, t, and u, the interchange $s \leftrightarrow u$ is made to yield the Bhabha cross section.[32] Both the Møller and Bhabha formulas, in terms of s, t, and u, are much more complicated than their spin-0 analogs. The result for the latter is given below

[30] Note that in the NR formula (4.102) v is the c.m. *relative* velocity, equal to $2\beta c$ in terms of the above β.

[31] For example, in the calculation of the energy loss rate of an electron going through matter.

[32] See Section 3.4 for the spin-0 case.

in terms of the c.m. scattering angle and, as in the Møller formula, in terms of the c.m. β and γ (of each particle):

$$\left(\frac{d\sigma}{d\Omega}\right)_B = \frac{\alpha^2 \Lambda^2}{16} \frac{1}{\gamma^2} \left[\left(\frac{1+\beta^2}{\beta^2} \right)^2 \csc^4 \frac{\theta}{2} - \frac{8\gamma^4 - 1}{\gamma^4 \beta^2} \csc^2 \frac{\theta}{2} \right.$$
$$\left. + \frac{12\gamma^4 + 1}{\gamma^4} - 4\beta^2(1+\beta^2) \sin^2 \frac{\theta}{2} + 4\beta^4 \sin^4 \frac{\theta}{2} \right]. \quad (4.208)$$

While the Møller formula (4.202) is most conveniently expressed in terms of θ, the Bhabha formula is more compact when written as a function of $\theta/2$.

The NR and ER limits of the general result (4.208) are readily obtained. We find

$$\left(\frac{d\sigma}{d\Omega}\right)_B \xrightarrow[\text{(NR)}]{} \frac{\alpha^2 \Lambda^2}{16\beta^4} \csc^4 \frac{\theta}{2}, \quad (4.209)$$

which is the same as the first term in the formula (4.102); that is, there is no exchange term and "annihilation" and magnetic moment effects are negligible in the NR limit. In the ER limit, the Bhabha cross section is simply $\cos^4(\theta/2)$ times the corresponding limit of the Møller formula:

$$\left(\frac{d\sigma}{d\Omega}\right)_B \to \left(\frac{d\sigma}{d\Omega}\right)_M \cos^4 \frac{\theta}{2} \qquad \text{(ER limit)} \quad (4.210)$$

As expected, for $\theta \ll 1$ and arbitrary energy, the limit of $(d\sigma/d\Omega)_B$ is the same as the result (4.207) for $(d\sigma/d\Omega)_M$ and the same as the limit of the spin-0 formula (4.151).

Bibliographical Notes

Classical Coulomb scattering is treated in just about every book on classical mechanics. It is also covered, along with the validity of both the classical and Born limits in

1. Bohm, D., *Quantum Theory*, New York: Prentice-Hall, Inc., 1951.

The general quantum theory of scattering (including Coulomb scattering) is developed in, for example, the following references, listed in order of increasing formality,

2. Mott, N. F. and Massey, H. S. W., *Theory of Atomic Collisions*, 3rd ed., Oxford, UK: Oxford University Press, 1965.
3. Landau, L. D. and Lifshitz, E. M., *Quantum Mechanics, Non-Relativistic Theory*, 2nd ed., Oxford, UK: Pergamon Press, 1965.
4. Goldberger, M. L. and Watson, K. M., *Collision Theory*, New York: John Wiley & Sons, Inc., 1964.

Scattering of relativistic charges is treated in many textbooks on quantum electrodynamics, such as References 3–8 listed at the end of Chapter 3; Reference 7 (Feynman) covers the (much simpler) spin-0 case, which is generally not included in most texts. An introduction to relativistic scattering, especially Møller scattering, is also provided in

5. Cheng, D. C. and O'Neill, G. K., *Elementary Particle Physics*, Reading, MA: Addison-Wesley Publ. Co., 1979.

Chapter Five

Compton Scattering

In Compton scattering, two photons are involved, and both the incoming and the outgoing one are of the radiation field type. Also, the process is of the same order (in α) as Coulomb scattering treated in the previous chapter, since two vertices are involved in diagrams associated with covariant perturbation theory for spin-$\frac{1}{2}$ charges. Actually, in the high-energy limit, the Compton cross section depends very much on the spin of the charged particle that scatters the photon. In this limit, the main contribution to the scattering is a result of interactions with the particle's magnetic moment. At high energies, then, the cross section for scattering by a spin-0 charge is much less than that for particles like electrons that have spin-$\frac{1}{2}$ (and an intrinsic magnetic moment).

At low energies, the situation is the other way around, and the principal interaction is with the electric charge rather than with the magnetic moment. The domain of low photon energies is of special interest since it is the region of applicability of the classical limit. This limit is quite interesting, and, as in the last chapter on Coulomb scattering, we first consider the Compton process in a purely classical formulation. The process will then be treated quantum mechanically, first in non-relativistic non-covariant QED, then in a relativistic covariant formulation for particles of spin-0 and spin-$\frac{1}{2}$. In this manner, the problem is discussed in increasing order of difficulty.

5.1 CLASSICAL LIMIT

5.1.1 Kinematics of the Scattering

Although the concept of the photon is foreign in a classical description of the process, it is convenient to have the basic relations for the energy of the scattered photon and electron in terms of the photon scattering angle, for we want to consider the validity of the classical limit and the conditions for which the particle nature of the photon must be brought in. The process is most conveniently described in a reference frame in which the charged particle is initially at rest. Also, it is convenient to employ units with $c = 1$; in the final formulas for energy or momentum, it is obvious where factors of c must be inserted to have the equations in, say, c.g.s. units.

If $\boldsymbol{k}_0$ and $\boldsymbol{k}$ are, respectively, the momenta of the photon before and after scattering, and $\boldsymbol{v}$ is the velocity of the charge after scattering, the energy- and momentum-conservation relations are ($k_0 = |\boldsymbol{k}_0|$; $k = |\boldsymbol{k}|$)

$$\begin{aligned} k_0 - k &= m(\gamma - 1), \\ \boldsymbol{k}_0 - \boldsymbol{k} &= m\gamma\boldsymbol{v}. \end{aligned} \tag{5.1}$$

These relations are exact and assume only that the photon is a massless particle; m is the mass of the charge and γ is the Lorentz factor at its scattering velocity. Squaring and subtracting the two relations (5.1), we have, since $\gamma^2 v^2 = \gamma^2 - 1$,

$$2k_0 k(1 - \cos\theta) = 2m^2(\gamma - 1) = 2m(k_0 - k), \tag{5.2}$$

where $\cos\theta = \mathbf{k}_0 \cdot \mathbf{k}/k_0 k$; that is, θ is the photon scattering angle. Solving for the energy (or momentum) of the scattered photon, yields

$$k = k_0/[1 + (k_0/m)(1 - \cos\theta)]. \tag{5.3}$$

The energy of the scattered charge is

$$T = k_0 - k = (k_0^2/m)(1 - \cos\theta)/[1 + (k_0/m)(1 - \cos\theta)]. \tag{5.4}$$

These relations can also be expressed in more compact form in terms of the half-angle for scattering, since $1 - \cos\theta = 2\sin^2(\theta/2)$. The fundamental relation (5.3) can also be written in terms of the photon wavelengths:

$$\lambda\!\!\!^{-} = \lambda\!\!\!^{-}_0 + \Lambda(1 - \cos\theta). \tag{5.5}$$

The wavelength change is then

$$\Delta\lambda\!\!\!^{-} = \Lambda(1 - \cos\theta) = 2\Lambda\sin^2(\theta/2), \tag{5.6}$$

being determined by θ and the "Compton wavelength" $\Lambda = \hbar/mc$, a characteristic wavelength that is basic in QED. We see that as $\hbar \to 0$, $\Delta\lambda\!\!\!^{-} \to 0$; for finite $\hbar$, $\Delta\lambda\!\!\!^{-} \sim \lambda\!\!\!^{-}_0$ for $\lambda\!\!\!^{-}_0 \sim \Lambda$ or for initial photon energy $\hbar\omega_0 \sim mc^2$. That is, photon characteristics are particularly important at high energies greater than or of the order of mc^2.

The above elementary results for, in particular, $\Delta\lambda\!\!\!^{-}$ were confirmed in experiments by Compton in 1922 and established the particle nature of the photon. The characteristic photon energy mc^2 is relevant in a discussion of the validity of the classical limit for the process.

5.1.2 Derivation of the Thomson Cross Section

The description of Compton scattering by classical theory (CED) is quite different from that in QED. Although the photon is not introduced, it is possible to derive a cross section for the process by computing the energy in the scattered radiation in terms of the incident energy flux. If J_E is this flux, the differential scattering cross section is evaluated from the rate of energy scattered per steradian:

$$J_E(d\sigma/d\Omega) = dE_{\text{sc}}/d\Omega\, dt. \tag{5.7}$$

The scattered energy rate is, in turn, computed from the corresponding magnitude of the Poynting vector S:

$$dE_{\text{sc}}/d\Omega\, dt = S(dA/d\Omega) = r^2 S, \tag{5.8}$$

dA being a differential area perpendicular to the scattering center at a distance r and outlined by the solid angle element $d\Omega$. In the classical description, the scattered component is radiated energy that is a result of the acceleration of the charge q by

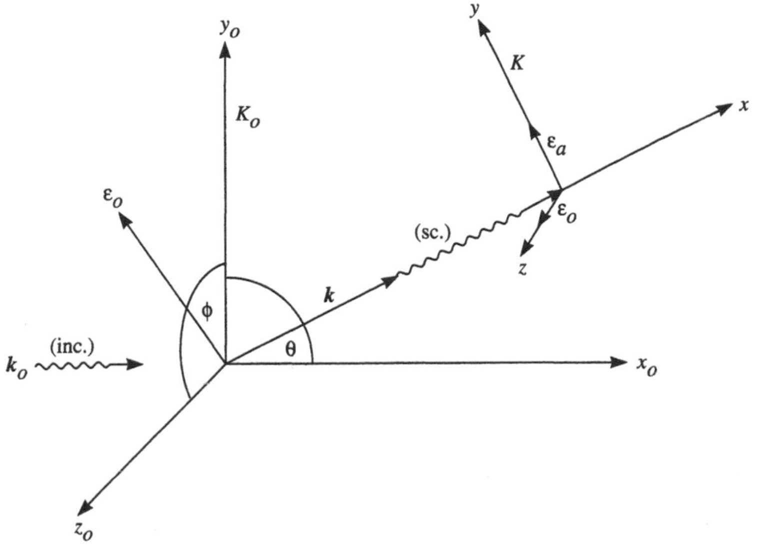

Figure 5.1 Photon scattering.

the field of the incident wave; that is, $dF_{sc} = dE_{rad}$. The wave is assumed to be sufficiently weak that only small velocities and accelerations are produced in the charge, which is initially at rest. In this limit, the emission from higher multipoles is negligible, and the radiation rate can be calculated from the classical dipole formula (see Chapter 2, Section 2.2).

For the description and evaluation of the process, it is convenient to employ the coordinate frames indicated in Figure 5.1. The charge motion is referred to the frame K_0 oriented such that the incident wave is propagating along the x-axis. We consider scattering of the radiation in a particular direction (defined by $d\Omega$) and orient the y_0- and z_0-axes of K_0 so that the scattered (or radiated) wave is in the x_0–y_0-plane at an angle θ with respect to the x_0-axis. A specific Fourier component of the incident wave having wave vector $\boldsymbol{k}_0$ is considered, and its polarization is described in terms of a unit vector $\boldsymbol{\varepsilon}_0$ in the y_0–z_0 plane at an angle ϕ_0 with respect to the y_0-axis. In terms of the arbitrary angles ϕ_0 and θ, the scattering process is then specified in complete generality. However, for the description of the polarization components of the scattered wave, it is convenient to introduce the second reference frame K' indicated in Figure 5.1. This frame has its x–y plane in the x–y plane of K_0 so that the z- and z_0-axes are parallel, but it is oriented such that the x-axis is along $\boldsymbol{k}$, the direction of the outgoing radiation. In the K frame, the scattered radiation will then have polarization components in the y- and z-directions. Thus, while the frame K_0 is convenient to describe the incident wave, K is more appropriate for the scattered radiation.

The electric field associated with the incident wave can be written in the form

$$\boldsymbol{E}_{\text{inc}} = \boldsymbol{\varepsilon}_0 E_0 \,\text{Re}\, e^{i(\boldsymbol{k}_0 \cdot \boldsymbol{r} - \omega_0 t)}, \tag{5.9}$$

with the unit polarization vector given by

$$\boldsymbol{\varepsilon}_0 = (0, \cos\phi_0, \sin\phi_0). \tag{5.10}$$

We neglect the magnetic force and find the acceleration of the charge to be

$$\dot{\boldsymbol{v}}_0 = (q/m)E_0(0, \cos\phi, \sin\phi)\cos\omega t. \tag{5.11}$$

The energy flux associated with the incident wave is

$$S_{\text{inc}} = J_E = (c/4\pi)E_0^2. \tag{5.12}$$

The acceleration (5.11) results in a radiated wave whose fields can be easily computed in the dipole approximation [see Equations (2.34)–(2.41) of Chapter 2]:

$$\begin{aligned} \boldsymbol{B} &= (1/c)hr\,\dot{\boldsymbol{A}}\times\boldsymbol{n}, \\ \boldsymbol{E} &= \boldsymbol{B}\times\boldsymbol{n}, \\ \dot{\boldsymbol{A}} &= (q/rc)\dot{\boldsymbol{v}}, \end{aligned} \tag{5.13}$$

where $\boldsymbol{n} = \boldsymbol{k}/k$ is a unit vector in the direction of propagation of the outgoing wave.

Actually, as already mentioned, it is more convenient to describe the fields of the outgoing wave with respect to the K frame for which the same type of relations are valid. In K, $\boldsymbol{n} = (1, 0, 0)$, and the components of the acceleration are

$$\dot{\boldsymbol{v}} = (\ddot{y}_0\sin\theta, \ddot{y}_0\cos\theta, \ddot{z}_0), \tag{5.14}$$

where $\ddot{y}_0$ and $\ddot{z}_0$ are the components of the (K_0-frame) acceleration (5.11). The magnetic and electric fields $\boldsymbol{B}$ and $\boldsymbol{E}$ are determined by

$$\begin{aligned} \dot{\boldsymbol{v}}\times\boldsymbol{n} &= (0, \ddot{z}_0, -\ddot{y}_0\cos\theta), \\ (\dot{\boldsymbol{v}}\times\boldsymbol{n})\times\boldsymbol{n} &= (0, -\ddot{y}_0\cos\theta, -\ddot{z}_0), \end{aligned} \tag{5.15}$$

the latter giving explicitly the $\boldsymbol{E}'$-field polarization components in the outgoing wave. Since $\ddot{y}_0$ and $\ddot{z}_0$ are proportional to $\cos\phi_0$ and $\sin\phi_0$, respectively [see Equation (5.11)], we see how the polarization components of the outgoing scattered wave depend on the scattering angle θ:

$$\boldsymbol{E} \propto (0, -\cos\phi_0\cos\theta, -\sin\phi_0). \tag{5.16}$$

The time-averaged Poynting vector associated with the scattered wave is in the direction of $\boldsymbol{n}$ and its magnitude is

$$\begin{aligned} S_{\text{sc}} &= (c/4\pi)(q^2/r^2c^4)(\ddot{y}_0^2\cos^2\theta + \ddot{z}_0^2) \\ &= (q^4/4\pi r^2c^3m^2)(\cos^2\phi_0\cos^2\theta + \sin^2\phi_0). \end{aligned} \tag{5.17}$$

The scattering cross section is, by Equations (5.7), (5.8), (5.12), and (5.17),

$$d\sigma/d\Omega = (q^2/mc^2)^2(\cos^2\phi_0\cos^2\theta + \sin^2\phi_0). \tag{5.18}$$

If $q = ze$, in terms of the classical electron radius $r_0 = \alpha \Lambda$, averaging over polarization directions $[\langle\cos^2\phi_0\rangle = \langle\sin^2\phi_0\rangle = \frac{1}{2}]$, we obtain the cross section for scattering of unpolarized radiation:

$$d\sigma/d\Omega = \tfrac{1}{2}z^4 r_0^2(1+\cos^2\theta). \tag{5.19}$$

Integrating over angles for the outgoing radiation, the total cross section is

$$\sigma = (8\pi/3)z^4 r_0^2 = z^4\sigma_{\mathrm{T}}, \tag{5.20}$$

where σ_{T} is the Thomson cross section. Both the total and differential classical cross sections are independent of frequency (or photon energy).

Before leaving the classical derivation to discuss its validity, let us look a little more closely at the process. Including the magnetic force from the incident wave and radiation reaction (see Section 6 of Chapter 2), the non-relativistic equation of motion for the charge is

$$m\ddot{\boldsymbol{r}} = q\left(\boldsymbol{E}_{\mathrm{inc}} + \frac{1}{c}\boldsymbol{v}\times\boldsymbol{B}_{\mathrm{inc}}\right) + \frac{2q^2}{3c^3}\dddot{\boldsymbol{r}}. \tag{5.21}$$

If $\boldsymbol{v} = \dot{\boldsymbol{r}}$ is sufficiently small, the magnetic force can be neglected, and the only correction to our previous treatment would come from an inclusion of the radiation-reaction force. This force can be regarded as a perturbation, and its magnitude can be evaluated by setting $\dddot{\boldsymbol{r}} = -i\omega_0\ddot{\boldsymbol{r}}$ and incorporating it with the term on the left which would then simply be multiplied by a factor $1 + i\varkappa$ with

$$\varkappa = 2q^2\omega_0/3mc^3. \tag{5.22}$$

The squared acceleration, the radiation rate, and the cross section would, as a result, be multiplied by a factor $(1+\varkappa^2)^{-1}$. This correction factor is very close to unity since $\varkappa$ can be written as

$$\varkappa = 2z^2\alpha\Lambda/3\lambda\!\!\!^{-}_0 \ll 1. \tag{5.23}$$

It is interesting to note[1] that the magnetic term $\boldsymbol{v}\times\boldsymbol{B}_{\mathrm{inc}}$ that we have neglected gives rise to a force in the direction of $\boldsymbol{k}_0$, since $\boldsymbol{v}$ is (approximately) in the direction of $\boldsymbol{E}_{\mathrm{inc}}$. Unlike the electric force, the magnetic force has a non-zero time average and, if allowed to act for sufficient time, results in a motion in the direction ($\boldsymbol{k}_0$) of the incident wave. In the classical description, this acceleration and velocity buildup is gradual, the latter being proportional to the time that the incident radiation has acted; the force is from the radiation pressure of the incident wave.

5.1.3 Validity of the Classical Limit

Generally, the classical limit[2] of an electromagnetic process is valid when the photon or particle character of the radiation is not important. The photon is "kinematically soft" if its energy and momentum are small compared with that of the charged particles involved in the process, and the effects of quantum mechanics do not manifest

---

[1] See the discussion in D. Bohm, *Quantum Theory*, New York: Prentice-Hall, Inc., 1951.

[2] A discussion of the classical limit of QED is given in Section 3.2.

themselves. Sometimes, however, classical formulas happen to hold outside their range of validity. For example, in Section 5 of Chapter 3, we saw how the formula for the probability of soft-photon limit emission applies even away from the soft-photon limit in the non-relativistic Born approximation; the basic formula is, moreover, identical to the classical one. In that case, which has an application to bremsstrahlung, the reason for the more general applicability had to do with the small momentum of the photon (even though its energy is comparable to that of the charges).

In Compton scattering, the important kinematic quantity is the photon energy. Even though, in an individual scattering, the charge undergoes a momentum change comparable to the momentum ($p_0 = \hbar c/\lambda_0$) of the incident photon, these recoils from individual scatterings do not invalidate a classical description of the process. The classical picture is that of the continuum limit corresponding to a flux of a large number of photons incident on the charge with their perturbations approximated by a total electric field. As long as the charge recoil velocity is $v \ll c$, the charge sees the same type of perturbing field from all the photons in the incident flux. On the other hand, when v is comparable to c, the charge would see various Doppler-shifted photons, and the classical description would not be possible. As is indicated in the results (5.3), (5.4), and (5.5), the charge receives such large recoil velocities and energies only when the incident photon energy is as large as mc^2. At these energies, the process can no longer be described in the classical continuum limit; in fact, when the recoil energy of the charge is relativistic, other quantum-mechanical effects must be brought in, such as the interactions associated with its magnetic moment.

The nature and validity of the classical limit will be seen a little more clearly when we give the quantum-mechanical formulation for the process in the following sections. Classical electrodynamics is, however, a valid limit for the process, and its domain of applicability is that of low energies such that $\hbar c k_0 \ll mc^2$.

5.2 QUANTUM-MECHANICAL DERIVATION: NON-RELATIVISTIC LIMIT

5.2.1 Interactions and Diagrams

Compton scattering computed by QED is considered a result of the action of perturbations associated with the interaction or coupling of the charge and photon. Strictly speaking, rather, it is the coupling of the charge and photon fields. The methods needed for this type of calculation were outlined in Chapter 3, and we give here the derivation of the scattering cross section by means of the non-relativistic theory.

In NR QED, there are two types of charge-photon couplings[3] corresponding to the single-photon vertex (Figure 3.2 or 3.3) and the two-photon vertex (Figure 3.4). The former coupling has an associated interaction Hamiltonian given by

[3]There is also a photon coupling to the particle's intrinsic magnetic moment. At NR energies, this is neglible compared with the charge-photon couplings. Photon scattering by a magentic moment is computed in NR theory in the following section.

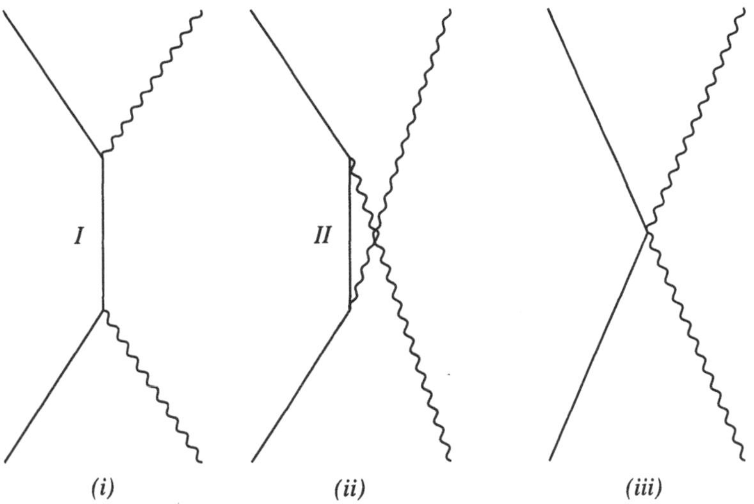

Figure 5.2 Diagrams for Compton scattering in a non-relativistic formulation. The same three types of diagrams would be associated with a covariant calculation for spin-0 charges. In these cases the diagram on the right is all-important. In a covariant spin-$\frac{1}{2}$ formulation the two diagrams on the left give the complete contribution.

the expression (3.20) of Chapter 3 (which we shall designate here as H') and the latter by formula (3.21) (here denoted by H''). That is, leaving out the factors involving the normalization volume L^3 (set equal to unity for convenience), the interaction Hamiltonian associated with the single-photon vertex is

$$H' = -\frac{i\hbar q}{m}\left(\frac{2\pi\hbar}{\omega_1}\right)^{1/2} \boldsymbol{\varepsilon}_1 \cdot \nabla, \tag{5.24}$$

where $\boldsymbol{\varepsilon}_1$ is the polarization unit vector for the photon (energy $\hbar\omega_1$). The two-photon vertex has an interaction Hamiltonian

$$H'' = \frac{\pi\hbar q^2}{m} \frac{\boldsymbol{\varepsilon}_1 \cdot \boldsymbol{\varepsilon}_2}{(\omega_1\omega_2)^{1/2}}, \tag{5.25}$$

in terms of the polarizations and frequencies of the two photons. Note that $H' \propto q$ while $H'' \propto q^2$, the latter being a (weaker) higher-order coupling.

In the scattering, there is a charge and photon in both the initial and final states, and the process takes place through the action of H' (twice) and H'', as indicated by the three diagrams shown in Figure 5.2. Let $\hbar\boldsymbol{k}_0$, $\hbar\omega_0$, and $\boldsymbol{\varepsilon}_0$ designate, respectively, the momentum, energy, and polarization of the incident photon, and $\hbar\boldsymbol{k}$, $\hbar\omega$, and $\boldsymbol{\varepsilon}$, $\hbar\boldsymbol{k}$, and $\boldsymbol{\varepsilon}$ be the quantities for the scattered photon. In H', the $\boldsymbol{\varepsilon}_1$ and ω_1 can refer to the "$\boldsymbol{\varepsilon}_0$–$\omega_0$" photon or the "$\boldsymbol{\varepsilon}$–$\omega$" one; that is, the "absorption" and "emission" can occur in either order as in diagrams (i) and (ii) of Figure 5.2. Similarly, in the perturbation H'', the "1" and "2" can refer to either the incident or scattered photon; effectively, then H'' acts twice [see Equation (3.19) of Chapter 3].

The effective perturbation Hamiltonian for the process in which the system makes a transition from an initial (0) state to a final (f) state is then

$$H'_{f0} = \sum_I H'_{fI} \frac{1}{E_0 - E_I} H'_{I0} + \sum_{II} H'_{fII} \frac{1}{E_0 - E_{II}} H'_{II0} + 2H''_{f0}. \tag{5.26}$$

In the result (5.26), the sums over the intermediate states I and II are associated with the diagrams (i) and (ii), respectively, of Figure 5.2.

5.2.2 Calculation of the Cross Section

It is convenient to compute the cross section in a frame in which the charge is initially at rest.[4] With a unit normalization volume, the photon flux incident on the charge is simply c, and the cross section is related to the transition probability per unit time for the process by

$$c\,d\sigma = \Delta W/\Delta t, \tag{5.27}$$

where [see Equation (3.43) of Chapter 3]

$$\frac{\Delta W}{\Delta t} = \frac{2\pi}{\hbar} \sum_f \left|H'_{f0}\right|^2 \delta(E_0 - E_f). \tag{5.28}$$

The sum over f in Equation (5.28) is over all coordinates not of interest in the final state. There are no interactions with the particle spins, so that the charge-plus-photon (field) states are,[5] in the position representation, the plane waves

$$u_{k_q k} = e^{i\boldsymbol{K}\cdot\boldsymbol{r}}, \tag{5.29}$$

where

$$\boldsymbol{K} = \boldsymbol{k}_q + \boldsymbol{k}, \tag{5.30}$$

$\hbar\boldsymbol{K}$ being the momentum of the charge and photon.

The matrix elements of the interactions yield, aside from other factors, momentum-conservation δ-functions; these can be ignored[6] and, in fact, do not appear in the more convenient and natural momentum representation. The single-photon vertex thus corresponds to the matrix element

$$\langle a|H'|b\rangle = H'_{ab} = (q/m)(2\pi\hbar^3/\omega)^{1/2}(\boldsymbol{\varepsilon}\cdot\boldsymbol{K}_b). \tag{5.31}$$

Since the photons are of the radiation field, $\boldsymbol{\varepsilon}\cdot\boldsymbol{k} = 0$ (that is, their fields are transverse), and

$$\boldsymbol{\varepsilon}\cdot\boldsymbol{K}_b = (\boldsymbol{\varepsilon}\cdot\boldsymbol{k}_q)_b. \tag{5.32}$$

We can now see that diagrams (i) and (ii) of Figure 5.2 do not contribute to the cross section. The matrix element H'_{I0} in Equation (5.26) is proportional to $\boldsymbol{\varepsilon}_0\cdot(\boldsymbol{k}_q)_0$,

---

[4]The charge is also assumed to be free, that is, not bound in an atom or molecule.

[5]A spin eigenfunction could multiply the state (5.29) but (spin) orthogonality conditions would yield, in the matrix elements, the result that the spin does not change. Thus, in this problem, it is best to simply ignore the spin coordinate.

[6]See Section 5.1 of Chapter 3 where the matrix element of H' was evaluated; note, in particular, Equations (3.118) and (3.119) therein.

where $\hbar(\boldsymbol{k}_q)_0$ is the initial momentum of the charge; but in the (convenient) reference frame chosen, $(\boldsymbol{k}_q)_0 = 0$. Similarly, H'_{fII} is proportional to $\boldsymbol{\varepsilon}_0 \cdot \boldsymbol{K}_{II}$, where[7] $\boldsymbol{K}_{II} = \boldsymbol{k}_{II} + \boldsymbol{k} + \boldsymbol{k}_0$, $\boldsymbol{k}_{II}$ being the charge wave vector in intermediate state II. But there is momentum conservation at the lower (and upper) vertex in diagram (II) of Figure 5.2, so that $0 = \boldsymbol{k} + \boldsymbol{k}_{II}$; thus, $\boldsymbol{\varepsilon}_0 \cdot \boldsymbol{K}_{II} = \boldsymbol{\varepsilon}_0 \cdot \boldsymbol{K}_0 = 0$, and this diagram also yields a zero amplitude.

The amplitude for the process is given completely by the matrix element of $2H''$ in terms of the coupling (5.25); that is, the two-photon vertex (iii) of Figure 5.2 gives the whole contribution. The cross section is determined by the contributions of the polarization components (of $\boldsymbol{\varepsilon}$) associated with the outgoing photon in $(\boldsymbol{\varepsilon} \cdot \boldsymbol{\varepsilon}_0)^2$. Referred to the frame K (see Figure 5.1), there are two possible polarizations; these are the two unit vectors

$$\boldsymbol{\varepsilon}_a = (0, 1, 0), \qquad \boldsymbol{\varepsilon}_b = (0, 0, 1). \tag{5.33}$$

Referred to the same frame K, the polarization components of the incoming photon are [see Equation (5.16)]

$$\boldsymbol{\varepsilon}_0 = (\cos\phi_0 \sin\theta, \cos\phi_0 \cos\theta, \sin\phi_0). \tag{5.34}$$

The squared scattering amplitudes from the two components are

$$\begin{aligned} A_a^2 &= (\boldsymbol{\varepsilon}_a \cdot \boldsymbol{\varepsilon}_0)^2 = \cos^2\phi_0 \cos^2\theta, \\ A_b^2 &= (\boldsymbol{\varepsilon}_b \cdot \boldsymbol{\varepsilon}_0)^2 = \sin^2\phi_0. \end{aligned} \tag{5.35}$$

Thus, summing over final polarizations and averaging over the initial polarization direction (angle ϕ_0), we have

$$\begin{aligned} \langle A_a^2 + A_b^2 \rangle &= \tfrac{1}{2}(1 + \cos^2\theta) \\ &= \tfrac{1}{2} \sum_{\boldsymbol{\varepsilon}_0} \sum_{\boldsymbol{\varepsilon}} (\boldsymbol{\varepsilon} \cdot \boldsymbol{\varepsilon}_0)^2. \end{aligned} \tag{5.36}$$

In the low-energy NR limit $\omega_1 = \omega_2 = \omega_0$, and the momentum of the outgoing photon is $p_f = \hbar\omega_0/c$. The cross section is given by

$$c\, d\sigma = \frac{2\pi}{\hbar} \cdot \frac{1}{2} \sum_{\boldsymbol{\varepsilon}_0} \sum_{\boldsymbol{\varepsilon}} |2H''|^2 \delta(E_0 - E_f), \tag{5.37}$$

in which the last factor corresponds to the density of final states for the outgoing photon:

$$\delta(E_0 - E_f) = \frac{dN_f}{dE_f} = \frac{p_f^2\, d\Omega}{c(2\pi\hbar)^3}. \tag{5.38}$$

Setting $q = ze$ in the coupling term H'', we then have the differential cross section

$$d\sigma/d\Omega = \tfrac{1}{2} z^4 \alpha^2 \Lambda^2 (1 + \cos^2\theta), \tag{5.39}$$

that is, a result identical to the formula (5.19) derived classically.

---

[7] Both photons are present in state II.

That only diagram (iii) of Figure 5.2 contributes in Compton scattering in the NR limit was mentioned earlier in Section 3.5 of Chapter 3. It is interesting that the other two diagrams (i) and (ii), which give nothing in the NR formulation, account for the whole amplitude in the covariant theory for spin-$\frac{1}{2}$, as we shall see in a later section. Diagrams in the covariant theory do not have precisely the same meaning as in the NR theory. They are, after all, only graphical reminders for perturbation theory formulations, and when the formulations are different, the diagrams carry different meanings.

5.3 SCATTERING BY A MAGNETIC MOMENT

As another application of NR QED, we can consider the scattering of a photon by a particle that has no charge but posesses an intrinsic magnetic moment. The neutron is an example of such a particle; the moment can be regarded as arising from its cloud of virtual pions. The cross section for the process is very small and its measurement would be extremely difficult; it is computed here as a further illustration of the basic theory. Only the specific case of spin-$\frac{1}{2}$ will be treated in detail, although it is clear how the problem for general spin would be handled in the same formulation.

In NR QED, the basic interaction between a photon and an intrinsic magnetic moment is expressed by the perturbation Hamiltonian (3.22) of Chapter 3. This interaction corresponds to a single-photon vertex in a diagrammatic representation, the vertex looking like that in Figure 3.3 or 3.4. The Hamiltonian is actually more conveniently expressed in terms of a photon polarization unit vector representing the magnetic field rather than the electric field. If $\boldsymbol{\varepsilon}$ is this quantity, the perturbation Hamiltonian is given by

$$H' = -i(2\pi\hbar\omega)^{1/2}\boldsymbol{\mu}\cdot\boldsymbol{\varepsilon}, \tag{5.40}$$

where $\boldsymbol{\mu}$ is the magnetic moment and $\hbar\omega$ is the photon energy. For spin-$\frac{1}{2}$, the magnetic-moment operator is given by

$$\boldsymbol{\mu} = \mu_0\boldsymbol{\sigma}, \tag{5.41}$$

in which μ_0 is the magnitude of the moment and $\boldsymbol{\sigma}$ is the Pauli spin operator. Calculated by perturbation theory, the process procedes as a result of the double action of the Hamiltonian (5.40). That is, the effective interaction would be given by an expression like the first two terms of Equation (5.26), corresponding to diagrams (i) and (ii) of Figure 5.2.

As in ordinary Compton scattering by a charge, we evaluate the process in a frame in which the particle is initially at rest. Since the interaction involves only the spin, this is the only kind of "motion" to be considered for the particle. The particle does, of course, receive a momentum recoil and a small kinetic energy as a result of the scattering [see Equations (5.3) and (5.4)]. There is momentum conservation at the vertices and, if $\hbar\omega_0 \ll mc^2$, the energy denominators are easily shown to be, to first order in $\hbar\omega_0/mc^2$,

$$\begin{aligned} E_0 - E_I &= \hbar\omega_0(1 - \hbar\omega_0/2mc^2), \\ E_0 - E_{II} &= -\hbar\omega_0[1 - (\hbar\omega_0/2mc^2)(1 - 2\cos\theta)]. \end{aligned} \tag{5.42}$$

In fact, to compute the cross section, we need only the denominators to lowest order (i.e., $\pm\hbar\omega_0$). Making use of the completeness relation

$$\sum_s |m_s\rangle\langle m_s| = I \quad \text{(unit matrix)} \tag{5.43}$$

for the spin eigenfunctions, the effective interaction Hamiltonian is then

$$H'_{f0} = -2\pi i\hbar c^2(\omega\omega_0)^{1/2}\mu_0^2 \times \left[\frac{\langle f|(\sigma\cdot\varepsilon)(\sigma\cdot\varepsilon_0)|0\rangle}{E_0 - E_I} + \frac{\langle f|(\sigma\cdot\varepsilon_0)(\sigma\cdot\varepsilon)|0\rangle}{E_0 - E_{II}}\right]. \tag{5.44}$$

The vectors ε and ε_0 refer to the scattered and incident photons, respectively. Figure 5.1 can again represent the scattering, and it is convenient (as in Section 5.2) to express the components of ε and ε_0 with respect to the frame K. Then[8] ε_0 is given by the form (5.34) and the two components of ε are given by the vectors (5.33). It is convenient to take the z-axis (or z_0-axis) as the polar direction of the spin coordinate. The initial and final states $|0\rangle$ and $\langle f|$ in the matrix elements (5.44) are then the basic spin-$\frac{1}{2}$ eigenstates $|\pm\rangle$ [see Chapter 3, Equation (3.140)] of the operator σ_z.

The lowest-order amplitudes A_I and A_{II} in the matrix element (5.44) do not cancel so that, as mentioned earlier, the zeroth-order energy denominators ($\pm\hbar\omega_0$) can be employed and ω can be set equal to ω_0. The square of H'_{f0} is then given by an expression of the form

$$|H'_{f0}|^2 = 4\pi^2\mu_0^4|A|^2. \tag{5.45}$$

For the production of a specific polarization (say, ε_a) in the scattered photon, the amplitude for the process between particular initial and final spin states would be $\langle f|A_a|0\rangle$, where

$$A_a = (\sigma\cdot\varepsilon_a)(\sigma\cdot\varepsilon_0) - (\sigma\cdot\varepsilon_0)(\sigma\cdot\varepsilon_a), \tag{5.46}$$

with a similar expression for polarization ε_b and amplitude A_b. Employing the identities satisfied by the spin operators that are given in Equations (3.143) and the expressions (5.33) and (5.34), we readily obtain the following non-vanishing squared amplitudes

$$\begin{aligned} |\langle\pm|A_a|\pm\rangle|^2 &= 4\cos^2\phi_0\sin^2\theta, \\ |\langle\pm|A_a|\mp\rangle|^2 &= 4\sin^2\phi_0, \\ |\langle\pm|A_b|\mp\rangle|^2 &= 4\cos^2\phi_0. \end{aligned} \tag{5.47}$$

Note that the polarization ε_a can result from interactions that leave the spin coordinate unchanged and also from spin changes; polarization ε_b involves only the latter. If the scattering moment is unpolarized, we can average over initial spin states, sum over final states, and also sum over final photon polarizations, to get

$$\begin{aligned} |A|^2 &= \tfrac{1}{2}\sum_{0,f}(|\langle f|A_a|0\rangle|^2 + |\langle f|A_b|0\rangle|^2) \\ &= 4(1 + \cos^2\phi_0\sin^2\theta). \end{aligned} \tag{5.48}$$

[8] Remember that these now refer to the photon's magnetic field.

For an unpolarized incident photon beam, we can average over ϕ_0 using $\langle\cos^2\phi_0\rangle = \frac{1}{2}$, to get

$$\left\langle |A|^2 \right\rangle_{\phi_0} = 2(2 + \sin^2\theta). \tag{5.49}$$

The scattering cross seciton is obtained, as in ordinary Compton scattering, from the relations (5.27) and (5.28), the energy-conservation δ-function yielding the density of final states $(dN/dE)_f = (p_0^2/c)d\Omega/(2\pi\hbar)^3$ on the summing over momentum states for the outgoing photon. Employing the results (5.45) and (5.49), we have

$$d\sigma/d\Omega = 2(\mu_0/\hbar)^4 p_0^2(2 + \sin^2\theta). \tag{5.50}$$

If we write

$$\mu_0 = (e\hbar/2mc)\varkappa, \tag{5.51}$$

where $\varkappa$ is the dimensionless anomalous moment,[9] in terms of $\Lambda_m = \hbar/mc$ and the fine-structure constant α,

$$d\sigma/d\Omega = \tfrac{1}{4}\alpha^2\Lambda_m^2\varkappa^4(\hbar\omega_0/mc^2)^2(1 + \tfrac{1}{2}\sin^2\theta). \tag{5.52}$$

The total cross section is

$$\sigma = \tfrac{4}{3}\pi\alpha^2\Lambda_m^2\varkappa^4(\hbar\omega_0/mc^2)^2. \tag{5.53}$$

We see that the cross section is smaller than the ordinary Thomson cross section (5.20) by a factor $\frac{1}{2}(\hbar\omega_0/mc^2)^2$ if $\varkappa = 1$, so it is very small.

The above results (5.52) and (5.53) can also be derived classically,[10] just as for ordinary Compton scattering. In fact, the elementary classical derivation is valid for arbitrary spin s, given the value for the corresponding $\varkappa$. The more general result is obtained from the spin-$\frac{1}{2}$ formulas by multiplying by $\frac{4}{3}s(s+1)$.

5.4 RELATIVISTIC SPIN-0 CASE

The basic theory for the QED of spin-0 charges was outlined in Section 4.2 of Chapter 3, and we have already applied it in the previous chapter. For the evaluation of Compton scattering, it is necessary to include effects of both the one- and two-photon vertices or interactions. The Feynman diagrams for the process are, in fact, the same[11] as is in the non-relativistic case (Figure 5.2). Moreover, it is again the case that only diagram (iii) contributes to the cross section. It is easy to show that diagrams (i) and (ii) do not contribute, from considerations of the form of the interaction associated with the single-photon vertex. For simplicity in the notation, let us use p_μ for the charge four-momentum and k_μ for that of the photon. Then for the single-photon vertex (Figure 3.2), the interaction amplitude is of the

---

[9] Equal to -1.913 for the neutron, for example. In the case of the neutron, the domain of validity of the scattering formula would be $\hbar\omega_0 \ll m_\pi c^2$, where m_π is the pion mass.

[10] See R. J. Gould, *Astrophys. J.* **417**, 12 (1993).

[11] As emphasized before, the diagrams have different meanings in NR and covariant QED.

form [see Equations (3.81) and (3.82) of Chapter 3] $M \propto \varepsilon \cdot (p_a + p_b)$, a four-dimensional scalar product in which ε_μ is the photon polarization four-vector and a and b designate the incoming and outgoing charges at the vertex. Through a gauge transformation (see Section 5.3 of Chapter 3), ε_μ can always be chosen to have only a space component, and

$$\varepsilon_\mu k_\mu = \boldsymbol{\varepsilon} \cdot \boldsymbol{k} = 0. \tag{5.54}$$

Also, since there is energy and momentum conservation at the vertices, the charge four-momentum for the intermediate state of diagram (i) of Figure 5.2 is given by $(p_I)_\mu = (p_0 + k_0)_\mu$. Associated with the lower vertex in this diagram, there is an amplitude of the form

$$M \propto (\varepsilon_0)_\mu (p_0 + p_I)_\mu = (\varepsilon_0)_\mu (2p_0 + k_0)_\mu = 2\boldsymbol{\varepsilon}_0 \cdot \boldsymbol{p}_0. \tag{5.55}$$

But in a frame where the charge is initially at rest, $\boldsymbol{p}_0 = 0$ and $M = 0$. Similarly, for diagram (ii), the lower vertex yields an amplitude

$$M' \propto \varepsilon_\mu (p_0 + p_{II})_\mu = \varepsilon_\mu (2p_0 - k)_\mu = 2\boldsymbol{\varepsilon} \cdot \boldsymbol{p}_0. \tag{5.56}$$

Thus, diagrams (i) and (ii) both have factors in their total amplitudes that are zero.

The whole contribution comes from the two-photon vertex diagram (iii), for which the interaction amplitude is of the form [see Equations (3.78) and (3.80) of Chapter 3]

$$M'' = \varepsilon \cdot \varepsilon_0 = \boldsymbol{\varepsilon} \cdot \boldsymbol{\varepsilon}_0, \tag{5.57}$$

in terms of the incoming charge and outgoing polarizations. Other factors involving the charge, etc., are left out for simplicity; as in the derivations in the previous chapter, the multiplying factor in the general scattering cross section will be fixed by a comparison with the formula (5.19) derived in the classical and non-relativistic limit. Since the invariant amplitude (5.57) is independent of kinematic quantities, the energy dependence of the cross section is determined from the four $1/E$ factors in the invariant normalizations and from the phase-space factor. The cross section is given by an expression of the form [see Chapter 4, Equation (4.126)]

$$c\,d\sigma \propto \int \frac{|\mathcal{M}|^2}{mk_0} \frac{d^3\boldsymbol{p}\, d^3\boldsymbol{k}}{Ek} \delta^{(4)}\left(\sum_{(f)} p_\mu - \sum_{(0)} p_\mu\right), \tag{5.58}$$

with $\mathcal{M} = M''$. The initial and final energies of the charge are, respectively, $m(c^2)$ and E; k_0 and k are, respectively, the initial and final photon energies and are related by formula (5.3). An integration over the momentum space $d^3\boldsymbol{p}$ of the outgoing charge can be performed using the momentum part of the four-dimensional $\delta^{(4)}$ to yield the two-particle phase-space factor in terms of a single integration over $d^3\boldsymbol{k}$:

$$\begin{aligned} \Phi &= \iint (d^3\boldsymbol{p}/E)(d^3\boldsymbol{k}/k)\, \delta(E + k - m - k_0)\, \delta^{(3)}(\boldsymbol{k} + \boldsymbol{p} - \boldsymbol{k}_0) \\ &= \int (d^3\boldsymbol{k}/Ek)\, \delta(E + k - m - k_0), \end{aligned} \tag{5.59}$$

in which

$$E = E(\boldsymbol{p} = \boldsymbol{k}_0 - \boldsymbol{k}) = [(\boldsymbol{k}_0 - \boldsymbol{k})^2 + m^2]^{1/2}. \tag{5.60}$$

With $d^3\boldsymbol{k} = k^2 dk\, d\Omega$, instead of integrating over dk, it is more convenient to use the δ-function argument

$$W = E + k - m - k_0, \tag{5.61}$$

in Equation (5.59) as integration variable. Then

$$\Phi = d\Omega \int (k/E)(dk/dW)\, \delta(W)\, dW = d\Omega\, [(k/E)(dk/dW)]_{W=0}. \tag{5.62}$$

By the relations (5.61) and (5.60),

$$dW/dk = 1 + dE/dk = 1 + (k - k_0 \cos\theta)/E, \tag{5.63}$$

and for $W = 0$, $E = m + k_0 - k$, yielding, very simply,

$$\Phi = k\, d\Omega / [m + k_0(1 - \cos\theta)]. \tag{5.64}$$

With k/k_0 given by the fundamental kinematic relation (5.3), we see that

$$\Phi = (k_0/m)(k/k_0)^2 d\Omega \tag{5.65}$$

and the cross section can be written

$$d\sigma/d\Omega \propto (k/k_0)^2 (\boldsymbol{\varepsilon} \cdot \boldsymbol{\varepsilon}_0)^2. \tag{5.66}$$

Again taking a polarization (initial) average and (final) sum [see Equation (5.36)] and establishing the proportionality constant through a comparison with the low-energy limit (5.19), we have

$$\frac{d\sigma}{d\Omega} = \frac{1}{2} z^4 \alpha^2 \Lambda^2 \frac{1 + \cos^2\theta}{[1 + (\hbar\omega_0/mc^2)(1 - \cos\theta)]^2}. \tag{5.67}$$

The above formula (5.67) is exact in that it is valid at any photon energy; that is, if

$$\nu = \hbar\omega_0/mc^2, \tag{5.68}$$

the formula holds for all ν from 0 to arbitrarily large values. It is the spin-0 analog of the more familiar Klein-Nishina formula, valid for spin-$\frac{1}{2}$, and given in the following section. The spin-0 case is very much simpler, as is clear from the easy derivation of the result. The comparison of the spin-0 and spin-$\frac{1}{2}$ formulas is quite interesting, especially in the limit $\nu \gg 1$, as we shall see.

The total cross section can be evaluated by integrating over θ from 0 to π, more readily performed over the variable $w = 1 + \nu(1 - \cos\theta)$, ranging from 1 to $1 + 2\nu$. We obtain

$$\sigma = \sigma(\nu) = \pi z^4 \alpha^2 \Lambda^2 f(\nu) \tag{5.69}$$

where

$$f(\nu) = \frac{2(\nu + 1)}{\nu^2} \left[\frac{2(\nu + 1)}{2\nu + 1} - \frac{1}{\nu} \ln(2\nu + 1) \right]. \tag{5.70}$$

In the low- and high-energy limits, this function has the forms

$$f(\nu) \to \begin{cases} \frac{8}{3}(1 - 2\nu + \cdots) & (\nu \ll 1), \\ 2/\nu - (1/\nu^2)(2\ln 2\nu - 3) + \cdots & (\nu \gg 1). \end{cases} \tag{5.71}$$

At low energies, the cross section approaches the Thomson-limit formula with a first-order relative correction -2ν due to the onset of relativistic effects. In the high-energy limit, the cross section varies inversely with the photon energy.

5.5 RELATIVISTIC SPIN-$\frac{1}{2}$ PROBLEM: KLEIN-NISHINA FORMULA

5.5.1 Formulation

Although in the covariant QED of spin-$\frac{1}{2}$ charges there is only one kind of vertex, the calculation of the cross section for Compton scattering is more complicated than that for the spin-0 case. In fact, in the original derviation by Klein and by Nishina by the older non-covariant methods, the calculation was quite lengthy and cumbersome. The modern methods provide an enormous simplification, as in other applications, but the detailed derivation of the expression still requires a fair number of algebraic manipulations. The basic steps in the problem will be laid down here, and the reader will be referred to the standard texts for the most lengthy part of the calculation, which is the evaluation of the trace. Moreover, in setting up the problem, we have to introduce and prove a few more elementary theorems involving Dirac matrices that have not been given previously in this book.

The Feynman diagrams for the process would be the two on the left of Figure 5.2; these will now give the whole contribution to the cross section, including effects associated with the photon interaction with the particle magentic moment. The basic single-photon vertex, corresponding to the interaction $\not{e}$ (see Chapter 3), includes both the charge and magnetic-moment coupling. The scattering process is a result of this type of coupling acting twice (involving the incoming and outgoing photons), and the amplitude for the process is given by an expression of the form

$$M = \not{e} P_F(I) \not{e}_0 + \not{e}_0 P_F(II) \not{e}, \tag{5.72}$$

where the P_F's are Feynman propagators, given by [see Chapter 3, Equations (3.106) and (3.107)]

$$P_F(p) = \frac{1}{\not{p} - imc} = \frac{\not{p} + imc}{p^2 + m^2c^2}. \tag{5.73}$$

In the two terms in Equation (5.72), corresponding to diagrams (i) and (ii), respectively, of Figure 5.2, multiplying numerical factors have been omitted; these will again be fixed in a comparison with the result in the NR limit, as was done for the spin-0 problem in the previous section. The notation is also the same, with k_ν and p_ν referring, respectively, to the photon and charge final four-momenta; a subscript zero designates the initial values (and the initial photon polarization). Because there is energy and momentum conservation at the vertices, the components of the four-momenta for the two intermediate states are given by

$$p_I = p_0 + k_0, \qquad p_{II} = p_0 - k. \tag{5.74}$$

Setting $c = 1$, since $p_0^2 = p^2 = -m^2$ and $k_0^2 = k^2 = 0$, the amplitude M is given by

$$M = \not{\varepsilon}\frac{\not{p}_0 + \not{k}_0 + im}{2(p_0 \cdot k_0)}\not{\varepsilon}_0 - \not{\varepsilon}_0\frac{\not{p}_0 - \not{k} + im}{2(p_0 \cdot k)}\not{\varepsilon}. \tag{5.75}$$

This expression can be simplified further. It is ultimately employed[12] between the two spinors $\overline{u}$ and u_0 to form

$$\mathcal{M} = \overline{u}Mu_0; \tag{5.76}$$

the operator $\not{p}_0$ can then be moved to the right of $\not{\varepsilon}_0$ and $\not{\varepsilon}$ with the help of the identity [see Chapter 3, Equation (3.103)] $\not{A}\not{B} + \not{B}\not{A} = 2A \cdot B$. Since u_0 is a free-particle Dirac spinor, it satisfies $\not{p}_0 u_0 = imu_0$, and M becomes

$$M = \frac{\not{\varepsilon}\not{k}_0\not{\varepsilon}_0 + 2\not{\varepsilon}(p_0 \cdot \varepsilon_0)}{2(p_0 \cdot k_0)} + \frac{\not{\varepsilon}_0\not{k}\not{\varepsilon} + 2\not{\varepsilon}_0(p_0 \cdot \varepsilon)}{2(p_0 \cdot k)}. \tag{5.77}$$

In a frame where $\boldsymbol{p}_0 = 0$, $(p_0 \cdot \varepsilon_0)$ and $(p_0 \cdot \varepsilon)$ are both zero because ε_ν and $(\varepsilon_\nu)_0$ have only space components (in a convenient gauge), and M is even more simple:

$$M = -\frac{1}{2m}\left(\frac{\not{\varepsilon}\not{k}_0\not{\varepsilon}_0}{k_0} + \frac{\not{\varepsilon}_0\not{k}\not{\varepsilon}}{k}\right) \tag{5.78}$$

(here k_0 and k are the incoming and outgoing photon energies).

Performing a spin (final) sum and (initial) average for the charge, we find that the quantity to be calculated is[12]

$$X = \frac{1}{2}\,\mathrm{Tr}\left(M'\frac{\not{p} + im}{2im}M\frac{\not{p}_0 + im}{2im}\right), \tag{5.79}$$

where

$$M' = \gamma_0 M^\dagger \gamma_0. \tag{5.80}$$

Now, for the case where M is, as in Equation (5.78), a string of operators of the form

$$M = \not{a}\not{b}\cdots\not{q}, \tag{5.81}$$

the expression (5.80) for M' can be simplified. It is further assumed that the a_μ are of the form $a_\mu = (ia, \boldsymbol{a})$ with a and $\boldsymbol{a}$ real, as is the case when they refer to momentum four-vectors or polarizations. With the help of the fundamental identity $\gamma_\mu\gamma_\nu + \gamma_\nu\gamma_\mu = 2\delta_{\mu\nu}$, we see that

$$\gamma_0\gamma_\mu\gamma_0 = \begin{cases} \gamma_\mu & (\mu = 0) \\ -\gamma_\mu & (\mu = 1, 2, 3). \end{cases} \tag{5.82}$$

---

[12]Many of the identities and expressions referred to here were introduced in Section 4.4.1.

We then obtain, since the γ_μ-matrices are Hermitian,

$$(\not{a})' = -\not{a},$$
$$(\not{a}\not{b})' = \not{b}\not{a}, \tag{5.83}$$
$$(\not{a}\not{b}\cdots\not{q})' = \pm\not{q}\cdots\not{b}\not{a} \qquad (+ \text{ for even, } - \text{ for odd number factors}).$$

Employing these identities, the trace (5.79) is found to be

$$X = \frac{1}{32m^4}\,\mathrm{Tr}\left[\left(\frac{\not{\varepsilon}_0\not{k}_0\not{\varepsilon}}{k_0} + \frac{\not{\varepsilon}\not{k}\not{\varepsilon}_0}{k}\right)(\not{p} + im)\right.$$
$$\left.\times\left(\frac{\not{\varepsilon}\not{k}_0\not{\varepsilon}_0}{k_0} + \frac{\not{\varepsilon}_0\not{k}\not{\varepsilon}}{k}\right)(\not{p}_0 + im)\right]. \tag{5.84}$$

5.5.2 Evaluation of the Cross Section

The calculation of trace (5.84) is straightforward but a little tedious; it is outlined in a number of texts on QED. The result is, however, very simple and is of the form

$$X \propto [k_0/k + k/k_0 - 2 + 4(\boldsymbol{\varepsilon}\cdot\boldsymbol{\varepsilon}_0)^2]. \tag{5.85}$$

The relationship between the invariant trace [which corresponds to the quantity $|\mathcal{M}|^2$ in the spin-0 formula (5.58)], and the cross section is the same as in the spin-0 formulation. That is, the same formulation now yields for the spin-$\frac{1}{2}$ case:

$$\frac{d\sigma}{d\Omega} \propto \left(\frac{k}{k_0}\right)^2 X \propto \left(\frac{k}{k_0}\right)^2\left[\frac{k_0}{k} + \frac{k}{k_0} - 2 + 4(\boldsymbol{\varepsilon}\cdot\boldsymbol{\varepsilon}_0)^2\right], \tag{5.86}$$

with k and k_0 again related by Equation (5.3). The multiplying constant in the result (5.86) can be determined through a comparison with the non-relativistic or low-energy limit for which $k \to k_0$. In that limit, before a summation over final-state polarization $\boldsymbol{\varepsilon}$ is made, the cross section is[13] [see Equations (5.36) and (5.39)]

$$d\sigma/d\Omega \to \alpha^2\Lambda^2(\boldsymbol{\varepsilon}\cdot\boldsymbol{\varepsilon}_0)^2. \tag{5.87}$$

Thus, the general spin-$\frac{1}{2}$ formula (5.86) must be

$$\frac{d\sigma}{d\Omega} = \frac{\alpha^2\Lambda^2}{4}\left(\frac{k}{k_0}\right)^2\left[\frac{k_0}{k} + \frac{k}{k_0} - 2 + 4(\boldsymbol{\varepsilon}\cdot\boldsymbol{\varepsilon}_0)^2\right]. \tag{5.88}$$

If we sum and average over final and initial polarizations, $(\boldsymbol{\varepsilon}\cdot\boldsymbol{\varepsilon}_0)^2$ would be replaced by $\frac{1}{2}(1 + \cos^2\theta)$ [see Equation (5.36)] and all other terms in brackets in Equation (5.88) would be multiplied by a factor[13] of 2. The expression then becomes

$$\left(\frac{d\sigma}{d\Omega}\right)_{\mathrm{KN}} = \frac{\alpha^2\Lambda^2}{2}\left(\frac{k}{k_0}\right)^2\left(\frac{k_0}{k} + \frac{k}{k_0} - \sin^2\theta\right), \tag{5.89}$$

known as the Klein-Nishina formula.

---

[13]The cross sections (5.87) and (5.88) are the values per final-state polarization.

The total cross section can be easily obtained through an integration over $d\Omega$, employing the same notation and integration variable as in the derivation of the spin-0 formula (5.69). The result can be written

$$\sigma = \pi\alpha^2\Lambda^2 f'(\nu), \tag{5.90}$$

where

$$f'(\nu) = \frac{1}{\nu}\left[\left(1 - \frac{2}{\nu} - \frac{2}{\nu^2}\right)\ln(2\nu+1) + \frac{1}{2} + \frac{4}{\nu} - \frac{1}{2(2\nu+1)^2}\right], \tag{5.91}$$

and, again, $\nu = \hbar\omega_0/mc^2$.

5.5.3 Invariant Forms

The cross section (5.89), which is summed and averaged over spins and polarizations, can be expressed in terms of the kinematic invariants introduced in Section 6.3 of Chapter 3. If the reaction $a+b \to c+d$ represents $e_0+\gamma_0 \to e+\gamma$, the invariants s, t, and u would be equal to

$$\begin{aligned} s &= (p_0 + k_0)^2 = (p+k)^2, \\ t &= (p_0 - p)^2 = (k_0 - k)^2, \\ u &= (p_0 - k)^2 = (p - k_0)^2. \end{aligned} \tag{5.92}$$

In the frame with the electron initially at rest,

$$\begin{aligned} s &= -2mk_0 - m^2, \\ t &= 2kk_0(1 - \cos\theta), \\ u &= 2mk - m^2. \end{aligned} \tag{5.93}$$

The three invariants are not independent, satisfying $s+t+u = -2m^2$ [see Chapter 3, Equation (3.214)], and an invariant expression like the trace (5.85) can be written in terms of s and u. In fact, the formula is more conveniently written in terms of the invariants

$$U = (m^2+u)^{-1}, \qquad S = (m^2+s)^{-1}. \tag{5.94}$$

In terms of these quantities, the factor in parentheses in Equation (5.89) resulting from the invariant trace summed over polarizations is given by

$$k_0/k + k/k_0 - \sin^2\theta = 4[(U+S)^2 + U + S - \tfrac{1}{4}(U/S + S/U)]. \tag{5.95}$$

The right-hand side of this relation is an explicit function of the kinematic invariants and the factor k/k_0 in the Klein-Nishina formula can also be expressed in terms of the invariants:

$$k/k_0 = -S/U. \tag{5.96}$$

Further, the angular element $d\Omega$ in the cross section can be written as a differential in dt, making use of the energy-momentum conservation expression $k_0/k = 1 + k_0(1 - \cos\theta)$. In this elementary manipulation k_0 can be regarded as a parameter, while k is a function of θ, so that

$$d\Omega/2\pi = 2U^2 dt. \tag{5.97}$$

The invariant form for the Klein-Nishina cross section is then

$$d\sigma = 8\pi\alpha^2\Lambda^2 S^2[(U+S)^2 + U + S - S/4U - U/4S]\,dt. \tag{5.98}$$

Sometimes it is convenient to express a cross section formula in terms of quantities in the c. m. frame (K_c, say) and sometimes in a frame in which one of the particles is at rest (K_r). The velocity of relative motion between these frames is in the direction of motion of the particles in K_c and K_r. The relationship between the cross section $d\sigma$, the "flux" j, and the transition probability dw is

$$dw = j\,d\sigma, \tag{5.99}$$

and all three quantities are invariants. The appropriate form for the invariant flux is

$$j = \frac{I}{VE_1E_2}, \tag{5.100}$$

where the denominator terms are associated with normalization. The invariant I is

$$I = [(p_1\cdot p_2)^2 - m_1^2 m_2^2]^{1/2}. \tag{5.101}$$

This form yields the correct results

$$I_c = (v_1+v_2)/V \quad \text{and} \quad I_r = v_{1\text{or}2}/V \tag{5.102}$$

for the two special cases. It is then the appropriate general invariant form for I.

5.5.4 Limiting Forms and Comparisons

Let us first consider the low-energy limit where $\nu \ll 1$. Specifically, in terms of ν and the scattering angle θ, the Klein-Nishina formula (5.89) can be written

$$\left(\frac{d\sigma}{d\Omega}\right)_{\text{KN}} = \frac{\alpha^2\Lambda^2}{2\big(1+\nu(1-\cos\theta)\big)^2} \times\left[1+\nu(1-\cos\theta)+\frac{1}{1+\nu(1-\cos\theta)} - \sin^2\theta\right]. \tag{5.103}$$

For a particle of the same mass and charge but spin 0, the formula is given by Equation (5.67), and the ratio is given by

$$\frac{(d\sigma/d\Omega)_{\text{KN}}}{(d\sigma/d\Omega)_{\text{spin-0}}} = 1+\nu^2\frac{(1-\cos\theta)^2}{1+\cos^2\theta} + O(\nu^3), \tag{5.104}$$

that is, the difference is second-order in ν. The KN formula in the low-energy limit can also be compared with the Thomson formula (5.19):

$$(d\sigma/d\Omega)_{\text{KN}} = \tfrac{1}{2}\alpha^2\Lambda^2\big(1+\cos^2\theta\big)\big(1-2\nu(1-\cos\theta)+O(\nu^2)\big); \tag{5.105}$$

here the lowest-order (in ν) relative correction is the same as in the spin-0 case. In other words, the effects fo spin are of relative order ν^2 in correcting the low-ν non-relativistic formula. Perhaps this is to be expected on the basis of the results

(5.52) and (5.53), which show that the cross section for scattering off a magnetic moment is of order ν^2 compared with the Thomson formula. Concerning the spin contribution to the KN formula, it is natural to ask if the magnetic-moment result (5.52) is obtained if the spin-0 formula (5.67) is subtracted from the KN formula when both are expanded to order ν^2. The answer is no; the reason is that the spin character of the charge affects the type of relativistic QED that applies, so that the spin effects cannot be totally removed from the relativistic corrections.

The total cross section for spin-$\frac{1}{2}$ can also be compared with the Thomson-limit formula (5.20) σ_T with a lowest-order correction. From an expansion of the expression (5.91) for small ν, we find

$$\sigma_{\text{KN}} = \sigma_T\big(1 - 2\nu + O(\nu^2)\big). \tag{5.106}$$

the relative correction to order ν being the same as in the spin-0 case [see Equation (5.71)].

The high-energy limit $\nu \gg 1$ is particularly interesting, especially when the spin-$\frac{1}{2}$ and spin-0 formulas are compared. Except at small scattering angles, the two formulas are given by

$$\left.\begin{aligned} \left(\frac{d\sigma}{d\Omega}\right)_{\text{KN}} &\to \frac{\alpha^2\Lambda^2}{2}\frac{1}{\nu(1-\cos\theta)} \\ \left(\frac{d\sigma}{d\Omega}\right)_{\text{spin-0}} &\to \frac{\alpha^2\Lambda^2}{2}\frac{1+\cos^2\theta}{\nu^2(1-\cos\theta)^2} \end{aligned}\right\} \quad \big(\theta \gg (2/\nu)^{1/2}\big). \tag{5.107}$$

We see that at small angles the first formula goes as $\nu^{-1}\theta^{-2}$ and the second as $\nu^{-2}\theta^{-4}$ with $\theta_{\min} = (2/\nu)^{1/2}$ being an effective minimum scattering angle. The total cross sections then have the forms

$$\begin{aligned} \sigma_{\text{KN}} &\propto \nu^{-1}\ln(1/\theta_{\min}) \sim \nu^{-1}\ln\nu, \\ \sigma_{\text{spin-0}} &\propto \nu^{-2}(1/\theta_{\min}^2) \sim \nu^{-1}. \end{aligned} \tag{5.108}$$

More precisely, the two formulas are [see Equations (5.71) and (5.91)]

$$\begin{aligned} \sigma_{\text{KN}} &\to \pi\alpha^2\Lambda^2\nu^{-1}(\ln 2\nu + \tfrac{1}{2}) \qquad \nu \gg 1, \\ \sigma_{\text{spin-0}} &\to \pi\alpha^2\Lambda^2(2/\nu) \qquad \nu \gg 1. \end{aligned} \tag{5.109}$$

The spin-$\frac{1}{2}$ cross section is larger by a logarithmic factor. Away from the small-angle limit, however, the spin-$\frac{1}{2}$ angular cross section $d\sigma/d\Omega$ is larger by a factor $\sim \nu$. Thus, at very high energies, the effects of spin are of great importance. Apparently, the principal contribution to the scattering comes from photon interactions with the intrinsic moment, in contrast to the low-energy limit where these interactions are negligible.

That magnetic-moment interactions are of great importance in the high-energy limit can perhaps be seen through considerations of the classical Hamiltonian for a relativistic charge (with also a magnetic moment) interacting with a radiation field.

With the gauge $\Phi = 0$, this Hamiltonian has the form

$$H_{\text{cl}} = [(\boldsymbol{p}c - q\boldsymbol{A})^2 + m^2c^4]^{1/2} - \boldsymbol{\mu} \cdot \text{curl } \boldsymbol{A}. \tag{5.110}$$

When $|\boldsymbol{p}| \gg mc$ and $|\boldsymbol{A}|$ is a small perturbation,

$$H_{\text{cl}} \to pc - q(\boldsymbol{A} \cdot \boldsymbol{p} + \boldsymbol{p} \cdot \boldsymbol{A})/2p + q^2A^2/2pc - \boldsymbol{\mu} \cdot \text{curl } \boldsymbol{A}. \tag{5.111}$$

If we designate the second term by H_q' and the last term by H_m', for $\mu \sim q\hbar/mc$ and $|\text{curl } \boldsymbol{A}| \sim k|\boldsymbol{A}|$,

$$H_m'/H_q' \sim \hbar\omega/mc^2, \tag{5.112}$$

and we can see how the magnetic interaction can dominate at large photon energies. For non-relativistic energies, the situation is the other way around, as is indicated in Equation (3.46) of Chapter 3.

5.6 RELATIONSHIP TO PAIR ANNIHILATION AND PRODUCTION

Two processes are closely related to Compton scattering. One is direct pair annihilation:

$$e^+ + e^- \to \gamma + \gamma'. \tag{5.113}$$

The other is pair production in photon-photon collisions:

$$\gamma + \gamma' \to e^+ + e^-, \tag{5.114}$$

which is just the reverse of pair annihilation. The Feynman diagrams for these processes are the same as in Compton scattering if the diagrams are turned on their sides, and correspond to a different "channel" for one basic process. Because of this, it is not necessary to do the difficult calculation of the trace for each of the processes (5.113) and (5.114), since it can be taken over from the Compton scattering calculation. Moreover, since the processes are the reverse of one another, the cross section for one can be gotten from the other by detailed balance. However, there is an important feature of the two processes that does not appear in the treatment of Compton scattering. This is the effect of "Coulomb focusing," which comes in because we have either an incoming or outgoing electron and positron. That is, at low energies in the rest frame of the $e^+ - e^-$ system the Born approximation is not valid and the charged-particle wave function cannot be represented by a plane wave. The low-energy correction to the Born approximation can be made easily, however, and it yields a different form for the cross sections. For pair annihilation there is also the competing process of positronium (Ps) formation:

$$e^+ + e^- \to \text{Ps} + \gamma, \tag{5.115}$$

in which the pair form a bound "atomic" state and there is a single outgoing photon. A high-energy positron passing through matter usually slows down by radiationless

Coulomb scattering off electrons before it annihilates, so that the low-energy limit for the processes is of great importance.[14]

A good treatment of the pair annihilation process (5.113) and pair production (5.114) may be found in Ref. 4, Ch. 3 (BLP) and we shall refer to that work for the detailed derivation of the general cross sections and their relationship to that for Compton scattering. The cross sections are most easily represented in terms of the velocity of the electron and positron in the c. m. frame: $\beta = v/c = \beta^+ = \beta^-$. The result for the annihilation cross section is (see BLP)

$$\sigma_{\text{an}} = \frac{1}{4}\alpha^2 \Lambda^2 \frac{1-\beta^2}{\beta^2} X(\beta), \tag{5.116}$$

where

$$X(\beta) = (3-\beta^2)\ln\frac{1+\beta}{1-\beta} - 2\beta(2-\beta^2). \tag{5.117}$$

In the low-energy limit ($\beta \ll 1$), $X \to 2\beta$, and

$$\sigma_{\text{an}} \to \tfrac{1}{2}\alpha^2 \Lambda^2 \beta^{-1}. \tag{5.118}$$

The low-energy result (5.118) can be corrected for Coulomb focusing by multiplying by the squared amplitude of a Coulomb wave function at zero separation of the e^+ and e^-. This procedure works because the interaction that causes the process is over a characteristic length scale of order Λ, while the Coulomb wave function amplitude varies over a much larger length scale $\sim a_0$ (Bohr radius). We have

$$|u(0)|^2 = w/(1-e^{-w}), \tag{5.119}$$

where

$$w = \pi\alpha/\beta. \tag{5.120}$$

At low energies, $|u(0)|^2 \to w$, and the annihilation cross section approaches

$$\sigma_{\text{an}} \to \tfrac{1}{2}\pi^2\alpha^3\Lambda^2\beta^{-2} \qquad \text{(very low energy).} \tag{5.121}$$

For the pair-production process (5.114) the cross section can be obtained from the reverse process (5.113) very simply, and the result is expressed as a function of the c. m. velocity β. If E_e is the electron or positron energy (including rest energy) and E_γ is the energy of one of the photons, these energies are equal. For annihilation, the "flux"is 2β and the cross section can be written in terms of the "interaction" $|H'|^2$ and the outgoing phase space factor is

$$\phi_{\text{an}} = \int \delta(2E_e - 2E_\gamma) d^3\mathbf{k}_\gamma/\sigma, \tag{5.122}$$

where σ is a symmetry number, equal to 2, to correct for counting the same identical pair of photons twice. In pair production the flux is 2 (times c) and the phase space factor involves an integration over the δ-function with a factor $d^3\mathbf{k}_e$. For the forward

---

[14]In a plasma the positrons thermalize before annihilating. The competing processes of direct annihilation and Ps formation are treated in, for example, R. J. Gould, *Astrophys. J.* **344**, 232 (1989).

and reverse processes the interaction is the same and we find, very simply,

$$\sigma_{\text{prod}}/\sigma_{\text{an}} = 2\beta^2. \tag{5.123}$$

That is, the cross section for pair production in photon-photon collisions will have the same function $X(\beta)$ given in (5.117). The low-energy Born result is

$$\sigma_{\text{prod}}(\text{Born}) \to \pi\alpha^2\Lambda^2\beta, \tag{5.124}$$

indicating a cross section going to zero at threshold. However, if we correct for Coulomb focusing[15] by multiplying by the factor (5.119), we obtain a cross section that is *finite* at threshold:

$$\sigma_{\text{prod}} \to \pi^2\alpha^3\Lambda^2 \qquad \text{(threshold)}. \tag{5.125}$$

5.7 DOUBLE COMPTON SCATTERING

The process

$$\gamma_0 + q \to q + \gamma + \gamma' \tag{5.126}$$

in which a photon scatters off a charge with the result that there are two photons in the final state is commonly called *double Compton scattering*. The name may be a little misleading, and a better designation might be "radiative Compton scattering," since it is just Compton scattering with an extra outgoing photon. In fact, in the evaluation of radiative corrections[16] to (ordinary) Compton scattering, it is necessary to consider the process of double Compton (DC) scattering in order to handle the infrared divergence. For arbitrary energy $\hbar\omega_0$ of the incident photon, the cross section for the DC process can be computed by the methods of modern QED, although the result cannot be expressed in terms of a simple analytic formula. Simple expressions can be obtained when, say, γ' is "soft," since DC is then closely related to ordinary Compton scattering. However, even for general $\hbar\omega'$, when $\hbar\omega_0 \ll mc^2$, a simple formula can be derived for the cross section.

5.7.1 Non-Relativistic Case. Soft-Photon Limit

When the "additional" outgoing photon is soft, the cross section for DC can be written very simply as

$$d\sigma_D(\text{soft}) = d\sigma_C\, dw', \tag{5.127}$$

in terms of the Compton cross section $d\sigma_C$ and a factor (dw') that is the probability that the ordinary Compton process is accompanied by this additional photon. The

---

[15] See R. J. Gould, *Astrophys. J.* **337**, 950 (1989).

[16] A radiative correction would be associated with the effect of a (virtual) photon being "emitted" and "reabsorbed" during the scattering process; that is, the virtual photon is present in an intermediate state. A brief discussion of the general topic of radiative corrections to electromagnetic processes has been given in Section 3.6.2.

process is separable in this sense, because the soft-photon emission does not affect the dynamics of the rest of the process, which is ordinary Compton scattering. Section 5 of Chapter 3 has been devoted to the general subject of soft-photon emission and expressions have been derived therein for the probability dw in various limiting cases.

Let us first consider the case where the incident photon has energy $\hbar\omega_0 \ll mc^2$, and evaluate the cross section in the frame where the charge is initially at rest. Then the kinetic energy of the charge after scattering is negligible, although its momentum is comparable to that of the photons involved in the process. In fact, if $\boldsymbol{\varepsilon}_0$, $\boldsymbol{\varepsilon}$, and $\boldsymbol{\varepsilon}'$ and $\boldsymbol{k}_0$, $\boldsymbol{k}$, and $\boldsymbol{k}'$ are, respectively, the polarizations and wave vectors of the photons involved in the process (5.126), the wave vector of the scattered charge is

$$\boldsymbol{p} = \boldsymbol{k}_0 - \boldsymbol{k} - \boldsymbol{k}'. \tag{5.128}$$

In the soft-photon limit where $k' = |\boldsymbol{k}'|$ is very small, the wave vectors k_0 and k are almost equal in magnitude [see Equation (5.3)] and

$$p^2 = 2k_0^2(1 - \cos\theta), \tag{5.129}$$

in terms of the scattering angle of the outgoing photon γ; that is, $\cos\theta = \boldsymbol{k}_0 \cdot \boldsymbol{k}/k_0^2$. We can then write down an expression for the cross section (5.127) from the Thomson-limit formula [see Equations (5.36) and (5.39)]

$$d\sigma_T = \alpha^2 \Lambda^2 z^4 (\boldsymbol{\varepsilon} \cdot \boldsymbol{\varepsilon}_0)^2 d\Omega. \tag{5.130}$$

and the expression (3.136) of Chapter 3 for the probability dw':

$$dw' = (\alpha/4\pi^2) z^2 (d\omega'/\omega')(\boldsymbol{\varepsilon}' \cdot \boldsymbol{\beta})^2 d\Omega', \tag{5.131}$$

where $\boldsymbol{\beta}c$ is the velocity of the scattered charge. The DC cross section is then given by

$$d\sigma_D = (\alpha^3/4\pi^2)\Lambda^2 z^6 (d\omega'/\omega')(\boldsymbol{\varepsilon} \cdot \boldsymbol{\varepsilon}_0)^2 (\boldsymbol{\varepsilon}' \cdot \boldsymbol{\beta})^2 d\Omega\, d\Omega', \tag{5.132}$$

and can be considered to be a very general expresion in that it gives the complete description of the process for arbitrary polarizations $\boldsymbol{\varepsilon}_0$, $\boldsymbol{\varepsilon}$, and $\boldsymbol{\varepsilon}'$ and gives the angular dependence on the directions of the outgoing photons. The velocity $\boldsymbol{\beta}$ is determined by the photon wave vectors; from the relation (5.128),

$$\boldsymbol{\beta} = \Lambda(\boldsymbol{k}_0 - \boldsymbol{k} - \boldsymbol{k}') \tag{5.133}$$

equal to, for small $\boldsymbol{k}'$, $\Lambda(\boldsymbol{k}_0 - \boldsymbol{k})$. If we are not interested in the direction or polarization of the outgoing soft photon, instead of the expression (5.131) for dw', we could employ the (polarization-summed and angular-integrated) expression (3.134):

$$dw' = (2\alpha/3\pi) z^2 \beta^2 d\omega'/\omega'. \tag{5.134}$$

Here, by equation (5.129), β^2 would be given by

$$\beta^2 = 2(\hbar\omega_0/mc^2)^2(1-\cos\theta). \tag{5.135}$$

We can also employ the cross section (5.39), which is the formula averaged over $\boldsymbol{\varepsilon}_0$ and summed over $\boldsymbol{\varepsilon}$:

$$\begin{aligned} d\sigma_{\mathrm{T}} &= \alpha^2 \Lambda^2 z^4 \left(\tfrac{1}{2}\sum_{\boldsymbol{\varepsilon}_0}\sum_{\boldsymbol{\varepsilon}}\right)(\boldsymbol{\varepsilon}\cdot\boldsymbol{\varepsilon}_0)^2 d\Omega \\ &= \tfrac{1}{2}\alpha^2\Lambda^2 z^4(1+\cos^2\theta)\,d\Omega. \end{aligned} \tag{5.136}$$

The DC cross section is then

$$\begin{aligned} d\sigma_D &= \frac{2\alpha^3}{3\pi}\Lambda^2 z^6 \left(\frac{\hbar\omega_0}{mc^2}\right)^2 \frac{d\omega'}{\omega'} \int (1+\cos^2\theta)(1-\cos\theta)\,d\Omega \\ &= (32/9)\alpha^3 z^6 \nu^2 d\omega'/\omega', \end{aligned} \tag{5.137}$$

expressed very simply in terms of $\nu = \hbar\omega_0/mc^2$ [see Equation (5.68)].

A more accurate expression for $d\sigma_D$ in the soft-photon limit may be obtained[17] by treating the kinematics of the process more accurately, using Equation (5.4) for the kinetic energy of the charge. Including a relative correction of order ν, instead of Equation (5.135), the more precise result is

$$\beta^2 = 2\nu^2(1-x)[1-\nu(1-x)+O(\nu^2)], \tag{5.138}$$

where

$$x = \cos\theta. \tag{5.139}$$

The more accurate expression for the Compton cross section, including a relativistic correction, would be the formula (5.105), valid for spin-0 or spin-$\frac{1}{2}$ charges (and presumably for other spin values):

$$d\sigma_C = \pi\alpha^2 z^4(1+x^2)[1-2\nu(1-x)+O(\nu^2)]\,dx. \tag{5.140}$$

The same formula (5.134) for the photon-emission probability can be used, since the relative correction to it is of order β^2 [see Equation (3.169) of Chapter 3], which is of order ν^2. Inserting the result (5.138) into Equation (5.134) and again computing $d\sigma_D$ by the formula (5.127), we now have

$$\begin{aligned} d\sigma_D &= \tfrac{4}{3}\alpha^3\Lambda^2 z^6\nu^2(d\omega'/\omega')\int_{-1}^{1}(1+x^2)(1-x)[1-3\nu(1-x)+O(\nu^2)]\,dx \\ &= (32/9)\alpha^3\Lambda^2 z^6\nu^2[1-(21/5)\nu]\,d\omega'/\omega'. \end{aligned} \tag{5.141}$$

The correction to the formula (5.137) is significant[15] because of the large value (21/5) of the coefficient of ν in brackets.

---

[17] See also R. J. Gould, *Astrophys. J.*, **230**, 967(1979).

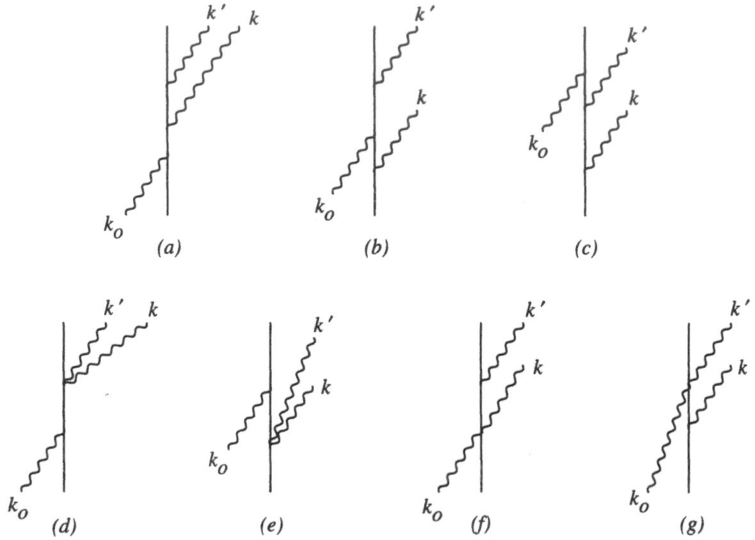

Figure 5.3 Perturbation diagrams for double Compton scattering. The bottom four are associated with the two-photon vertex.

5.7.2 Non-Relativistic Case. Arbitrary Energy

When one of the outgoing photons in DC scattering is not soft, the process must be treated in a more general manner. For example, since there are two photons in the final state, exchange effects must be included; that is, the photons are no longer essentially distinguishable by virtue of their vastly different energies. In a quantum-mechanical formulation of the problem, the process (5.126) is described as a result of the actions of the electromagnetic perturbations or photon-charge couplings, as in the case of ordinary Compton scattering. For the latter, there are three types of perturbation amplitudes corresponding to the diagrams in Figure 5.2. In DC scattering, we can consider the additional outgoing photon to "come from" or be coupled to any one of the charged-particle lines of these diagrams, and $3+3+2 = 8$ diagrams could be indicated plus as many exchange diagrams. However, three of these diagrams would be duplications of others in the same group, as is easily seen. Also, more general perturbations must be considered involving the diagrams with the two-photon vertex.

To see the perturbation diagrams necessary for inclusion in the total amplitude, it is best to indicate the charge states as a straight vertical line with the incoming photon (state) indicated on the left and the two outgoing photons going to the right and up from the charge line. We see then that there are seven diagrams and these are shown in Figure 5.3; there would be, of course, seven additional exchange diagrams. However, of the seven diagrams associated with the direct amplitude, only two contribute, along with their exchange amplitudes. The contributing diagrams are (e) and (f) of Figure 5.3, being the only ones where the "first" vertex is of the two-photon type; all of the diagrams with a single-photon vertex acting first do not contribute. The reason for this has to do with the form of the interaction Hamiltonians for the two vertices; these are given in Chapter 3 as Equations (3.20) and (3.21) and quoted in Equations (5.24) and (5.25) of this chapter. However, let us exhibit them again, this time with H' in the momentum representation, with both expressed in terms of the wave vectors ($k = \omega/c$) of the photons and with $\boldsymbol{p}$ designating the wave vector of the charge. For unit normalization volume, the one- and two-photon vertices then have interaction Hamiltonians of the form [see Equations (5.24) and (5.25)]

$$H' = \left(\frac{2\pi\hbar^3}{c}\right)^{1/2} \frac{q}{m} \frac{\boldsymbol{\varepsilon}\cdot\boldsymbol{p}}{k^{1/2}}, \tag{5.142}$$

$$H'' = \frac{(2)\pi\hbar\, q^2}{c} \frac{}{m} \frac{\boldsymbol{\varepsilon}_1\cdot\boldsymbol{\varepsilon}_2}{(k_1 k_2)^{1/2}}. \tag{5.143}$$

Note that in H' the charge wave vector $\boldsymbol{p}$ is that *entering* the vertex; $\boldsymbol{\varepsilon}$ is the polarization unit vector associated with the photon involved in the vertex and k ($= |\boldsymbol{k}| = \omega/c$) is the magnitude of its wave vector. When we compute the cross section in a frame where the charge is initially at rest, we then see why the lowest or first vertex cannot be of the single-photon type since the corresponding $\boldsymbol{p}$ would be zero. This is why only diagrams (e) and (f) can contribute, for which H'', acts first.[18] Concerning the expression (5.143) for H'', it should be noted that the result has been quoted with, when applicable, an additional factor of 2. The reason for inserting the factor has been discussed in Section 5.2; very simply, only when one photon is incoming and the other is outgoing is the factor 2 to be included.

Let us designate the intermediate state of diagram (f) by II and that for diagram (e) by I. The effective perturbation Hamiltonian for the total direct amplitude would then have a matrix element between initial (0) and final (f) state:

$$H_{f0} = \sum_I \frac{H'_{fI} H''_{I0}}{E_0 - E_I} + \sum_{II} \frac{H'_{fII} H''_{II0}}{E_0 - E_{II}}. \tag{5.144}$$

The sums over intermediate states in Equation (5.8) are trivial in the momentum representation, being sums over momentum or wave-vector states ($\boldsymbol{p}_I$ and $\boldsymbol{p}_{II}$) of the charge. But since there is momentum conservation at the vertices, only a single state contributes for each diagram, corresponding to

$$\boldsymbol{p}_I = \boldsymbol{p} + \boldsymbol{k}', \qquad \boldsymbol{p}_{II} = \boldsymbol{p} - \boldsymbol{k}_0, \tag{5.145}$$

where $\boldsymbol{p}$ is the wave vector for the outgoing charge [see Equation (5.128)]. In the energy denominators in H_{f0}, the contribution from the kinetic energy of the charge is negligible, so that only the energies in photons need be included. For state I in diagram (f), only the photon γ is present while, for state II in (e), all three photons are present. By energy conservation for the whole process, since the outgoing kinetic energy of the charge is negligible,

$$k_0 = k + k'. \tag{5.146}$$

[18] In ordinary Compton scattering (see Section 5.2.1), this is the same reason why only the two-photon vertex is effective.

The energy denominators are then given by

$$E_0 - E_I = \hbar c(k_0 - k) = \hbar c k',$$
$$E_0 - E_{II} = \hbar c(k_0 - k_0 - k - k') = -\hbar c k_0. \qquad (5.147)$$

In the matrix elements in the amplitude (5.144), we can make use of the identities associated with the transverse nature of the three photons:

$$\boldsymbol{\varepsilon}_0 \cdot \boldsymbol{k}_0 = 0, \qquad \boldsymbol{\varepsilon} \cdot \boldsymbol{k} = 0, \qquad \boldsymbol{\varepsilon}' \cdot \boldsymbol{k}' = 0. \qquad (5.148)$$

The matrix element H'_{fI} would be proportional to $\boldsymbol{\varepsilon} \cdot \boldsymbol{p}_I$, but because of the relations (5.145) and (5.148), this would be equal to $\boldsymbol{\varepsilon} \cdot \boldsymbol{p}$. Similarly, H'_{fII} would be proportional to $\boldsymbol{\varepsilon}_0 \cdot \boldsymbol{p}_{II} = \boldsymbol{\varepsilon}_0 \cdot \boldsymbol{p}$. In the H'' matrix elements H''_{I0} would include the additional factor of 2 in parentheses in the form (5.143), and $\boldsymbol{\varepsilon}_1 \cdot \boldsymbol{\varepsilon}_2/(k_1k_2)^{1/2}$ would equal $\boldsymbol{\varepsilon}_0 \cdot \boldsymbol{\varepsilon}/(k_0k)^{1/2}$, while H''_{II0} would not include the 2, and $\boldsymbol{\varepsilon}_1 \cdot \boldsymbol{\varepsilon}_2/(k_1k_2)^{1/2}$ would equal $\boldsymbol{\varepsilon} \cdot \boldsymbol{\varepsilon}'/(kk')^{1/2}$. To have included the extra factor of 2 in H''_{II0} would amount to adding the exchange amplitude, which, in fact, is done eventually to give a total amplitude; we leave it out of the direct amplitude, however. The total direct amplitude, including both terms in Equation (5.144), corresponding to diagrams (f) and (e), is then given by

$$H_{f0}(\text{direct}) = C(k_0kk')^{-1/2}(M - \tfrac{1}{2}M'), \qquad (5.149)$$

where

$$C = (2\pi)^{3/2}(\hbar/c^5)^{1/2}(q^3/m^2), \qquad (5.150)$$

and

$$M = (\boldsymbol{\varepsilon}' \cdot \boldsymbol{p})(\boldsymbol{\varepsilon} \cdot \boldsymbol{\varepsilon}_0)/k', \qquad (5.151)$$
$$M' = (\boldsymbol{\varepsilon}_0 \cdot \boldsymbol{p})(\boldsymbol{\varepsilon} \cdot \boldsymbol{\varepsilon}')/k_0. \qquad (5.152)$$

The exchange amplitude is obtained by replacing

$$\boldsymbol{\varepsilon}' \leftrightarrow \boldsymbol{\varepsilon}, \qquad \boldsymbol{k}' \leftrightarrow \boldsymbol{k}. \qquad (5.153)$$

The result is an exchange amplitude

$$M_e = (\boldsymbol{\varepsilon} \cdot \boldsymbol{p})(\boldsymbol{\varepsilon}' \cdot \boldsymbol{\varepsilon}_0)/k, \qquad (5.154)$$

while M'_e is identical to M'. Including the exchange contribution, the total amplitude is

$$\mathcal{M} = M + M_e - M', \qquad (5.155)$$

and its square is given by

$$\mathcal{M}^2 = M^2 + M_e^2 + M'^2 + 2MM_e - 2M'(M + M_e). \qquad (5.156)$$

It is this quantity that determines the cross section. In the various terms M, M_e, and M', the final wave vector ($\boldsymbol{p}$) of the charge is given by Equation (5.128) in terms of $\boldsymbol{k}_0$, $\boldsymbol{k}$, and $\boldsymbol{k}'$. The terms in $\mathcal{M}^2$ can then be written in terms of the photon polarizations and wave vectors.

For the evaluations of the various scalar products in M, M_e, M', it is necessary to adopt a convenient set of reference axes to represent the polarizations and wave vectors. The most convenient convention is to take $\boldsymbol{k}_0$ along the x_0-axis of a frame

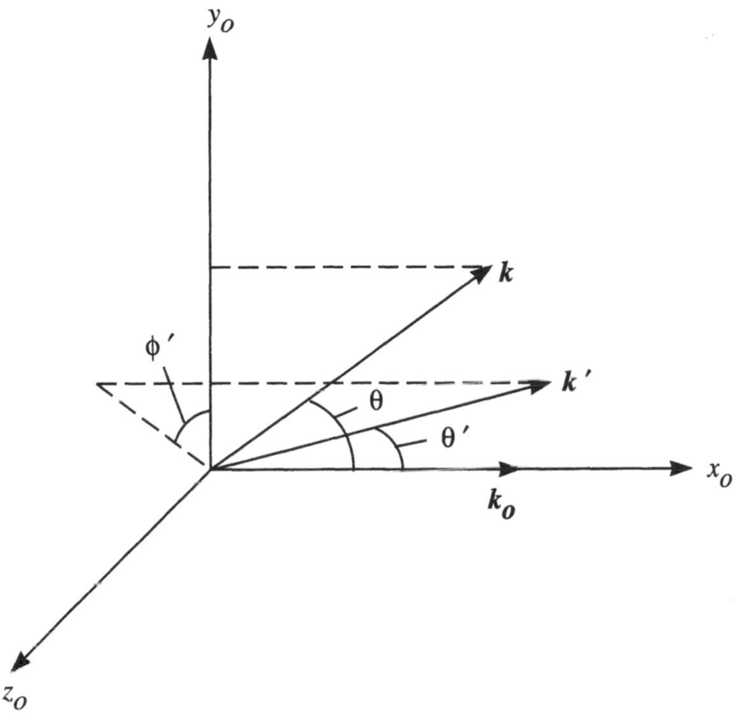

Figure 5.4 Angles involved in the description of outgoing photons in double Compton scattering.

(K_0), with the two independent linear polarization states of $\boldsymbol{\varepsilon}_0$ along the y_0- and z_0-axes. The orientation of these axes is such that $\boldsymbol{k}$ is in the x_0–y_0 plane at an angle θ with respect to $\boldsymbol{k}_0$. The wave vector $\boldsymbol{k}'$ is at angle θ' with respect to $\boldsymbol{k}_0$, but at an azimuthal angle ϕ' with respect to the y_0-axis when projected on the y_0–z_0 plane. Figure 5.4 indicates the angles involved. With respect to the frame K_0, the various photon linear polarization states, designated by a and b, are[19]

$$\begin{aligned} &\boldsymbol{\varepsilon}_{0a} = (0, 1, 0) \qquad \boldsymbol{\varepsilon}_{0b} = (0, 0, 1) \\ &\boldsymbol{\varepsilon}_a = (-\sin\theta, \cos\theta, 0) \qquad \boldsymbol{\varepsilon}_b = (0, 0, 1) \\ &\boldsymbol{\varepsilon}'_a = (-\sin\theta', \cos\theta' \cos\phi', \cos\theta' \sin\phi') \\ &\boldsymbol{\varepsilon}'_b = (0, -\sin\phi', \cos\phi') \end{aligned} \tag{5.157}$$

[19]These relations can be obtained through considerations of transformations of vectors represented in different frames. In particular, the polarization components $\boldsymbol{\varepsilon}'$ are represented in the frame K_0 by transformations employing rotation matrices associated with (see Figure 5.4) (i) a rotation through an angle θ' around the z_0-axis and (ii) a rotation through an angle ϕ' around the x_0-axis, the matrix for the combined operation being the product $M(\phi')M(\theta') = M(\phi', \theta')$. In a frame ($K'$, say) with the x'-axis along $\boldsymbol{k}'$, the polarization components are $\boldsymbol{\varepsilon}'_a = (0, 1, 0)$ and $\boldsymbol{\varepsilon}'_b = (0, 0, 1)$, and the values quoted in Equation (5.157) are obtained from these through multiplication by the matrix $M(\phi', \theta')$.

The photon wave vectors are given by

$$\begin{aligned} \boldsymbol{k}_0 &= k_0(1, 0, 0), \\ \boldsymbol{k} &= k(\cos\theta, \sin\theta, 0), \\ \boldsymbol{k}' &= k'(\cos\theta', \sin\theta'\cos\phi', \sin\theta'\sin\phi'). \end{aligned} \tag{5.158}$$

The amplitude $\mathcal{M}$ and its square (5.156) can then be written, for each polarization state, as a function of k_0, k, and k' and the angles θ, θ', and ϕ'.

The cross section for the process is computed from [see Equations (5.27) and (5.28)]

$$c\,d\sigma = (2\pi/\hbar)\sum_f |H_{f0}|^2\delta(E_0 - E_f), \tag{5.159}$$

in which the matrix element H_{f0} would correspond to the total amplitude, including the exchange contribution; that is, it would be given by the form (5.149) with $M - \frac{1}{2}M'$ replaced by $\mathcal{M}$. If the incident photon is unpolarized, we can average over its components ($\boldsymbol{\varepsilon}_0$) and sum over the polarizations of the outgoing photons. We also integrate over the outgoing photon momentum states, but, if we integrate over all of the space of $\boldsymbol{k}$ and $\boldsymbol{k}'$, a factor of $\frac{1}{2}$ must be inserted to avoid counting the same scattering twice. The total cross section is then

$$d\sigma = \frac{2\pi}{\hbar c}\left(\frac{1}{2}\sum_{\varepsilon_0}\sum_{\varepsilon}\sum_{\varepsilon'}\right)\iint \frac{1}{2}\frac{d^3\boldsymbol{k}\,d^3\boldsymbol{k}'}{(2\pi)^6}C^2\frac{\mathcal{M}^2}{k_0 k k'}\frac{\delta(k_0 - k - k')}{\hbar c}. \tag{5.160}$$

The outgoing-photon wave-vector phase space volume element can be written

$$d^3\boldsymbol{k}\,d^3\boldsymbol{k}' = (2\pi)^2 k^2 dk\,k'^2 dk' dx\,dx' d\phi'/2\pi, \tag{5.161}$$

in which $x = \cos\theta$ and $x' = \cos\theta'$. Integration over dk with the δ-function results in simply setting $k' = k_0 - k'$, yielding a cross section differential in dk'.

The polarization sums and angular integrations are trivial but a little tedious. Although the cross section before summation and integration is complicated and contains many terms, the expression differential in dk' is very simple. The result is

$$\frac{d\sigma}{dw} = \frac{8}{9}\alpha^3\Lambda^2 z^6\left(\frac{\hbar\omega_0}{mc^2}\right)^2 F(w), \tag{5.162}$$

in terms of z $(= q/e)$, the dimensionless

$$w = k'/k_0 = \omega'/\omega_0, \tag{5.163}$$

and the function

$$F(w) = w(1-w)\left[\frac{1+(1-w)^2}{w^2} + \frac{1+w^2}{(1-w)^2} + w^2 + (1-w)^2\right]. \tag{5.164}$$

There is an interesting feature of this calculation: there are no contributions from the interference or cross terms in the squared total amplitude (5.156). The first and second terms in brackets in Equation (5.164) come from M^2 and M_e^2, respectively, and the last two terms come from M'^2. Although the interference terms are present in the angular cross section, there is a cancellation on summation over polarizations and integration over angles for the outgoing photons. Perhaps this feature of the result is connected with the restriction to the limit $\hbar\omega_0 \ll mc^2$, which is also the domain of validity of the classical limit for normal Compton scattering. Although the cross section for the latter can be derived classically in this limit, the double Compton formula seems to require a quantum-mechanical derivation.

Another characteristic of the formula (5.25) should be noted. The function $F(w)$ is symmetric around $w = \frac{1}{2}$, as it should be, since a specification of the energy of one outgoing photon determines that of the other, their total being fixed. It is clear from its form that $F(w) = F(1 - w)$, and that the direct and exchange amplitudes contribute in a symmetric way. Note also that the cross section exhibits an infrared divergence both at $w = 0$ and $w = 1$, the latter limit corresponding to the case where the "other" photon is infinitesimally soft. Actually, it is probably better to agree to count photons of energy only up to $w = \frac{1}{2}$ and multiply the cross section (5.162) by a factor of 2. The expression then agrees with the formula (5.137) already derived for the soft-photon limit. Moreover, it is interesting that the amplitude M' [diagram (e) in Figure 5.3] does not contribute in this limit.

5.7.3 Extreme Relativistic Limit

The cross section for double Compton scattering also simplifies in the limit $\nu \gg 1$ for which the scattered electron is highly relativistic. The probability that ordinary Compton scattering is accompanied by an additional soft photon can then be computed from the limit of the relation [see Chapter 2, Equation (2.149) or Chapter 3, Equation (3.168)]

$$dw' = \frac{\alpha}{\pi} z^2 \left(\frac{1}{\beta} \ln \frac{1+\beta}{1-\beta} - 2 \right) \frac{d\omega'}{\omega'}. \tag{5.165}$$

For $\beta = (1 - 1/\gamma^2)^{1/2} \to 1$ ($\gamma \gg 1$), this expression becomes [see Chapter 3, Equation (3.190)]

$$dw' = (2/\pi)\alpha z^2 (\ln 2\gamma - 1)\, d\omega'/\omega'. \tag{5.166}$$

If the DC cross section is written, as is commonly done,

$$d\sigma_D = d\sigma_C\, dw' = (\alpha/\pi)\, \delta_D\, d\sigma_C, \tag{5.167}$$

the dimensionless δ_D is simply

$$\delta_D = 2z^2 (\ln 2\gamma - 1)\, d\omega'/\omega'. \tag{5.168}$$

Actually, a more accurate formula could be obtained by employing the exact expression (5.165) for dw' and, for spin-$\frac{1}{2}$ particles (electrons: $z = 1$), the exact KN cross section (5.89) for $d\sigma_C$. The β of the electron could be computed from Equation (5.4), giving $\beta(\theta)$, and an integration over $d\Omega = 2\pi \sin\theta\, d\theta$ performed with the KN formula. This would be a complicated integration, and we shall, instead, derive a simpler formula valid to logarithmic accuracy [that is, to relative accuracy $\sim (\ln \nu)^{-1}$].

In the Compton cross section in Equation (5.167), we simply take the total σ_{KN} given in Equation (5.109), and, in the expression (5.168) for δ_D, the logarithmic factor is replaced by $\ln \nu$. The result is then

$$d\sigma_D \approx 2\alpha^3 \Lambda^2 \nu^{-1} \ln^2 \nu \, d\omega'/\omega'; \tag{5.169}$$

note that terms of order unity compared with $\ln \nu$ have been neglected. Integrating over $d\omega'$, we have a total cross section

$$\sigma_D \approx 2\alpha^3 \Lambda^2 \nu^{-1} \ln^2 \nu \, \ln(\nu'/\nu'_{\mathrm{m}}), \tag{5.170}$$

where ν'_{m} is some minimum soft-photon energy. The result (5.170) corresponds to

$$\delta_D = 2 \ln \nu \, \ln(\nu'/\nu'_{\mathrm{m}}). \tag{5.171}$$

Because the product of the two logarithms can become large, we see that $\sigma_D/\sigma_C = (\alpha/\pi)\delta_D$ can become greater than unity. That is, even though DC is a higher-order process, its cross section can become greater than that for ordinary Compton scattering. Of course, δ_D could be made arbitrarily large by simply taking ν'_{m} sufficiently small. Concerning the applicability of the result at a *maximum* ν', because of the very weak logarithmic dependence of δ_D on ν', we can expect that the formula is still (logarithmically) accurate even for ν' of the order ν. In fact, $\nu' \sim \nu$ will be the characteristic maximum energy of the second photon since the sum of the energies of the two outgoing photons will be $\approx \nu$ (minus the energy of the scattered electron). Suppose we arbitrarily take $\nu'_{\mathrm{min}} \sim 1$; then $\delta_D \approx 2 \ln^2 \nu$, and

$$\sigma_D/\sigma_C = (\alpha/\pi)\delta_D > 1, \tag{5.172}$$

for

$$\nu > e^{(\pi/2\alpha)^{1/2}} \approx 2 \times 10^6. \tag{5.173}$$

Above this energy, the cross section for double Compton scattering is larger than that for ordinary Compton scattering. This does not necessarily mean that DC is of prime importance in, for example, some general Compton process. For example, it is also necessary to evaluate the radiative corrections to ordinary Compton scattering.

The radiative correction is of the same order in α as the DC cross section; the corrected cross section can be written

$$(d\sigma_C)_{\mathrm{corr}} = d\sigma_C[1 + (\alpha/\pi)\delta_r], \tag{5.174}$$

in terms of a dimensionless relative correction δ_r. The correction defined here—a coefficient of α—is the lowest-order radiative correction and, in various limits, it has been evaluated by Brown and Feynman[20] and others. Characteristic of all radiative correction effects, δ_r possesses an infrared divergence; the divergence phenomenon is, of course, well-understood in quantum electrodynamics (see Section 6.2 of Chapter 3). In fact, we can even call upon our complete understanding of the phenomenon to relate δ_r to δ_D by contending that δ_r *must be*, in the high-energy limit, of the form

$$\delta_r = -2 \ln \nu \, \ln(\nu''_{\mathrm{char}}/\nu''_{\mathrm{m}}). \tag{5.175}$$

[20] L. M. Brown and R. P. Feynman, *Phys. Rev.*, **85**, 231 (1952).

In this expression, $\nu''_{\rm m}$ is a "low-energy cutoff" associated with the infrared divergence and can be identified with the $\nu'_{\rm m}$ in Equation (5.170). The value of $\nu''_{\rm char}$ is some characteristic energy in the problem and is of order ν. Now, if we ask for the probability of single Compton scattering with, in the latter, a ν' up to some $\nu'_{\rm max}$, the resulting probability will be proportional to

$$d\sigma_C + d\sigma_D = d\sigma_C[1 + (\alpha/\pi)(\delta_r + \delta_D)]. \tag{5.176}$$

Then in $\delta_r + \delta_D$, both $\nu'_{\rm m}$ and $\nu''_{\rm m}$ can go to zero, and the divergence cancels with

$$\delta_r + \delta_D \sim (\pm) \ln \nu, \tag{5.177}$$

which is appreciably less than α/π up to an extremely high energy. In this manner, we can see how the effects of radiative corrections and infinitesimally soft additional soft proton production can cancel. As a result, even though there are an infinite number of these soft photons, their presence does not necessarily greatly modify the effects of the lower-order process. For a more specific discussion of some of these points, the reader is referred to the reference in Footnote 15.

Bibliographical Notes

Derivations of the Thomson formula in a purely classical treatment are given in the well-known texts by Jackson and Panofsky and Phillips (References 4 and 5 at the end of Chapter 1).

One of the earliest quantum-mechanical derivations of the Compton cross section in the non-relativistic limit was given in the famous review by Fermi (Reference 2 of Chapter 3). It was noted therein, for example, that the two-photon interaction Hamiltonian gives the whole contribution to the cross section, while the double action of the single-photon interaction does not contribute. Fermi's derivation employed standard field-theory techniques.

A relativistic covariant treatment of Compton scattering by spin-0 charges is given in the lecture notes by Feynman (Reference 7 of Chapter 3). The spin-$\frac{1}{2}$ problem is also treated there in detail, including the evaluation of the trace in the problem.

Double Compton scattering is discussed at some length in the book by Jauch and Rohrloch (Reference 3 of Chapter 3).

Chapter Six

Bremsstrahlung

Bremsstrahlung is *the* basic radiative or photon-producing mechanism. The process is especially important when it takes place in the Coulomb field of an ion (z) and where the scattered particle is an electron; this is electron-ion (e–z) bremsstrahlung for which the emission is "dipole" in nature at non-relativistic energies. At these energies, e–e bremsstrahlung is quadrupole and the corresponding cross section is much smaller. At relativistic energies, the e–z and e–e cross sections are comparable and, in fact, are closely related in the extreme relativistic limit. All of these cases and energy domains will be considered in this chapter. We begin with a brief treatment of the process at low energies, which is the domain of the classical limit. Then we treat the problem in the Born limit but still at non-relativistic energies. The general domain between the classical and Born limit presents a complex quantum-mechanical problem, but there are various ways of expressing the cross section in a simple approximate form in this intermediate region. At relativistic energies, the Born approximation is completely valid but only the e–z cross section can be expressed in terms of an analytic (Bethe-Heitler) formula. The $e^{\pm}$–$e^{\pm}$ and e^{+}–e^{-} cross sections simplify only at ultra-relativistic energies and cannot be expressed in closed form for $E \sim mc^2$.

6.1 CLASSICAL LIMIT

The general problem of classical bremsstrahlung in a (fixed) Coulomb field is quite complicated mathematically and will not be treated in complete detail here. The classical domain does have important applications[1] corresponding to the case where the electron energies are small compared to 1 Ry.

6.1.1 Soft-Photon Limit

In this limit, the classical problem simplifies with the help of a convenient trick.[2] We evaluate the cross section by considering two domains of impact parameter (b) in which we make two different approximations in computing the contributions to the total cross section. The domains are

$$\begin{aligned} 0 < b < b_0, \\ b_0 < b < \infty \end{aligned} \tag{6.1}$$

[1]An example would be the ionized interstellar gas where $kT \approx 0.5$ eV and $kT/\text{Ry} \approx 0.04$.

[2]The treatment here will follow that in the author's paper: *Am. J. Phys.* **38**, 189 (1970).

and the differential cross section for producing a photon of frequency within $d\omega$ is

$$d\sigma = d\sigma_{<} + d\sigma_{>}, \tag{6.2}$$

with contributions coming from the domains $b < b_0$ and $b > b_0$ ("close" and "distant" encounters). In the former domain, we make the approximation that $\omega\tau_{\text{coll}} \ll 1$, where τ_{coll} is the characteristic collision or scattering time, roughly equal to b/v in terms of the incident velocity of the scattered charge. In fact, this criterion yields an effective maximum impact parameter

$$b_{\text{max}} \sim v/\omega; \tag{6.3}$$

actually, we integrate over b to ∞ and, as we shall see, the contribution falls off rapidly for $b > b_{\text{max}}$. For the domain of distant encounters, the assumption $\omega \ll 1/\tau_{\text{coll}}$ is not made, but in the expression for $d\sigma_{>}$ we approximate the particle orbit by a straight line; this is the approximation of small-angle-scattering (see Section 1.1 of Chapter 4). The dividing line b_0 between the close- and distant-encounter domains is such that its precise value need not be specified, since in adding $d\sigma_{<}$ and $d\sigma_{>}$ the terms involving b_0 cancel. Basically, this happens because the domains of applicability of the approximation overlap. Cancellations of this type are common in treatments of various physical processes. In quantum-mechanical formulations in the Born approximation, instead of an integration over impact parameters, usually the integration variable is the momentum transfer $\hbar q$. In that case, sometimes an intermediate-value q_0 is employed that does not appear in a final formula.

For small b, the contribution to the soft-photon bremsstrahlung cross section is

$$d\sigma_{<} = 2\pi \int_0^{b_0} b\,db\,dw_\omega(b), \tag{6.4}$$

where $dw_\omega(b)$ is the soft-photon emission probability. If the charge of the scattered particle is $q = ze$, this probability is given by[3]

$$dw_\omega = (2\alpha/3\pi)z^2(\Delta\mathbf{v}/c)^2 d\omega/\omega, \tag{6.5}$$

and can be expressed in terms of the scattering angle θ_{sc}, since

$$(\Delta\mathbf{v})^2 = 2v^2(1 - \cos\theta_{\text{sc}}). \tag{6.6}$$

If the charge (mass m) is scattered by a heavy charge Ze, the impact parameter b is given explicitly in terms of θ_{sc} [see Chapter 4, Equation (4.21)]:

$$b(\theta_{\text{sc}}) = (zZe^2/mv^2)\cot(\theta_{\text{sc}}/2). \tag{6.7}$$

The integral (6.4) can then be transformed in terms of θ_{sc} as the integration variable runs from the small value $\theta_{\text{sc}}(b_0) = 2zZe^2/mv^2b_0 \ll 1$ to the maximum value of π. We then find

$$d\sigma_{<} = (16/3)\alpha^3\Lambda^2(z^4Z^2/\beta^2)(d\omega/\omega)\ln(mv^2b_0/zZe^2), \tag{6.8}$$

where $\beta = v/c$ and $\Lambda = \hbar/mc$.

[3]See Equation (2.143) of Chapter 2. This is the probability integrated over directions of emission and summed over polarizations; it is differential only in the photon frequency.

For large b, the contribution to the bremsstrahlung cross section is computed from

$$d\sigma_{>} = 2\pi \int_{b_0}^{\infty} b\, db\, dw'_\omega(b), \tag{6.9}$$

where $dw'_\omega(b)$ is a more general expression than Equation (6.5) and is given by [see Chapter 2, Equation (2.77)]

$$dw'_\omega = (8\pi z^2 e^2/3\hbar c^3)|\boldsymbol{a}_\omega|^2(d\omega/\omega). \tag{6.10}$$

Here $\boldsymbol{a}_\omega$ is the Fourier transform of the acceleration, and if the charge motion is in the x–y plane, its square is, in turn, given by the Fourier transforms of the force components:

$$|\boldsymbol{a}_\omega|^2 = (|F_{x\omega}|^2 + |F_{y\omega}|^2)/m^2. \tag{6.11}$$

The general expression (6.10) for dw'_ω could have been used for the small-b domain as well and, in fact, the formula (6.10) is valid for arbitrary ω; that is, it holds even away from the soft-photon limit. This more general expression, which does *not* assume $\omega t \ll 1$ in the Fourier transforms, provides a natural convergence in the upper limit of the integral (6.9).

The simplifying assumption that *is* made in the distant-encounter domain is to approximate the particle path by a straight line in evaluating the Fourier transforms of the force components. We set $x(t) \approx vt$, $y(t) \approx b$, and obtain[4]

$$\begin{aligned} F_{x\omega} &= \frac{zZe^2}{2\pi} \int_{-\infty}^{\infty} \frac{vt}{(v^2t^2 + b^2)^{3/2}} e^{i\omega t}\, dt, \\ F_{y\omega} &= \frac{zZe^2}{2\pi} \int_{-\infty}^{\infty} \frac{b}{(v^2t^2 + b^2)^{3/2}} e^{i\omega t}\, dt. \end{aligned} \tag{6.12}$$

These expressions must be squared and added to obtain the (squared) Fourier transform in Equation (6.11), to be substituted into the probability (6.10) and then to be integrated over b to compute the cross section (6.9). Rather than b, a more convenient integration variable is the dimensionless

$$\rho \equiv \omega b/v \tag{6.13}$$

with $\rho_0 = \omega b_0/v \to 0$ in the soft-photon limit. The Fourier transforms (6.12) can be expressed in terms of ρ and are actually integral representations of modified Bessel functions. However, for the evaluation of the integral (6.9), the properties of the functions required can be derived from the expressions (6.12) themselves; that is, it is not necessary to employ any theorems on Bessel functions. The mathematics is thus elementary and is given in the reference in Footnote 2; the result is

$$d\sigma_{>} = (16/3)\alpha^3 \Lambda^2 (z^4 Z^2/\beta^2)(d\omega/\omega) \ln(2v/\Gamma\omega b_0), \tag{6.14}$$

where $\Gamma = e^\gamma = 1.781\ldots$ (γ =Euler's constant).

When the results (6.8) and (6.14) are combined, the total cross section is then independent of b_0:

$$d\sigma = d\sigma_{<} + d\sigma_{>} = (16/3)\alpha^3 \Lambda^2 (z^4 Z^2/\beta^2)(d\omega/\omega) \ln(2mv^3/\Gamma z Z e^2 \omega). \tag{6.15}$$

[4]We have already seen integrals of this type in Chapter 2 [Equations (2.165) and (2.166)] in treating the Weizsäcker-Williams method.

This is basically a classical formula; if we were to multiply by $\hbar\omega$ to get an "energy emission" cross section, the result would be independent of $\hbar$. The argument of the logarithm is large. Later in this chapter, we shall derive the corresponding formula for $d\sigma$ for soft-photon bremsstrahlung in the Born approximation. That formula is the same except for the argument of a logarithm, which is, instead, $2mv^2/\hbar\omega$. We shall also give the result in the general case encompassing both the classical and Born limits and reducing to each's correct formula.

6.1.2 General Case: Definition of the Gaunt Factor

As mentioned earlier, the general classical problem of bremsstrahlung in a Coulomb field is mathematically somewhat complex. The final result for the cross section for arbitrary ω cannot be expressed in terms of elementary functions, but involves a Hankel function and its derivative. Away from the soft-photon limit, the characteristic photon energy produced in the scattering of an electron of velocity v is the maximum value

$$\hbar\omega_{\text{char}} \sim \tfrac{1}{2}mv^2. \tag{6.16}$$

Although, in a purely classical treatment of the problem, photon frequencies much larger that ω_{char} are predicted, imposition of the finite value of $\hbar$ leads us to ignore these higher frequencies as unphysical. Of course, the classical calculation neglects effects of the emission of radiation on the electron orbit in the Coulomb field of the ion; classical radiation reaction is always small for non-relativistic particles. However, the classical calculation is valid for $mv^2 \ll Z^2\text{Ry}$ (see Section 1.4 of Chapter 4) and the result can be applied if we impose, in the end, the condition

$$\omega < mv^2/2\hbar. \tag{6.17}$$

There is, in the problem, a characteristic classical frequency given by $\omega_{\text{cl}} \sim v/r_v$, where $Ze^2/r_v \sim mv^2$; that is,

$$\omega_{\text{cl}} \sim mv^3/Ze^2. \tag{6.18}$$

We then see from the results (6.16) and (6.18) that

$$\omega_{\text{char}}/\omega_{\text{cl}} \sim Ze^2/\hbar v = Z\alpha/\beta \gg 1. \tag{6.19}$$

In other words, the characteristic photon energy (6.16) corresponds to the high-frequency limit in the classical problem. Employment of this limit in simplifying the classical problem from the beginning would seem a useful approach, but the problem is still really quite difficult. Therefore, the completely general formulation will be outlined here and the reader is referred to the treatment by Landau and Lifshitz[5] for further details.

For the general classical bremsstrahlung problem, the cross section is again computed from the photon-emission probability formula (6.10), with $d\sigma$ computed by an integration over all impact parameters:

$$d\sigma = 2\pi \int_0^\infty b\,db\,dw'_\omega(b). \tag{6.20}$$

[5] L. D. Landau and E. M. Lifshitz, *The Classical Theory of Fields*, 4th ed., New York: Pergamon Press, 1975. See also their volume on *Mechanics*, 2nd ed. (Pergamon 1969).

However, instead of evaluating the Fourier transform

$$\boldsymbol{a}_\omega = \frac{1}{2\pi}\int_{-\infty}^{\infty} \boldsymbol{a}(t)\, e^{i\omega t}\, dt, \tag{6.21}$$

in terms of the exact acceleration $\boldsymbol{a}(t) = \ddot{\boldsymbol{r}}$, it is more convenient to integrate by parts twice[6] and obtain the result

$$\boldsymbol{a}_\omega = -\omega^2 \boldsymbol{r}_\omega. \tag{6.22}$$

The transform $\boldsymbol{r}_\omega$ can be most easily evaluated by expressing $\boldsymbol{r}(t)$ in parametric form[5]

$$\begin{aligned} x &= a(\epsilon - \cosh\xi), \\ y &= a(\epsilon^2 - 1)^{1/2}\sinh\xi, \\ r &= a(\epsilon\cosh\xi - 1), \\ t &= t_0(\epsilon\sinh\xi - \xi), \end{aligned} \tag{6.23}$$

where

$$a = Ze^2/mv^2, \quad t_0 = a/v, \quad \epsilon^2 = 1 + (b/a)^2. \tag{6.24}$$

The (hyperbolic) orbit description (6.23) corresponds to motion in an attractive Coulomb potential $U = -Ze^2/r$; that is, that of an electron in the field of a positive ion. When the Coulomb field is repulsive, the description is different and the bremsstrahlung cross section is very different[5] (much smaller in magnitude). The velocity v in the expressions (6.24) is that of the electron incident on the heavy fixed ion; a, t_0 and ϵ are constants for the classical orbit and are fixed by v and b. Note that $1/t_0 \sim \omega_{\rm cl}$, given by the expression (6.18). The variable ξ provides the parametric description and varies from $-\infty$ to $+\infty$. Note further that the coordinate axes are chosen such that the orbit is in the x–y-plane and is symmetric about the x-axis, having a closest-approach distance

$$r_{\rm min} = a(\epsilon - 1) \tag{6.25}$$

on the x-axis.

The Fourier transforms of the position coordinates are expressed in terms of integrals over the variable ξ by employing the parametric relations (6.23):

$$x_\omega = \frac{at_0}{2\pi}\int_{-\infty}^{\infty} (\epsilon - \cosh\xi)\, e^{i\nu(\epsilon\sinh\xi - \xi)}(\epsilon\cosh\xi - 1)\, d\xi, \tag{6.26}$$

$$y_\omega = \frac{at_0}{2\pi}\int_{-\infty}^{\infty} (\epsilon^2 - 1)^{1/2}\sinh\xi\, e^{i\nu(\epsilon\sinh\xi - \xi)}(\epsilon\cosh\xi - 1)\, d\xi, \tag{6.27}$$

where

$$\nu = \omega t_0 \tag{6.28}$$

is a dimensionless frequency parameter. The tranforms (6.26) and (6.27) can be simplified further by integrating by parts, neglecting the oscillatory integrated terms;

[6]The integral (6.21) can be taken between the limits $-T$ to T with $T \to \infty$. The integrated terms are highly oscillatory, since $\dot{\boldsymbol{r}}(\pm T) = \text{const}$, having the factors $\exp(\pm i\omega T)$; these terms can be neglected.

x_ω is integrated twice and y_ω once to give

$$x_\omega = -\frac{ia}{2\pi\omega}\int_{-\infty}^{\infty} \sinh\xi \, e^{i\nu(\epsilon \sinh\xi - \xi)} d\xi, \tag{6.29}$$

$$y_\omega = \frac{ia(\epsilon^2 - 1)^{1/2}}{2\pi\epsilon\omega}\int_{-\infty}^{\infty} e^{i\nu(\epsilon \sinh\xi - \xi)} d\xi. \tag{6.30}$$

The integral (6.30) happens to be a Hankel function (H) and that in Equation (6.29) is the derivative (H') of a Hankel function (see Footnote 5). The bremsstrahlung cross section is then computed from the integral (6.20) in terms of x_ω and y_ω through the relations (6.10) and (6.22); the result has the form

$$d\sigma \propto (d\omega/\omega)\omega^4 \int_0^\infty b\, db(|x_\omega|^2 + |y_\omega|^2) = \omega^3 d\omega\, a^2 \int_1^\infty \epsilon\, d\epsilon(|x_\omega|^2 + |y_\omega|^2). \tag{6.31}$$

The last integral can be evaluated and expressed in terms of a product HH' of the Hankel functions with the help of identities involving these functions (see Landau and Lifshitz, Footnote 5, for details). The result is an exact (in the classical limit) explicit expression in terms of ω and v and $i\nu$ as an argument of the Hankel functions.

Rather than going through the mathematics described above, let us concentrate on the more physical aspects of a very important limiting form for the classical bremsstrahlung cross section. This is the frequency domain away from the soft-photon limit[7] for which ω has the characteristic value (6.16) of the order $mv^2/\hbar$. From the definition (6.28) this corresponds to the limit

$$\nu \sim Ze^2/\hbar v = Z\alpha/\beta \gg 1, \tag{6.32}$$

in the region of classical scattering. We then see that the integrals (6.29) and (6.30)—the functions H' and H—have highly oscillatory integrands except for ϵ near 1 and ξ near 0. Therefore, for $\nu \gg 1$, the main contribution to the integral (6.31) comes from the lower limit around $\epsilon = 1$; this corresponds to the limit $b \to 0$ in the impact parameter and a closest-approach distance $r \to 0$. In other words, the main contribution to the photon emission for $\omega \sim mv^2/\hbar$ comes from the zero-b orbits that pass very near the nucleus Z. Asymptotic forms for the Hankel functions can be taken, since only the region around $\xi = 0$ contributes to the integral representations [essentially the integrals (6.29) and (6.30)]. The cross section in this limit is given by

$$d\sigma_0 = \frac{16\pi}{3\sqrt{3}} Z^2\alpha^3\Lambda^2 \frac{1}{\beta^2}\frac{d\omega}{\omega}, \tag{6.33}$$

where, because of its simplicity and importance, a subscript 0 has been added to the formula. In fact, the form (6.33) is so convenient that it is common to express the cross section outside of the range of validity of $d\sigma_0$ in terms of this formula and a correction factor:

$$d\sigma = d\sigma_0 g. \tag{6.34}$$

[7]The general formula for $d\sigma$ in terms of the Hankel functions reduces to the result (6.15) when the small-ν asymptotic forms of the functions are taken (see Footnote 5).

The correction factor g is called a "Gaunt factor," and the procedure is useful because it turns out that g is a very slowly varying function of v and ω and not appreciably different from unity even well outside the classical domain.[8]

Often the quantity of interest is the "luminosity" or photon-energy integrated cross section defined by

$$\Omega = \int \hbar\omega (d\sigma/d\omega)\, d\omega, \tag{6.35}$$

where the limits on the integral would be 0 and $mv^2/2\hbar$. For the cross section (6.33), the result is extremely simple:

$$\Omega_0 = (8\pi/3\sqrt{3}) Z^2 \alpha^3 \Lambda^2 mc^2; \tag{6.36}$$

that is, Ω_θ is a constant. Outside the range of validity of the classical formula (6.33) it would be necessary to include the correction factor $g(v, \omega)$ in the integral (6.35) and the energy-integrated cross section could be written

$$\Omega = \Omega_0 \bar{g}, \tag{6.37}$$

where $\bar{g}$ is a frequency-averaged Gaunt factor and would be a (weak) function of v.

Finally, concerning classical Coulomb bremsstrahlung, it is well to emphasize again that the cross section away from the soft-photon limit is very different when the Coulomb field is repulsive rather than attractive. The cross section for a repulsive potential is much smaller, and there is a simple physical reason for this. For scattering in a repulsive Coulomb potential there is a finite "distance of closest approach" even for zero impact parameter. On the other hand, as we have seen, for the bremsstrahlung problem with an attractive Coulomb potential, the main contribution to the emission process comes from the very small b-values for which the classical orbit passes near the scattering center at which the acceleration is very large. The repulsive-potential problem is outlined briefly in Landau and Lifshitz (see Footnote 5) and a result is given for the cross section, showing it to be *exponentially small* in the limit (6.32). Later we shall see that in the Born approximation limit, there is no difference between the bremsstrahlung cross sections for the attractive and repulsive cases. The Born limit is fundamentally different from the classical limit in that the scattering potential is treated as a (basically weak) perturbation.

6.2 NON-RELATIVISTIC BORN LIMIT

6.2.1 General Formulation for Single-Particle Bremsstrahlung

We consider the problem in which a particle of charge ze is scattered by an "external potential" with a photon being produced in the process. The overall process is evaluated by pertubation theory as a result of the combined action of two perturbations: (i) the scattering potential V and (ii) the photon-coupling interaction H'. The basic

[8]We shall have more to say about Gaunt factors in treating bremsstrahlung quantum mechanically. The terminology and use of "g" factors to fix up formulas is widespread. One speaks, for example, of photoionization and recombination Gaunt factors, and there the g is again a factor to correct a particularly simple (but not quite correct) formula.

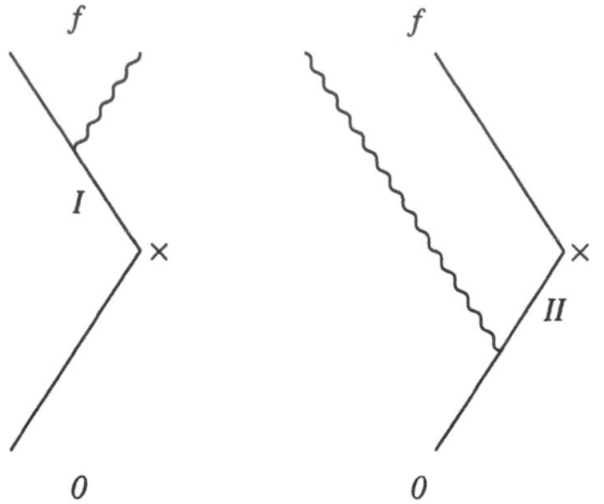

Figure 6.1 Perturbation diagrams for bremsstrahlung in the scattering by a fixed external potential.

assumption is that each perturbation "acts once" (Born approximation), and the total amplitude for the process is computed by second-order perturbation theory. This amplitude is the effective perturbation matrix element

$$H_{f0} = \sum_{I} H'_{fI} \frac{1}{E_0 - E_I} V_{I0} + \sum_{II} V_{fII} \frac{1}{E_0 - E_{II}} H'_{II0}, \tag{6.38}$$

represented by the diagrams in Figure 6.1. These diagrams are the same as those in Figure 3.7 and, in fact, our treatment of bremsstrahlung here will parallel the derivation of results for soft-photon emission in Section 5 of Chapter 3. As we have seen for the general soft-photon problem, the formula for the photon emission probability holds even *away from* the soft-photon limit when the Born approximation holds. However, instead of simply taking over that result and applying it to the bremsstrahlung problem, we derive it again, expressing the cross section formula in the form

$$d\sigma_b = \int d\sigma_{\text{sc}}\, dw_\omega. \tag{6.39}$$

Here $d\sigma_{\text{sc}}$ is the (radiationless) scattering cross section and dw_ω is a probability for photon emission within frequency $d\omega$. In the formula of the form (6.39), the convenient integration variable will be seen to be the momentum transfer to the scattering center.

The problem is most easily formulated in the momentum representation. Since in the sums over the intermediate states I and II in the matrix element (6.38) only single terms contribute,[9] and there being momentum conservation at the vertices, the amplitude H_{f0} can be written more simply as

$$H_{f0} = H' \frac{1}{\Delta E(I)} V + V \frac{1}{\Delta E(II)} H'. \tag{6.40}$$

[9] In the position-space representation, the intermediate-state sums would just yield momentum-conservation δ-functions [see Section 5.1 of Chapter 3, especially Equation (3.123)].

In this expression, it is inherently implied that momentum (but not energy) conservation is to be applied at each vertex and $\Delta E(I)$ and $\Delta E(II)$, the analogs of the Feynman propagators in the covariant theory, are the energy denominators for the respective diagrams.

Let us employ a notation whereby $\boldsymbol{k}_0$ and $\boldsymbol{k}_f$ are, respectively, the initial and final wave vectors of the charge; $\boldsymbol{k}_I$ and $\boldsymbol{k}_{II}$ are the values in the intermediate states, while $\boldsymbol{k}$ (no subscript) is the wave vector for the outgoing photon (polarization $\boldsymbol{\varepsilon}$). The momentum or wave-vector conservation relations are

$$\begin{aligned} \boldsymbol{k}_I &= \boldsymbol{k}_f + \boldsymbol{k}, \\ \boldsymbol{k}_{II} &= \boldsymbol{k}_0 - \boldsymbol{k}. \end{aligned} \tag{6.41}$$

The momentum transferred to the scattering center is $\hbar\boldsymbol{q}$ so that

$$\boldsymbol{q} = \boldsymbol{k}_0 - \boldsymbol{k}_I = \boldsymbol{k}_{II} - \boldsymbol{k}_f = \boldsymbol{k}_0 - \boldsymbol{k}_f - \boldsymbol{k}. \tag{6.42}$$

Although the scattering center or external potential absorbs momentum, it does not carry off energy so that

$$\hbar\omega = \hbar ck = (\hbar^2/2m)(k_0^2 - k_f^2). \tag{6.43}$$

The intermediate-state energies are

$$\begin{aligned} E_I &= (\hbar^2/2m)k_I^2, \\ E_{II} &= (\hbar^2/2m)k_{II}^2 + \hbar ck. \end{aligned} \tag{6.44}$$

Application of the momentum-conservation relations (6.41) and the (total) energy-conservation relation (6.43) yields for the energy denominators:

$$\begin{aligned} \Delta E(I) &= \hbar ck - (\hbar^2/2m)[2(\boldsymbol{k}_f \cdot \boldsymbol{k}) + k^2], \\ \Delta E(II) &= -\hbar ck + (\hbar^2/2m)[2(\boldsymbol{k}_0 \cdot \boldsymbol{k}) - k^2]. \end{aligned} \tag{6.45}$$

Since the particles are non-relativistic, the first term on the right of both of these expressions gives the main contribution and

$$\Delta E(I) \approx -\Delta E(II) \approx \hbar ck. \tag{6.46}$$

Taking a unit normalization volume ($L^3 = 1$), the matrix element for the photon-emission vertex[10] is, in the momentum representation [see Chapter 5, Equations (5.24) and (5.142)],

$$H'_{ba} = \left(\frac{2\pi\hbar^3}{c}\right)^{1/2} \frac{ze}{m} \frac{\boldsymbol{\varepsilon} \cdot \boldsymbol{k}_a}{k^{1/2}}, \tag{6.47}$$

where $\boldsymbol{k}_a$ is the wave number of the charge entering the vertex. The matrix elements of the scattering potential are

$$V_{ba} = \int V(\boldsymbol{r})e^{i(\boldsymbol{k}_a - \boldsymbol{k}_b)\cdot\boldsymbol{r}} d^3\boldsymbol{r} = \int V(\boldsymbol{r})e^{i\boldsymbol{q}\cdot\boldsymbol{r}} d^3\boldsymbol{r} = V(\boldsymbol{q}), \tag{6.48}$$

that is, just the Fourier transform of the scattering potential—a familiar result [see Chapter 4, Equation (4.42)]. Note that it has been assumed that the scattering

[10]This is the vertex exhibited in Figure 3.2.

potential is velocity-independent. We have also ignored the spin coordinate, since all of the interactions do not involve it; if we include the spin coordinate in the formulation, the orthogonality of the associated eigenfunctions would simply yield the result that the spin coordinate does not change.

In the first term on the right for the amplitude (6.40), the matrix element of H' involves the factor $\boldsymbol{\varepsilon} \cdot \boldsymbol{k}_I$, but since $\boldsymbol{\varepsilon} \cdot \boldsymbol{k} = 0$, from the first relation (6.41), $\boldsymbol{\varepsilon} \cdot \boldsymbol{k}_I = \boldsymbol{\varepsilon} \cdot \boldsymbol{k}_f$. Making the approximation (6.44) for the energy denominators, the amplitude (6.40) becomes

$$H_{f0} = \left(\frac{2\pi\hbar^3}{c}\right)^{1/2} \frac{ze}{m\hbar c} \frac{\boldsymbol{\varepsilon} \cdot (\boldsymbol{k}_f - \boldsymbol{k}_0)}{k^{3/2}} V(\boldsymbol{q}). \tag{6.49}$$

The wave numbers are related to $\boldsymbol{q}$ by

$$\boldsymbol{k}_0 = \boldsymbol{k}_f + \boldsymbol{k} + \boldsymbol{q}, \tag{6.50}$$

but the photon wave vector k is made much smaller in magnitude than k_0, k_f, or q [see Equation (6.43)]; thus,

$$\boldsymbol{q} \approx \boldsymbol{k}_0 - \boldsymbol{k}_f, \tag{6.51}$$

and

$$H_{f0} \approx -\left(\frac{2\pi\hbar^3}{c}\right)^{1/2} \frac{ze}{m\hbar c} \frac{\boldsymbol{\varepsilon} \cdot \boldsymbol{q}}{k^{3/2}} V(\boldsymbol{q}). \tag{6.52}$$

It is significant that H_{f0} is directly proportional to the Fourier transform $V(\boldsymbol{q})$, since the Born cross section for radiationless scattering is proprtional to $|V(\boldsymbol{q})|^2$. The probability for bremsstrahlung is determined by $|H_{f0}|^2$, so that $(\Lambda = \hbar/mc)$

$$\frac{|H_{f0}|^2}{|V(\boldsymbol{q})|^2} = W_{\text{ph}} = 2\pi\alpha\Lambda^2 z^2 \frac{(\boldsymbol{\varepsilon} \cdot \boldsymbol{q})^2}{k^3} \tag{6.53}$$

represents the probability per photon state that the scattering will be accompanied by photon production. Per polarization state, the differential number of photon states is $dN_{\text{states}} = d^3\boldsymbol{k}/(2\pi)^3 = (2\pi)^{-3}k^2 dk\, d\Omega$, and if we multiply this factor by W_{ph}, we obtain

$$dw = \frac{1}{4\pi^2}\alpha z^2 \Lambda^2 (\boldsymbol{\varepsilon} \cdot \boldsymbol{q})^2 \frac{dk}{k} d\Omega = dw_k(\boldsymbol{q}) \tag{6.54}$$

as the probability for photon production within dk and $d\Omega$ in polarization state $\boldsymbol{\varepsilon}$. The expression (6.54) is essentially the result (3.131) [and (3.137)] of Chapter 3, and again it should be emphasized that it has *not* been assumed that the emitted photon is soft. It is only the photon momentum that has been taken to be small, and this is guaranteed by the energy relation (6.43).

Let us now evaluate the bremsstrahlung cross section and see, as mentioned earlier in connection with the formula (6.39), how the momentum transfer is the natural integration variable in determining the cross section differential in dk and $d\Omega$. For unit normalization volume, the particle flux incident on the scattering center is v_0 and the cross section is computed from

$$v_0 d\sigma = (2\pi/\hbar)|H_{f0}|^2 \delta(E_0 - E_f)\, dN_f. \tag{6.55}$$

With an outgoing photon and charge, the number of final states is

$$dN_f = \frac{k_f^2\, dk_f\, d\Omega_f}{(2\pi)^3} \frac{k^2\, dk\, d\Omega}{(2\pi)^3}, \tag{6.56}$$

and, if required, a summation over photon polarization states would also be performed. For given photon energy $\hbar ck$, since

$$E_0 - E_f = (\hbar^2/2m)(k_0^2 - k_f^2) - \hbar ck, \tag{6.57}$$

the integration over the δ-function in Equation (6.55) can be performed by writing $(\hbar^2/m)k_f dk_f = dE_f$. Also, if θ_f is the angle $\boldsymbol{k}_f$ makes with $\boldsymbol{k}_0$, since $d\Omega_f = 2\pi \sin\theta_f d\theta_f$, and since

$$\boldsymbol{q}^2 \approx (\boldsymbol{k}_0 - \boldsymbol{k}_f)^2 = k_0^2 + k_f^2 - 2k_0 k_f \cos\theta_f, \tag{6.58}$$

we can write

$$2q\, dq = 2k_0 k_f \sin\theta_f d\theta_f. \tag{6.59}$$

Then

$$k_f d\Omega_f = (2\pi/k_0) q\, dq, \tag{6.60}$$

and, with the additional factor $1/v_0 = m/\hbar k_0$ in $d\sigma$,

$$(k_f/k_0) d\Omega_f = (2\pi/k_0^2) q\, dq \tag{6.61}$$

is seen to be the convenient differential variable for integration. That is, from the basic expression (6.55) with the result (6.52) for H_{f0}, the bremsstrahlung cross section can be written as in Equation (6.39):

$$d\sigma = \int d\sigma_{\text{sc}}(\boldsymbol{q})\, dw_k(\boldsymbol{q}), \tag{6.62}$$

with

$$d\sigma_{\text{sc}} = \frac{2\pi}{k_0^2} \left(\frac{m}{2\pi\hbar^2}\right)^2 |V(\boldsymbol{q})|^2 q\, dq, \tag{6.63}$$

and the factor $dw_k(\boldsymbol{q})$ given by the result (6.54).

The formula (6.62) is very general, holding for any scattering potential as long as the Born approximation is valid, and it is not restricted to the soft-photon limit. However, in the limits q_{min} and q_{max}, it is necessary to consider the finite energy carried away by the bremsstrahlung photon. If we write

$$k_f/k_0 = \xi, \tag{6.64}$$

the photon energy is

$$\hbar\omega = (\hbar^2/2m)k_0^2(1 - \xi^2) = E_0(1 - \xi^2), \tag{6.65}$$

and

$$\frac{dk}{k} = \frac{2\xi\, d\xi}{1 - \xi^2}, \tag{6.66}$$

while

$$q_{\max} = k_0(1+\xi), \quad q_{\min} = k_0(1-\xi). \tag{6.67}$$

If the photon polarization is not of interest, a sum over the two transverse components can be performed (see also Section 5.1 of Chapter 3):

$$\sum_{\boldsymbol{\varepsilon}} (\boldsymbol{\varepsilon} \cdot \boldsymbol{q})^2 = q^2 - (\boldsymbol{k} \cdot \boldsymbol{q})^2/k^2 = q^2 \sin^2\theta, \tag{6.68}$$

where θ is the angle between $\boldsymbol{k}$ and $\boldsymbol{q}$; that is, it is now convenient to refer the photon direction to $\boldsymbol{q}$. Integration of $\sin^2\theta$ over $d\Omega$ yields a factor $8\pi/3$, and the photon emission probability is

$$dw_k = (2/3\pi)\alpha z^2 \Lambda^2 (dk/k) q^2. \tag{6.69}$$

The general cross section differential in $d\Omega_f dk\, d\Omega$ and for arbitrary polarization $\boldsymbol{\varepsilon}$ can be readily obtained from Equation (6.55) and the subsequent developments. However, often the cross section differential in dk and summed over polarizations is the quantity of interest, and this cross section would be computed from the integral (6.62) with the expression (6.69) for dw_k. In terms of the de Broglie wavelength $\lambda\!\!\!^{-}_0 = 1/k_0$ of the charge incident on the scattering center, the differential bremsstrahlung cross section is then, for an arbitrary scattering potential,

$$d\sigma = \frac{1}{3}\alpha\left(\frac{z\lambda\!\!\!^{-}_0}{\pi\hbar c}\right)^2 \frac{dk}{k} \int |V(\boldsymbol{q})|^2 q^3 dq. \tag{6.70}$$

Note that in this general formula, valid in the Born approximation, aside from the factor dk/k, the photon energy comes in only in the limits of the integral in the value

$$\xi = (1 - \hbar ck/E_0)^{1/2}, \tag{6.71}$$

determining $q_{\max}$ and $q_{\min}$ in Equation (6.67).

6.2.2 Coulomb (and Screened-Coulomb) Bremsstrahlung

The most important application of the general bremsstrahlung cross section (6.70) is to the problem where the scattering is due to a Coulomb potential. Moreover, the case where the scattered particle is an electron ($z = -1$) is of prime importance. Contained within this special case is the more general problem where the scattering potential is that of a "Yukawa" or screened-Coulomb field

$$V(\boldsymbol{r}) = (zZe^2/r)e^{-\varkappa r}, \tag{6.72}$$

for which [Equation (4.51), Chapter 4]

$$V(\boldsymbol{q}) = 4\pi zZe^2/(\varkappa^2 + q^2). \tag{6.73}$$

The bremsstrahlung cross section is then

$$d\sigma = \frac{16}{3}\alpha^3 z^4 Z^2 \lambda\!\!\!^{-2}_0 \frac{dk}{k} \int \frac{q^3 dq}{(\varkappa^2 + q^2)^2}, \tag{6.74}$$

and the integral can be evaluated easily:

$$I = \int_{q_{\min}}^{q_{\max}} \frac{q^3 dq}{(\varkappa^2 + q^2)^2} = \frac{1}{2}\left(\ln \frac{\varkappa^2 + q_{\max}^2}{\varkappa^2 + q_{\min}^2} + \frac{\varkappa^2}{\varkappa^2 + q_{\max}^2} - \frac{\varkappa^2}{\varkappa^2 + q_{\min}^2}\right). \quad (6.75)$$

For pure Coulomb bremsstrahlung ($\varkappa = 0$)

$$I \underset{\varkappa \to 0}{\longrightarrow} \ln \frac{q_{\max}}{q_{\min}} = \ln \frac{1+\xi}{1-\xi}, \quad (6.76)$$

while in the limit of strong screening

$$I \underset{\varkappa \gg k_0}{\longrightarrow} (q_{\max}^4 - q_{\min}^4)/4\varkappa^4 = 2(k_0/\varkappa)^4 \xi(1+\xi^2). \quad (6.77)$$

The results in the two limits are very different in their dependence on the energy of the incident charge and on the energy of the outgoing photon [see Equations (6.66) and (6.71)]. The Born formula for Coulomb bremsstrahlung can be expressed in terms of a Gaunt factor [see Equation (6.34)]:

$$g_{\text{Born}} = \frac{\sqrt{3}}{\pi} \ln \frac{1+\xi}{1-\xi}. \quad (6.78)$$

The Born photon-energy integrated cross section (6.35) is easily obtained:

$$\Omega_{\text{Born}} = (16/3)\alpha^3 z^4 Z^2 \Lambda^2 mc^2, \quad (6.79)$$

like the result (6.36) valid in the classical[11] limit, independent of energy; the ratio is $\Omega_{\text{Born}}/\Omega_0 = 2\sqrt{3}/\pi = 1.103$.

A final remark should be made concerning the Born bremsstrahlung formulas. They are independent of the signs of the scattered charge; this is not so in the classical limit.

6.2.3 Born Correction: Sommerfeld-Elwert Factor

There is a simple procedure for correcting the Born formula for Coulomb bremsstrahlung that is due to Elwert.[12] By the results (6.74) and (6.76), the Born formula is

$$d\sigma_{\text{Born}} = \frac{16}{3}\alpha^3 z^4 Z^2 \lambda_0^2 \frac{dk}{k} \ln \frac{1+\xi}{1-\xi}. \quad (6.80)$$

The Elwert prescription is

$$d\sigma = d\sigma_{\text{Born}} f_S, \quad (6.81)$$

where $d\sigma$ is the corrected cross section and,[13] for the case $z = -1$ (electron-ion bremsstrahlung),

$$f_S = \frac{\beta_0}{\beta_f} \frac{1 - e^{-2\pi\alpha Z/\beta_0}}{1 - e^{-2\pi\alpha Z/\beta_f}}. \quad (6.82)$$

---

[11] In the classical formula, however, a quantum-mechanical cutoff $\hbar\omega < E_0$ has been applied for the photon spectrum.

[12] G. Elwert, *Ann. Phys.* **34**, 178 (1939); *Z. Naturforsch.* **3a**, 477 (1948).

[13] The S in f_S stands for Sommerfeld and the factor f_S is commonly called the "Sommerfeld factor." It is related to another factor employed in the theory of β-decay to correct for the effects of the Coulomb field on the outgoing electron or positron; that factor is often referred to as a "Fermi factor."

The correction factor f_S is an approximate one, but there is a theoretical or physical basis for its use. Unfortunately, most textbooks give little explanation as to why the form (6.82) should correct the Born formula, so we shall try to provide some motivation for its introduction.

A point to make is that the Coulomb field has a large effective range such that the electron incident on the ion begins to feel the scattering potential far from the smaller-r region that is most important in contributing to the photon-emission amplitude. In the classical treatment of bremsstrahlung in the previous section, it can also be seen clearly how the main contribution to emission comes from the part of the orbit near the scattering center where the acceleration is largest. Now let us recall the result (4.76) of Chapter 4 giving the square of the Coulomb wave function at the origin when the incident amplitude is normalized to unity at infinity:

$$|u(0)|^2 = 2\pi n/(e^{2\pi n} - 1), \tag{6.83}$$

where

$$n = zZe^2/\hbar v, \tag{6.84}$$

v being the velocity of the charge at infinity. The factor f_S is then, simply,

$$f_S = |u_f(0)|^2/|u_0(0)|^2, \tag{6.85}$$

and it remains to provide arguments to understand why this ratio should correct the Born cross section.

Perhaps we should start with the Born formula (6.70), which has the form

$$d\sigma_{\text{Born}} \propto (dk/k) \int |V(\boldsymbol{q})|^2 q^3 dq, \tag{6.86}$$

and note some features of the result. First, the factor λ_0^2 has been left out; this factor arises only from the overall kinematics of the process [see Equation (6.61)] and not from the details of the scattering or emission perturbations. Second, the integration variable is the momentum transfer to the scattering center and involves the variable k and the parameter λ_0 only in the limits. The approximate expression (6.85) can be seen as essentially a Coulomb-field focusing factor that accounts for the effects of the Coulomb field on the incident and outgoing charge. That it should be the *ratio* of the squared wave functions at the origin may be understood as follows. Suppose we take the focused u_0 as the unperturbed state with normalization $|u_0(0)|^2 L^3 = 1$ (rather than $|u_0(\infty)|^2 L^3 = 1$); the normalization volume L^3 does not appear in the final $d\sigma$. The most important spatial region for the process is near the origin, at which we can write

$$u_0(\text{small } \boldsymbol{r}) \approx L^{-3/2} e^{-i\boldsymbol{K}_0 \cdot \boldsymbol{r}}, \tag{6.87}$$

where $\boldsymbol{K}_0$ is related to but does not equal $\boldsymbol{k}_0$. Similarly, the outgoing wave function is of the form

$$u_f(\text{small } \boldsymbol{r}) \approx L^{-3/2}\big(u_f(0)/u_0(0)\big) e^{-i\boldsymbol{K}_f \cdot \boldsymbol{r}}, \tag{6.88}$$

which has the normalization

$$|u_f(0)|^2 L^3 = f_S. \tag{6.89}$$

Note that the wave functions (6.87) and (6.88) are plane wave approximations to Coulomb wave functions in the neighborhood of the origin with their relative amplitudes corrected for Coulomb focusing by the necessary factor. Employing the plane-wave u's to evaluate the bremsstrahlung cross section amounts to the Born approximation, except that the amplitude u_f has been corrected by the factor $f_S^{1/2}$. The effective wave vectors $\boldsymbol{K}_0$ and $\boldsymbol{K}_f$ are related by

$$\boldsymbol{K}_0 = \boldsymbol{K}_f + \boldsymbol{q}, \tag{6.90}$$

where $\hbar\boldsymbol{q}$ is the same physical momentum transfer to the scattering center. The modified wave functions (6.87) and (6.88) would then be used in a calculation of the cross section in the same manner as in the pure Born approximation, yielding the result (6.81). Thus, it is seen how the correction factor is basically the square of the effective amplitude for the outgoing charge at $\boldsymbol{r} = 0$ when expressed in terms of that for u_0. The factor differs from unity because the focusing is greater for the (lower-energy) outgoing charge, and this must be taken into account in applying a consistent normalization for u_0 and u_f. The choice $|u_0(0)|^2 L^3 = 1$ is a natural one since it implies an L^3 independent of the variable k (although dependent on the parameter λ_0). If, on the other hand, $|u_f(0)|^2 L^3$ had been set equal to unity, the normalization volume would be a function of k.

It should be emphasized that the factor f_S is an approximate one, and we shall discuss its accuracy later in Section 6.4 where the "intermediate" (classical-to-Born) case is considered and where the problem must be treated using exact Coulomb wave functions. We can note here, for example, that as $k \to 0$ (soft-photon limit), $f_S \to 1$, suggesting that the Born formula is exact in this limit. This is not correct, as can be seen from the classical-limit formula (6.15), which differs from the Born formula.

The Elwert factor is particularly convenient as a correction when the relative correction is small, that is, when the Born formula is a good approximation and we want the lowest-order relative correction. In this limit, n_0 and n_f [defined by Equation (6.84)] are both small and

$$f_S = \frac{n_f}{n_0} \frac{e^{2\pi n_0} - 1}{e^{2\pi n_f} - 1} \to 1 + \pi(n_0 - n_f) + \cdots . \tag{6.91}$$

In terms of ξ [Equation (6.64)], $n_f = n_0/\xi$, and

$$f_S \approx 1 - \pi n_0(1/\xi - 1), \tag{6.92}$$

which provides a simple correction to the Born formula. For example, it is then easy to compute the correction to the Born integrated cross section (6.35) which can be expressed in terms of the limit (6.79):

$$\Omega \approx \Omega_{\text{Born}}(1 - \omega), \tag{6.93}$$

where

$$\omega = \pi n_0(2\ln 2 - 1). \tag{6.94}$$

In terms of the Rydberg energy Ry $= e^2/2a_0$ and the kinetic energy (E_0) of the incident charge,

$$n_0 = zZ(\text{Ry}/E_0)^{1/2}. \tag{6.95}$$

For the important case of electron-ion bremsstrahlung, $z = -1$ and ω is negative, yielding a positive relative correction in the result (6.93).

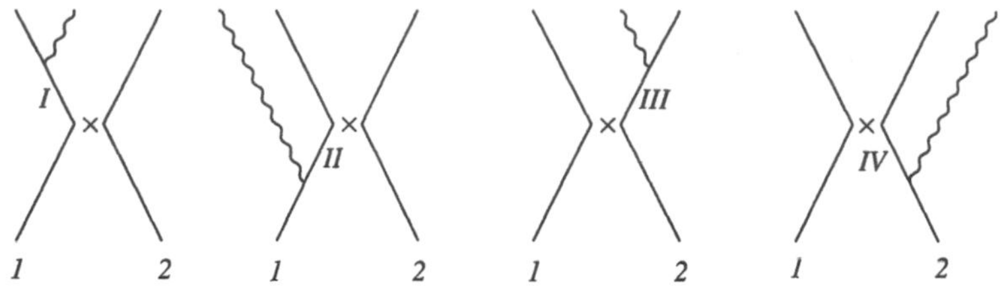

Figure 6.2 Diagrams for bremsstrahlung in two-body scattering.

6.2.4 Electron-Positron Bremsstrahlung

The problem of bremsstrahlung in electron-positron scattering is closely related to that for electron-ion bremsstrahlung. Photon production is again a result of an electromagnetic perturbation, that is, a photon-charge coupling. However, now, in contrast to the case of single-particle bremsstrahlung, as in the electron–ion problem, the photon coupling to both charges must be included. The effect for e^+-e^- bremsstrahlung is to double the amplitude for photon production and thus to quadruple the photon-emission probability. This can also be seen classically from the dipole-emission formulas (see Section 2.2 and 2.3) for which both charges contribute to the dipole moment. The classical treatment of e^+-e^- bremsstrahlung will not be given here, since it is the same as the problem of electron-ion (e–Z) bremsstrahlung. The Born limit for the e^+-e^- problem also requires a rather simple modification of the result for the e–Z case, but it will be considered here, in part because it provides some introduction to the more difficult problem of electron-electron bremsstrahlung treated in the following section.

In the Born approximation, the process is evaluated as a result of the actions of the perturbation V associated with the scattering and the photon-emission perturbation H', each acting once. There are four perturbation diagrams, exhibited in Figure 6.2, corresponding to the effective perturbation [see Equation (6.40) for the single-particle case]

$$\begin{aligned} H_{f)} &= H'\frac{1}{\Delta E(I)}V + V\frac{1}{\Delta E(II)}H' + H'\frac{1}{\Delta E(III)}V + V\frac{1}{\Delta E(IV)}H' \\ &= A_I + A_{II} + A_{III} + A_{IV}. \end{aligned} \tag{6.96}$$

The perturbation H' is the same as that in the matrix element (6.47), and the cross section is evaluated by the methods developed in this section for e–Z bremsstrahlung. Here we have three outgoing particles, but again the photon momentum or wave number is small compared to that of the charges e^+ and e^-. In the c.m. frame, then, the initial and final wave number of the charges will be $\pm\boldsymbol{k}_0$ and $\pm\boldsymbol{k}_f$, respectively.[14] If we define $\xi = k_f/k_0$ as in the e–Z problem, the photon energy will be

$$\hbar\omega = (\hbar^2/m)(k_0^2 - k_f^2) = (\hbar^2 k_0^2/m)(1-\xi^2), \tag{6.97}$$

[14]In terms of the inital c.m. velocity of the e^+ and e^-, $\beta_r = 2\beta_0$.

that is, k_0 is the magnitude of the initial wave number of either charge in the c.m. frame. Since the charge motion is non-relativistic, the c.m. photon energy $\hbar ck$ will also be equal, in lowest order, to the photon energy (say, $\hbar ck'$) in, for example, a "lab" system where one charge is initially at rest.

The essential feature of the calculation of electron-positron bremsstrahlung is the result that, to lowest order,

$$A_I + A_{II} = A_{III} + A_{IV}, \tag{6.98}$$

and it is very easy to see how this equality comes about. The energy denominators will satisfy the (lowest-order) relations of the form (6.46):

$$\Delta E(I) = -\Delta E(II) = \Delta E(III) = -\Delta E(IV) = \hbar ck. \tag{6.99}$$

Moreover, the matrix elements of the photon-emission perturbation will be given by [see Equation (6.47)]

$$H'_{ba} \propto zk^{-1/2}\boldsymbol{\varepsilon} \cdot \boldsymbol{k}_a, \tag{6.100}$$

where ze is the charge of the particle and $\boldsymbol{k}_a$ is its wave vector entering the vertex. Since the photon wave number $\boldsymbol{k}$ is small, there is a common momentum transfer in the four amplitudes (6.96): $\boldsymbol{q}_1 = -\boldsymbol{q}_2 = \boldsymbol{q}$, say. The matrix elements of the scattering potential will then be the same for the four amplitudes: $V_{ba} \propto 1/q^2$. Since in A_{III} and A_{IV} *both* z and $\boldsymbol{k}_a$ will have the same corresponding values: $H'(I) = H'(III)$; $H'(II) = H'(IV)$. Thus, we see the reason for the basic identity (6.98).

As in the e–Z problem, it is convenient to evaluate the bremsstrahlung cross section $d\sigma$ by integrating over the momentum transfer (rather than, say, the scattering angle) [see Equation (6.60)]. The $d\sigma$ is computed as in the expression (6.62) from the scattering cross section $d\sigma_{\text{sc}}$ and the photon-production probability dw_k. Differential in q, the Born scattering cross section can be obtained from, for example, the first term on the right of equation (4.95) of Chapter 4 and is, very simply,

$$d\sigma_{\text{sc}} = 8\pi(\alpha/\beta_r)^2 dq/q^3, \tag{6.101}$$

where $\beta_r c$ is the initial relative velocity.[14] In terms of q, the probability dw_k is now, because of the identity (6.98), four times the expression (6.69):

$$dw_k(\boldsymbol{q}) = (8/3\pi)\alpha\Lambda^2(dk/k)q^2 \quad (e^+\text{–}e^-). \tag{6.102}$$

The bremsstrahlung cross section is then

$$\begin{aligned} d\sigma &= \int d\sigma_{\text{sc}}(\boldsymbol{q})dw_k(\boldsymbol{q}) \\ &= \frac{64}{3}\alpha^3\Lambda^2\frac{1}{\beta_r^2}\frac{dk}{k}\int\frac{dq}{q}. \end{aligned} \tag{6.103}$$

The limits on q are determined by the fractional energy carried away (ξ) [see Equation (6.97)]: $q_{\max} = k_0 + k_f = k_0(1+\xi)$; $q_{\min} = k_0 - k_f = k_0(1-\xi)$. The factor

Λ^2/β_r^2 is just λ_r^2 where λ_r is the inital relative de Broglie wavelength and $d\sigma$ is

$$d\sigma = \frac{64}{3}\alpha^3\lambda_r^2\frac{dk}{k}\ln\frac{1+\xi}{1-\xi}. \tag{6.104}$$

Differential in ξ, dk/k could again be written in the form (6.66) if desired.

The Born formula (6.103) can be corrected by an application of an appropriate Sommerfeld-Elwert factor. From the relationship between the two-body and equivalent one-body problems (see Section 2.3 of Chapter 4), the $f_S(e^+\!-e^-)$ can be obtained from the e–Z formula (6.82) by setting $Z = 1$ and by making the replacements $\beta_0 \to \beta_r$; $\beta_f \to \xi\beta_r$.

6.3 ELECTRON-ELECTRON BREMSSTRAHLUNG. NON-RELATIVISTIC

In emission in e–e scattering, the amplitude terms that add in the e^+–e^- problem now cancel in lowest order and must be handled more carefully. When an expansion is made in k/k_e (photon/electron momenta), there results a finite term in the amplitude for the process corresponding to quadrupole emission, and an expression for the cross section can be derived. At non-relativistic energies, $d\sigma_{ee}$ is much smaller than the electron-ion formula $d\sigma_{eZ}$ [see Equation (6.80)]. The derivation is outlined in this section and the various special features of the problem are emphasized, especially those not present in single-particle bremsstrahlung.

6.3.1 Direct Born Amplitude

The basic perturbation diagrams for e–e bremsstrahlung are the same as those in Figure 3.2 for the $e^+\!-e^-$ problem. The photon-coupling perturbation (H') can be a part of any of the four legs of the diagram, with the other perturbation (V) being associated with the scattering potential. Both H' and V act once (Born approximation) and the effective perturbation Hamiltonian for the combined process would again be given by an expression of the form (6.96). However, in this case, A_I is almost cancelled by A_{III} and a similar near cancellation occurs between A_{II} and A_{IV}. The finite cross section is due to the effects of the incompleteness of the cancellation and, in particular, to the role of the finite momentum $\hbar\mathbf{k}$ of the outgoing photon. This will be exhibited very explicitly in the formulation that follows. The effect has an analog in classical theory, corresponding to the cancellation of dipole fields, requiring the evaluation of retardation. Because of the more critical role played by the photon momentum, quantum mechanics is more important in the e–e problem than in e–z bremsstrahlung.

Now let us evaluate the fundamental amplitude for the process, comparing the result to that computed from Equation (6.96) for e^+–e^- bremsstrahlung. This H_{f0} for the e–e problem would be the "direct" amplitude (A) for the process, and, since there are two outgoing particles, there must also be an exchange amplitude A_e. As in the $e^+\!-e^-$ problem, it is convenient to calculate the cross section in terms of the c.m. variables. The particle and photon wave vectors will satisfy the (momentum) conservation relation

$$\boldsymbol{k}_1 + \boldsymbol{k}_2 = 0 = \boldsymbol{k}_1' + \boldsymbol{k}_2' + \boldsymbol{k}. \tag{6.105}$$

The initial wave vectors are thus related by

$$\boldsymbol{k}_1 = -\boldsymbol{k}_2 \equiv \boldsymbol{k}_0, \tag{6.106}$$

and, since $\boldsymbol{k}$ is small, the final values are given by

$$\boldsymbol{k}_1' = \boldsymbol{k}_f = -(\boldsymbol{k}_2' + \boldsymbol{k}) \approx -\boldsymbol{k}_2'. \tag{6.107}$$

We can now see why the amplitudes A_I and A_{III} as well as A_{II} and A_{IV} nearly cancel. The matrix elements H_{ba}', given by formulas (6.47) and (6.100) are proportional to $\boldsymbol{k}_a$, which are, by Equations (6.106) and (6.107), simply minus the values in the corresponding amplitudes.[15] The lowest-order matrix elements V_{ba} will all be equal, again because $\boldsymbol{k}$ is small, and since the energy denominators will satisfy the identities (6.99) (to lowest order),

$$A_I \approx -A_{III}, \quad A_{II} \approx -A_{IV}. \tag{6.108}$$

To obtain a finite amplitude, the matrix elements and energy denominators must be evaluated more carefully. In terms of the intermediate-state particle wave vectors, the latter are

$$\begin{aligned}
\Delta E(I) &= E_0 - (\hbar^2/2m)(k_I^2 + k_2'^2),\\
\Delta E(III) &= E_0 - (\hbar^2/2m)(k_{III}^2 + k_1'^2),\\
\Delta E(II) &= E_0 - (\hbar^2/2m)(k_{II}^2 + k_2{}^2) - \hbar ck,\\
\Delta E(IV) &= E_0 - (\hbar^2/2m)(k_{IV}^2 + k_1{}^2) - \hbar ck,
\end{aligned} \tag{6.109}$$

where

$$E_0 = (\hbar^2/2m)(k_1^2 + k_2^2) = (\hbar^2/m)k_0^2. \tag{6.110}$$

There is momentum conservation at the vertices H', so that

$$\begin{aligned}
\boldsymbol{k}_I &= \boldsymbol{k}_1' + \boldsymbol{k},\\
\boldsymbol{k}_{III} &= \boldsymbol{k}_2' + \boldsymbol{k},\\
\boldsymbol{k}_{II} &= \boldsymbol{k}_1' - \boldsymbol{k},\\
\boldsymbol{k}_{IV} &= \boldsymbol{k}_2' - \boldsymbol{k}.
\end{aligned} \tag{6.111}$$

There is also conservation of total energy in the process:

$$\hbar ck = E_0 - (\hbar^2/2m)(k_1'^2 + k_2'^2). \tag{6.112}$$

The expressions (6.111) cna be substituted into the energy denominators (6.109) with terms in k^2 neglected; employing also the relations (6.106) and (6.107), we then have

$$\begin{aligned}
\Delta E(I) &= \hbar ck(1 - \boldsymbol{k}\cdot\boldsymbol{k}_f/\varkappa k),\\
\Delta E(III) &= \hbar ck(1 + \boldsymbol{k}\cdot\boldsymbol{k}_f/\varkappa k),\\
\Delta E(II) &= -\hbar ck(1 - \boldsymbol{k}\cdot\boldsymbol{k}_0/\varkappa k),\\
\Delta E(IV) &= -\hbar ck(1 + \boldsymbol{k}\cdot\boldsymbol{k}_0/\varkappa k),
\end{aligned} \tag{6.113}$$

[15] Recall that in $e^+ - e^-$ bremsstrahlung, the additional factors of z then reversed the sign again, and the corresponding amplitudes added instead of subtracted.

where

$$\varkappa = mc/\hbar = \Lambda^{-1}. \tag{6.114}$$

Since the electrons are non-relativistic, the second terms in parentheses in the expression (6.113) are always small, and we can define the (small) quantities

$$\eta_0 = \boldsymbol{k} \cdot \boldsymbol{k}_0/\varkappa k, \quad \eta_f = \boldsymbol{k} \cdot \boldsymbol{k}_f/\varkappa k. \tag{6.115}$$

These can be used as expansion parameters in the energy denominators to write $(1 \pm \eta)^{-1} = 1 \mp \eta + O(\eta^2)$ for η_0 or η_f.

With the matrix elements H'_{ba} given by the form (6.47), the fundamental factor is $\boldsymbol{\varepsilon} \cdot \boldsymbol{k}_a$, where $\boldsymbol{k}_a = \boldsymbol{k}_I, \boldsymbol{k}_{III}, \boldsymbol{k}_1$, and $\boldsymbol{k}_2$ in the four amplitudes I, II, III, and IV, respectively. But because of the identities (6.106), (6.107), and (6.111) and the transversality condition $\boldsymbol{\varepsilon} \cdot \boldsymbol{k} = 0$, in these amplitudes the matrix element has the values, say, H'_f, $-H'_f$, H'_0, and $-H'_0$, respectively, with

$$H'_f = \left(\frac{2\pi\hbar^3}{c}\right)^{1/2} \frac{ze}{m} \frac{\boldsymbol{\varepsilon} \cdot \boldsymbol{k}_f}{k^{1/2}}, \tag{6.116}$$

$$H'_0 = \left(\frac{2\pi\hbar^3}{c}\right)^{1/2} \frac{ze}{m} \frac{\boldsymbol{\varepsilon} \cdot \boldsymbol{k}_0}{k^{1/2}}. \tag{6.117}$$

Perhaps the greatest subtlety in the *e–e* bremsstrahlung problem concerns the scattering potential and its matrix element V_{ba}. This is the matrix element between the two-particle plane-wave states

$$u(\boldsymbol{r}_1, \boldsymbol{r}_2) = \exp[i(\boldsymbol{k}_1 \cdot \boldsymbol{r}_1 + \boldsymbol{k}_2 \cdot \boldsymbol{r}_2)]. \tag{6.118}$$

The matrix element

$$V_{ba} = \iint d^3\boldsymbol{r}_1\, d^3\boldsymbol{r}_2\, V(\boldsymbol{r}_1, \boldsymbol{r}_2) \exp[i(\boldsymbol{k}_{1a} - \boldsymbol{k}_{1b}) \cdot \boldsymbol{r}_1 + i(\boldsymbol{k}_{2a} - \boldsymbol{k}_{2b}) \cdot \boldsymbol{r}_2] \tag{6.119}$$

is most conveniently evaluated, when V is a central potential $V(\boldsymbol{r})$, by transforming to c.m. and relative coordinates:

$$\boldsymbol{R} = \tfrac{1}{2}(\boldsymbol{r}_1 + \boldsymbol{r}_2), \quad \boldsymbol{r} = \boldsymbol{r}_1 - \boldsymbol{r}_2. \tag{6.120}$$

Then, since $d^3\boldsymbol{r}_1\, d^3\boldsymbol{r}_2 = d^3\boldsymbol{R}\, d^3\boldsymbol{r}$, the $\boldsymbol{R}$-integration yields total momentum conservation and the $\boldsymbol{r}$-integration yields the Fourier transform of $V(\boldsymbol{r})$ in terms of the momentum-transfer[16] variable $\boldsymbol{q}$. That is, if we take 1-to-2 as the momentum transfer direction,

$$\boldsymbol{k}_{1a} = \boldsymbol{k}_{1b} + \boldsymbol{q}, \quad \boldsymbol{k}_{2b} = \boldsymbol{k}_{2a} + \boldsymbol{q}. \tag{6.121}$$

The matrix element (6.119) is then simply

$$V_{ba} = \int d^3\boldsymbol{r}\, V(\boldsymbol{r})\, e^{i\boldsymbol{q}\cdot\boldsymbol{r}} = V(\boldsymbol{q}), \tag{6.122}$$

[16]Rather, $\hbar\boldsymbol{q}$ is the momentum transfer.

but the $\boldsymbol{q}$-variable is slightly different in the four amplitudes. The values are (see Figure 6.2)

$$\begin{aligned} \boldsymbol{q}_I &= \boldsymbol{k}_0 - \boldsymbol{k}_I = \boldsymbol{k}_0 - \boldsymbol{k}'_1 - \boldsymbol{k} \equiv \boldsymbol{q}, \\ \boldsymbol{q}_{III} &= \boldsymbol{k}_0 - \boldsymbol{k}'_1 = \boldsymbol{q} + \boldsymbol{k}, \\ \boldsymbol{q}_{II} &= \boldsymbol{k}_{II} - \boldsymbol{k}'_1 = \boldsymbol{k}_0 - \boldsymbol{k}'_1 - \boldsymbol{k} = \boldsymbol{q}, \\ \boldsymbol{q}_{IV} &= \boldsymbol{k}_0 - \boldsymbol{k}'_1 = \boldsymbol{q} + \boldsymbol{k}. \end{aligned} \tag{6.123}$$

Since $\boldsymbol{k}$ is small, the matrix elements with the argument $\boldsymbol{q} + \boldsymbol{k}$ can be written

$$\begin{aligned} V(\boldsymbol{q} + \boldsymbol{k}) &= V(\boldsymbol{q}) + \frac{\partial V}{\partial \boldsymbol{q}} \cdot \boldsymbol{k} + \cdots \\ &= V(\boldsymbol{q}) + \frac{1}{q}\frac{\partial V}{\partial q}\boldsymbol{q} \cdot \boldsymbol{k} + \cdots, \end{aligned} \tag{6.124}$$

and we can introduce another small parameter η_V defined by

$$\eta_V = \frac{\partial}{\partial q}(\ln V)\frac{\boldsymbol{q} \cdot \boldsymbol{k}}{q}. \tag{6.125}$$

The matrix element (6.124) is then

$$V(\boldsymbol{q} + \boldsymbol{k}) = V(\boldsymbol{q})(1 + \eta_V). \tag{6.126}$$

We can now write down the basic direct amplitude for the process in terms of the expansion parameters η_0, η_f, and η_V [see Equation (6.96)]:

$$\begin{aligned} H_{f0} &= \frac{V(\boldsymbol{q})}{\hbar c k}\left[\frac{H'_f}{1 - \eta_f} - \frac{H'_f}{1 + \eta_f}(1 + \eta_V) - \frac{H'_0}{1 - \eta_0} + \frac{H'_0}{1 + \eta_0}(1 + \eta_V)\right] \\ &\approx \left(2V(\boldsymbol{q})/\hbar c k\right)\left[\eta_f H'_f - \eta_0 H'_0 - \tfrac{1}{2}\eta_V (H'_f - H'_0)\right]. \end{aligned} \tag{6.127}$$

For characteristic $\boldsymbol{k}$ in the bremsstrahlung problem, the three η-parameters are of the same order ($\sim v/c$ for the charges). The result (6.127) shows explicitly the (linear) dependence of the basic amplitude for the process on the expansion parameters associated with the energy denominators and the form of the scattering potential. The variable $\boldsymbol{q}$ in the amplitude (6.127) is essentially [see Equations (6.105), (6.106), (6.123)]

$$\boldsymbol{q} \approx \boldsymbol{k}_0 - \boldsymbol{k}_f, \tag{6.128}$$

and, substituting in the perturbations (6.116) and (6.117), we obtain

$$H_{ba} = (8\pi\alpha)^{1/2} z V(\boldsymbol{q}) k^{-3/2} a, \tag{6.129}$$

where a is the dimensionless factor[17]

$$\begin{aligned} a(\boldsymbol{k}, \boldsymbol{\varepsilon}; \boldsymbol{k}_0, \boldsymbol{k}_f) = \frac{1}{\varkappa^2 k}\bigg[&(\boldsymbol{k} \cdot \boldsymbol{k}_0)(\boldsymbol{\varepsilon} \cdot \boldsymbol{k}_0) - (\boldsymbol{k} \cdot \boldsymbol{k}_f)(\boldsymbol{\varepsilon} \cdot \boldsymbol{k}_f) \\ &- \frac{k\varkappa}{2q}\frac{\partial}{\partial q}(\ln V)(\boldsymbol{k} \cdot \boldsymbol{q})(\boldsymbol{\varepsilon} \cdot \boldsymbol{q})\bigg]. \end{aligned} \tag{6.130}$$

[17] The result (6.129) was obtained by R. J. Gould, *Phys. Rev.* **A23**, 2851 (1981).

For a Coulomb scattering potential $V = V(\boldsymbol{q}) = 4\pi z^2 e^2/q^2$, and

$$a_{\text{Coul}} = \frac{1}{\varkappa^2 k}\left[(\boldsymbol{k}\cdot\boldsymbol{k}_0)(\boldsymbol{\varepsilon}\cdot\boldsymbol{k}_0) - (\boldsymbol{k}\cdot\boldsymbol{k}_f)(\boldsymbol{\varepsilon}\cdot\boldsymbol{k}_f) + \frac{k\varkappa}{q^2}(\boldsymbol{k}\cdot\boldsymbol{q})(\boldsymbol{\varepsilon}\cdot\boldsymbol{q})\right], \quad (6.131)$$

which is an explicit expression in terms of the photon parameters $\boldsymbol{k}$ and $\boldsymbol{\varepsilon}$ and the scattering parameters $\boldsymbol{k}_0$ and $\boldsymbol{k}_f$. The factor $k\varkappa$ is [see Equation (6.112)]

$$\varkappa k = k_0^2 - k_f^2. \quad (6.132)$$

6.3.2 Photon-Emission Probability (without Exchange)

The amplitude or effective perturbation Hamiltonian (6.129) can be compared with the single-particle expression (6.49), which is also proportional to the Fourier transform $V(\boldsymbol{q})$. Significantly, in the former result the amplitude has a dependence, in addition to its proportionality to V, on the shape of the scattering potential. Given the form of V, as in the case of the Coulomb potential with the amplitude (6.131), it is possible to introduce a photon-emission probability as in single-particle bremsstrahlung:

$$W_{\text{ph}} = |H_{ba}|^2/|V(\boldsymbol{q})|^2 = 8\pi z^2 \alpha^2 k^{-3} a^2. \quad (6.133)$$

This is the probability per photon state; multiplying by the differential number of photon states per polarization $dN = d^3\boldsymbol{k}/(2\pi)^3 = k^2 dk\, d\Omega/(2\pi)^3$, we obtain a differential probability for emission within dk and $d\Omega$ with polarization $\boldsymbol{\varepsilon}$:

$$dw_k = W_{\text{ph}}\, dN = z^2(\alpha/\pi^2)a^2(dk/k)d\Omega. \quad (6.134)$$

It should be emphasized that this is the expression neglecting exchange. It would be relevant if, for example, despite their same z and m, the charges were somehow distinguishable.

If we wish to sum over photon polarizations, in the squared amplitude a^2 terms of the type $(\boldsymbol{\varepsilon}\cdot\boldsymbol{i}_\alpha)(\boldsymbol{\varepsilon}\cdot\boldsymbol{i}_\beta)$ appear, where $\boldsymbol{i}_\alpha$ and $\boldsymbol{i}_\beta$ are particle momentum unit vectors $(\boldsymbol{i}_\alpha = \boldsymbol{k}_\alpha/|\boldsymbol{k}_\alpha|)$. The associated polarization sums are

$$\sum_{\varepsilon}(\boldsymbol{\varepsilon}\cdot\boldsymbol{i}_\alpha)(\boldsymbol{\varepsilon}\cdot\boldsymbol{i}_\beta) = \boldsymbol{i}_\alpha\cdot\boldsymbol{i}_\beta - (\boldsymbol{i}\cdot\boldsymbol{i}_\alpha)(\boldsymbol{i}\cdot\boldsymbol{i}_\beta), \quad (6.135)$$

where $\boldsymbol{i} = \boldsymbol{k}/k$ is the unit vector in the direction of the photon momentum. An integration over the solid angle of emission direction can also be made and the identities (easily proved)

$$\begin{aligned} \int(\boldsymbol{i}\cdot\boldsymbol{i}_\alpha)(\boldsymbol{i}\cdot\boldsymbol{i}_\beta)\, d\Omega/4\pi &= \tfrac{1}{3}\boldsymbol{i}_\alpha\cdot\boldsymbol{i}_\beta, \\ \int(\boldsymbol{i}\cdot\boldsymbol{i}_\alpha)^2(\boldsymbol{i}\cdot\boldsymbol{i}_\beta)^2 d\Omega/4\pi &= \tfrac{1}{15} + \tfrac{2}{15}(\boldsymbol{i}_\alpha\cdot\boldsymbol{i}_\beta)^2 \end{aligned} \quad (6.136)$$

can be employed. If this summation and integration is performed on the result (6.134) with a_{Coul}, the differential probability is

$$dw_{\text{Coul(no exch.)}} = (8\alpha/15\pi)z^2\beta_0^4(dk/k)\left[4(1-\xi^2)^2 + 3\xi^2\sin^2\theta_f\right], \quad (6.137)$$

where θ_f is the c.m. scattering angle ($\cos\theta_f = (\boldsymbol{k}_0 \cdot \boldsymbol{k}_f)/k_0 k_f$) and ξ is again defined as in Equation (6.97).

There is a case where a photon emission probability is relevant for scattering of identical charges even when exchange effects are to be included. It is when the scattering potential has a short range, and let us see why this is so. Again for the case where spin interactions are negligible, the exchange amplitude (A_e) is obtained from the direct amplitude (A) by interchanging the final-state values $\boldsymbol{k}_1'$ and $\boldsymbol{k}_2'$ or, equivalently,

$$A_e = A(\boldsymbol{k}_f \to -\boldsymbol{k}_f). \tag{6.138}$$

From the basic expression (6.129) for the direct amplitude, we see that in the factor (6.130) only the last term in brackets is affected by the transformation $\boldsymbol{k}_f \to -\boldsymbol{k}_f$. Moreover, this term is small for a short-range potential; for example, for the screened-Coulomb potential (6.72) with a range parameter $\varkappa_V$,

$$\frac{\partial}{\partial q} \ln V = -\frac{2q}{\varkappa_V^2 + q^2}. \tag{6.139}$$

Then, as long as

$$q \sim k_0 \ll \varkappa_V, \tag{6.140}$$

the last term in the brackets (6.130) is negligible. The rest of the expression is such that

$$a_e = a \quad \text{(for short-range potential)} \tag{6.141}$$

and the total amplitude (6.129) is a "separable" function (product) of the scattering potential $V(\boldsymbol{q})$ and the factor (a) associated with the photon production. The process is then similar to the single-particle or dipole case for which there is a photon-emission probability independent of the form of the scattering potential. Exchange effects still do come into the problem, but only in the factor $V(\boldsymbol{q})$, which has an exchange value $V(\boldsymbol{q}_e)$ with

$$\boldsymbol{q}_e = \boldsymbol{k}_0 + \boldsymbol{k}_f. \tag{6.142}$$

The bremsstrahlung cross section would again be computed from a product of the type (6.62) with $d\sigma_{sc}$ including the exchange and interference terms, and, because of the identity (6.141), dw_k given by the form (6.134). In the short-range case for which

$$a_{\text{short}} = (\varkappa^2 k)^{-1}\big[(\boldsymbol{k}\cdot\boldsymbol{k}_0)(\boldsymbol{\varepsilon}\cdot\boldsymbol{k}_0) - (\boldsymbol{k}\cdot\boldsymbol{k}_f)(\boldsymbol{\varepsilon}\cdot\boldsymbol{k}_f)\big] \tag{6.143}$$

is a more simple expression, the probability summed over polarizations and integrated over $d\Omega$ is now

$$dw_k(\text{short}) = (8\alpha/15\pi)z^2\beta_0^4(dk/k)\big[(1-\xi^2)^2 + 3\xi^2\sin^2\theta\big], \tag{6.144}$$

that is, the same as the expression (6.137) for a Coulomb field except without the coefficient 4 in the first term in brackets.

The short-range-potential function (6.144) does have a restriction in its applicability other than the Born-approximation requirement. The inequality (6.140) provides an upper limit to the energy of the incident charge and at higher energies the photon emission involves the intricacies of the scattering potential to a greater degree.

6.3.3 Cross Section (with Exchange)

The cross section for two-particle bremsstrahlung is computed from

$$2v_0\, d\sigma = \frac{2\pi}{\hbar}\left(\int \sum_{\boldsymbol{\varepsilon}}\right)|H_{f0}|^2\delta(E_0 - E_f)\frac{d^3\boldsymbol{k}_f d^3\boldsymbol{k}}{(2\pi)^6}. \qquad (6.145)$$

In this expression the factor $2v_0$ is the particle flux factor from relative motion expressed in c.m. variables. The last factor on the right is the differential number of final (charge and photon) states and the $\boldsymbol{k}$-space element can be written

$$d^3\,\boldsymbol{k}_f\, d^3\boldsymbol{k} = k_f^2\, dk_f\, d\Omega_f k^2 dk\, d\Omega, \qquad (6.146)$$

differential in the magnitude of the wave vectors and in the directional solid-angle elements. Summing over polarization (if desired) and integrating over other final-state quantities not of interest can then yield a cross section differential only in dk. The argument of the δ-function is

$$E_0 - E_f = (\hbar^2/m)(k_0^2 - k_f^2) - \hbar ck, \qquad (6.147)$$

and since

$$\boldsymbol{q} \approx \boldsymbol{k}_0 - \boldsymbol{k}_f, \qquad (6.148)$$

the differential $d^3\boldsymbol{k}_f = 2\pi k_f^2\, dk_f \sin\theta_f d\theta_f$ can be rewritten as

$$d^3\boldsymbol{k}_f = (\pi/k_0)q\, dq\, d(k_f^2), \qquad (6.149)$$

by employing $q^2 = k_0^2 + k_f^2 - 2k_0k_f\cos\theta_f$. The δ-function can then be used in an integration over $d(k_f^2)$ and, since $v_0 = \hbar k_0/m$, the cross section is expressed in the following convenient form

$$d\sigma = \frac{m^2}{8\pi\hbar^4}\frac{1}{k_0^2}\int q\, dq \sum_{\boldsymbol{\varepsilon}} |H_{f0}|^2\frac{k^2 dk\, d\Omega}{(2\pi)^3}. \qquad (6.150)$$

The squared amplitude $|H_{f0}|^2$ must contain an appropriate mixture of direct, exchange, and interference terms. With the direct amplitude given by [see Equation (6.129)]

$$H_{f0}(\text{direct}) = A = (8\pi\alpha)^{1/2} zV(\boldsymbol{q})k^{-3/2}a, \qquad (6.151)$$

a combination of terms $(A + A_e)^2$ and $(A - A_e)^2$ must be taken with weighting factors determined by the relative numbers of two-particle spin states having the right spin-exchange symmetry (see Section 2.4 of Chapter 4). For unpolarized spin-$\frac{1}{2}$ particles, the combination is

$$W_{1/2} = \tfrac{1}{4}(A + A_e)^2 + \tfrac{3}{4}(A - A_e)^2 = A^2 + A_e^2 - AA_e, \qquad (6.152)$$

while for general spin

$$W_s = A^2 + A_e^2 + \frac{2(-1)^{2s}}{(2s+1)}AA_e. \qquad (6.153)$$

If the scattering potential is provided by a Coulomb field, $A \propto a/q^2$ and $A_e \propto a_e/q_e^2$, in which the exchange terms are obtained by the replacement $\boldsymbol{k}_f \rightarrow -\boldsymbol{k}_f$ [see Equations (6.138) and (6.142)]. The combination (6.152) for spin-$\frac{1}{2}$ or, more generally, the expression (6.153) for arbitrary spin, is to be employed in place of $|H_{f0}|^2$ in the cross section (6.150). Summation over $\boldsymbol{\varepsilon}$ and integration over $d\Omega$ is then performed using the identities (6.135) and (6.136). Then an integration over dq is performed between the limits $q_{\min} = k_0 - k_f$ and $q_{\max} = (k_0^2 + k_f^2)^{1/2}$, corresponding, respectively, to $\theta_f = 0$ and $\pi/2$. The result for Coulomb scattering of unpolarized particles of arbitrary spin can be written

$$d\sigma = \tfrac{4}{15} z^6 \alpha^3 \Lambda^2 (dk/k) S_s(\xi), \tag{6.154}$$

with $\Lambda = \hbar/mc$ and

$$S_s(\xi) = \Phi(\xi) + F_s(\xi), \tag{6.155}$$

where

$$\Phi(\xi) = 6(1+\xi^2)\ln\frac{1+\xi}{1-\xi} + 20\xi, \tag{6.156}$$

$$F_s(\xi) = \frac{2(-1)^{2s}}{2s+1}\left[\frac{7(1-\xi^4)^2 + 3(1-\xi^2)^4}{2(1+\xi^2)^3}\ln\frac{1+\xi}{1-\xi} + \frac{6\xi(1+\xi^4)}{(1+\xi^2)^2}\right]. \tag{6.157}$$

The parameter ξ is determined by k_0 and k by the relation (6.97) and $|dk/k| = 2\xi\, d\xi/(1-\xi^2)$.

The integrated cross section is, for spin-$\frac{1}{2}$,

$$\Omega_{1/2}(E_0) = \int \hbar ck(d\sigma/dk)\, dk = \tfrac{5}{9}(44 - 3\pi^2) z^6 \alpha^3 \Lambda^2 E_0, \tag{6.158}$$

in terms of the total initial c.m. energy $E_0 = \hbar^2 k_0^2/m$. This is in the Born approximation, like the cross section (6.154), and a Born correction can be made. The simplest procedure is to apply a Sommerfeld-Elwert factor to multiply the cross section. For electron-electron bremsstrahlung, this factor would be the same as the electron-ion formula (6.91) with $z = Z = -1$ and the initial and final velocities equal to relative values. In terms of the initial c.m. velocity v_0 (half the relative velocity $v_{0r} = 2v_0$), the corresponding n_0 is

$$(n_0)_{ee} = e^2/2\hbar v_0 = (\mathrm{Ry}/2E_0)^{1/2}. \tag{6.159}$$

That is, if $\frac{1}{2}$Ry ($= \mathrm{Ry}_\mu$) is the reduced-mass Rydberg constant and E_0 the total initial c.m. energy, the Sommerfeld-Elwert factor has the same form in terms of Ry_μ/E_0 as the one-particle formula (6.92):

$$f_{S(ee)} \approx 1 - \pi(\mathrm{Ry}/2E_0)^{1/2}(1/\xi - 1). \tag{6.160}$$

The correction [of the form (6.93)] to the integrated cross section can be evaluated, with Ω_{Born} given by the result (6.158). To do this, as well as other computations, it is convenient to have a simpler representation of the function (6.157). A useful approximation is

$$F_{1/2}(\xi) \approx -16\xi + c_2\xi^2, \tag{6.161}$$

with $c_2 = 14.630$. The form (6.161) has the right limit and derivative at $\xi = 0$ and yields the correct integrated cross section (6.158). Then with

$$\Omega_{ee} = \Omega_{\text{Born}}(1 - \omega_{ee}), \tag{6.162}$$

we obtain

$$\omega_{ee} = 0.738(\text{Ry}/E_0)^{1/2}. \tag{6.163}$$

6.4 INTERMEDIATE ENERGIES

For electron energies $E_e \sim Z^2\text{Ry}$ neither the classical nor the Born approximation holds for the scattering of an electron by a heavy ion Z (or, for $Z = 1$, in e–e scattering). At this "intermediate" energy, only the problem of electron-ion bremsstrahlung has been solved—and only with the help of some clever mathematical manipulations. The calculation of the cross section is a rather complex task and, instead of reproducing the derivation here, reference will be made to the original work by Sommerfeld and his collaborators. However, we shall try to exhibit some useful general formulas and show their relationship to results in the classical and Born limits.

6.4.1 General Result. Gaunt Factor

In the "exact" treatment of Coulomb bremsstrahlung[18] the electromagnetic perturbation H' associated with photon emission is regarded as a perturbation, but its matrix element is computed using exact Coulomb wave functions for the incoming and outgoing electron. These wave functions were introduced and discussed briefly in Section 2.2. of Chapter 4; they are hypergeometric functions and, although they cannot be represented in simple analytic form, the functions have the appearance of oscillatory trigonometric functions with variable amplitude, wavelength, and phase:

$$R_{nl}(r) = A(r)\sin[k(r)r + \phi(r)]. \tag{6.164}$$

Sommerfeld and others have shown how the Coulomb functions can be represented in terms of integrals in a complex plane, and these expressions are convenient for computing the matrix elements of H' in the bremsstrahlung problem. The cross section is evaluated in the usual way from the Fermi Golden Rule formula and, on integration over angles for the outgoing electron and photon and summation over polarizations, the result can be expressed differential in the photon energy. We do not even exhibit the result[19] here; it involves a derivative of the square of a hypergeometric function.

The only assumption or approximation made in the calculation is that retardation is negligible (dipole approximation). This is a very good approximation in the

[18] Only electron-ion bremsstrahlung has been given an exact quantum mechanical treatment; e–e bremsstrahlung has been computed (previous section) in the Born approximation (and with a Sommerfeld-Elwert correction factor) but not in an exact formulation.

[19] See H. A. Bethe and E. E. Salpeter, *Quantum Mechanics of One- and Two-Electron Atoms*, Berlin: Springer-Verlag, 1957; herein references may be found to the relevent original papers.

intermediate-energy domain and corresponds to

$$k\Delta x \ll 1, \tag{6.165}$$

where k is the photon wave number and Δx is some characteristic spatial extent of the range of integration in the matrix element for the process. Let us assume that $\hbar ck \sim E_0$ = electron initial energy and, for simplicity, take $Z = 1$. In the classical limit, ($E_0 \ll$ Ry), $\Delta x \sim e^2/E_0 \sim$ distance of closest approach, and $k\Delta x \sim e^2/\hbar c = \alpha \approx 1/137$, so that the dipole approximation is well satisfied. In the Born limit ($E_0 \gg$ Ry), $\Delta x \sim \lambda\!\!\!^{-}_e \sim \hbar(mE_0)^{-1/2}$, and $k\Delta x \sim (E_0/mc^2)^{1/2}$, which is small until E_0 is as large as, say, ~ 10 keV. Actually, in the angular-integrated cross section, the neglect of retardation can be expected to introduce a relative error[20] of the order $(k\Delta x)^2$. We should note that the dipole approximation will fail at energies that are an appreciable fraction of mc^2. However, then the Born approximation will be valid, being more accurate at the higher energies. At the characteristic energy $E_0 \sim$ Ry, $k\Delta x \sim \alpha$ as in the classical domain, so the dipole approximation is quite good throughout the low and intermediate domains.

As mentioned briefly in Section 6.1.2, it is convenient to express the general bremsstrahlung cross section in terms of the relation (6.34): $d\sigma = d\sigma_0 g$. Let us also remember that the cross section $d\sigma_0$ is a *limit*[21] of the general classical formula. From an exact calculation of $d\sigma$ using Coulomb wave functions, the ratio $d\sigma/d\sigma_0$ ($\equiv g$) can be computed, and the result is a function of, say, the initial electron energy E_0 and the bremsstrahlung energy $\hbar\omega$:

$$g = g(E_0, \omega). \tag{6.166}$$

The most significant property of the Gaunt factor (6.166) is its weak dependence on E_0 and ω. The function $g(E_0, \omega)$ has been computed and exhibited graphically[22] as a function of E_0 for various values of ω and as a function of ω for various E_0. The Karzas-Latter (K-L) curves in their Figures 1 and 2 may be a little misleading, however. The K-L Figures 1 and 2 are plots of the Gaunt factor for *absorption*, that is, the process inverse to bremsstrahlung. Therefore, their E_i is actually our final electron energy E_f, so that for (emission) bremsstrahlung the initial electron energy is $E_0 = \hbar\omega + E_f = \hbar\omega + E_i$ (with the K-L E_i). The most important domain is $\hbar\omega \sim E_0$, and g is then close to unity, while the lower portion of the K-L Figure 1 (for which $g < 1$) corresponds to the very high-frequency end of the emission spectrum ($E_f \ll E_0$). In fact, for $\hbar\omega/E_0 \sim 0.1$ to 1, the factor g is in the narrow range 1 to 2, and it would be useful if a more precise exhibition of g were made in this range. This would relieve the difficulty of reading from the K-L curves, which are log-log plots spread over a factor of 10^5 in g.

Away from the soft-photon limit, it is possible to obtain a simple approximation for $g(E_0, \omega)$ by applying the Sommerfeld-Elwert factor (6.82) to multiply the Born cross section (6.80). Comparison with the K-L calculations will then give us some idea of the accuracy factor f_S. From the definition (6.81) for f_S, the corresponding

---

[20] See Section 2.2.

[21] Often in the literature, the formula for $d\sigma_0$ is referred to as the "classical" expression.

[22] W. J. Karzas and R. Latter, *Astrophys. J. Suppl.* **6**, 167 (1961).

approximation to the Gaunt factor is

$$g \approx g_{\mathrm{B}} f_{\mathrm{S}}, \tag{6.167}$$

with

$$g_{\mathrm{B}} = \frac{\sqrt{3}}{\pi} \ln \frac{1+\xi}{1-\xi}, \tag{6.168}$$

where

$$\hbar\omega = E_0 - E_f = E_0(1 - \xi^2). \tag{6.169}$$

Further, for $E_0 \sim E_f \lesssim Z^2\mathrm{Ry}$, the exponential terms in the expression (6.82) for f_{S} are small ($\sim 10^{-3}$) and

$$f_{\mathrm{S}} \approx 1/\xi, \tag{6.170}$$

yielding

$$g_{\mathrm{S}} \approx \frac{\sqrt{3}}{\pi\xi} \ln \frac{1+\xi}{1-\xi}. \tag{6.171}$$

As a measure of the accuracy of this expression, we can, for example, take $\xi = 2^{-1/2}$ corresponding to $\hbar\omega = \frac{1}{2}E_0$; the expression (6.171) then yields $g_{\mathrm{S}} \approx 1.37$. The value at $\hbar\omega = Z^2\mathrm{Ry}$ from the K-L Figure 1 seems to be quite close to 1.37 (within $\sim 3\%$ on the basis of a rough reading of the K-L curve).

For small ξ, that is, near the high-frequency end of the photon spectrum, the expression (6.171) can be expanded:

$$g_{\mathrm{S}} = \frac{2\sqrt{3}}{\pi}\left(1 + \tfrac{1}{3}\xi^2 + \tfrac{1}{5}\xi^4 + \cdots\right), \tag{6.172}$$

giving a value $g_{\mathrm{S}} \to 2\sqrt{3}/\pi = 1.10$ at $\xi = 0$, again close to the extrapolation of the K-L curve (for absorption) at $\hbar\omega = Z^2\mathrm{Ry}$. The forms (6.171) and (6.172) are particularly important in yielding a finite (and accurate) value for the cross section at the high-frequency limit. On the other hand, the Born formula (6.168) without the factor f_{S} gives (incorrectly) a zero value for the cross section in this limit.

Another quantity of interest is the integrated cross section Ω [see Equation (6.35)], which can be computed easily using the approximate form (6.171) to correct the cross section (6.33). The result can be written in terms of the factor Ω_0 [see Equation (6.36)]:

$$\Omega_{\mathrm{S}} = \Omega_0 \bar{g}_{\mathrm{S}}, \tag{6.173}$$

where

$$\bar{g}_{\mathrm{S}} = 2\int_0^1 g_{\mathrm{S}}(\xi)\,\xi\,d\xi = 4\sqrt{3}\ln 2/\pi = 1.5286 \tag{6.174}$$

is an averaged Sommerfeld-Elwert factor. The result can also be compared with the Born value (6.79):

$$\Omega_{\mathrm{S}}/\Omega_{\mathrm{B}} = 24\ln 2/\pi^2 = 1.6855, \tag{6.175}$$

which again indicates the magnitude of the f_S correction when applied. It is well, at the same time, to emphasize the limitations or shortcoming of both results. The expressions apply only in the intermediate domain where $E_0 \sim z^2\text{Ry}$ and, more precisely, the factor $\overline{g}_S$ should be a function of E_0 and have the limiting value unity for $E_0 \ll Z^2\text{Ry}$. This shortcoming is simply a reflection of the approximate nature of the expression (6.82) for the factor f_S; as is clear from the analysis [see Equation (6.85)] that yielded the formula, the (approximate) correction is specifically designed for the domain $E_0 \sim Z^2\text{Ry}$. At this energy, the result (6.174) can be compared with the peak value (1.43) of the K-L Figure 6; the approximate value is about 7% too high.

6.4.2 Soft-Photon Limit

In the limit $\hbar\omega \ll E_0$, the general bremsstrahlung problem simplifies somewhat in that the exact expression for the cross section for arbitrary E_0, although it involves a special transcendental function, can be evaluated easily. The cross section was derived by Elwert[23] and the result is

$$d\sigma = (16/3)\alpha^3 \Lambda^2 Z^2 \beta_0^{-2} (d\omega/\omega) L_E, \tag{6.176}$$

where

$$L_E = L_B - \Phi, \tag{6.177}$$

in which L_B is the Born logarithm [see Equation (6.80) in the limit $\hbar\omega \ll E_0$]

$$L_B = \ln(2mv_0^2/\hbar\omega), \tag{6.178}$$

and

$$\Phi = \Phi(n) = \gamma + \text{Re}\,\psi(in). \tag{6.179}$$

Here (for simplicity the subscript 0 has been left off n)

$$n = Ze^2/\hbar v_0, \tag{6.180}$$

and ψ is the digamma function[24] or logarithmic derivative of the gamma function; γ (= 0.5772...) is Euler's constant.

The function $\Phi(n)$ is tabulated[24] but much more interesting are its asymptotic series expansions in powers of n and $1/n$. Since the classical Born limits correspond to, respectively, $n \gg 1$ and $n \ll 1$, the series in powers of n would then be a Born series and that in powers of $1/n$, would be a WKB (Wentzel-Kramers-Brillouin) series. The former series is[24]

$$\Phi = \sum_{k=1}^{\infty} (-1)^{k-1} \zeta(2k+1) n^{2k}, \tag{6.181}$$

but a better way of writing this series is in the form

$$\Phi = n^2/(1+n^2) + \sum_{k=1}^{\infty} (-1)^{k-1} [\zeta(2k+1) - 1] n^{2k}. \tag{6.182}$$

---

[23] G. Elwert, *Ann. Phys.* **34**, 178 (1939); *Z. Naturforsch.* **3a**, 477 (1948).

[24] The properties of this function, including asymptotic expansions that will be employed, are summarized by M. Abramowitz and J. A. Stegun, *Handbook of Mathematical Functions*, New York: Dover Publ. Inc., 1965.

In these series, in which ζ is the Riemann ζ–function, the second (6.182) is actually more useful although it appears to be more complicated; this is because the coefficients $(\zeta - 1)$ become increasingly very small for the larger k. As a result, although it is not precisely a power series in n like the result (6.181), the series (6.182) is more convenient for computing Φ when n is small.

The WKB series is[24]

$$\Phi = \ln n + \gamma + \frac{1}{12n^2}\left(1 + \frac{1}{10n^2} + \frac{1}{21n^4} + \cdots\right). \tag{6.183}$$

Although, in terms exhibited, the coefficients are small, this is actually an asymptotic series; that is, it does not converge and must eventually be cut off. The result (6.183) yields, in the classical limit $n \gg 1$, the limit

$$L_{\mathrm{E}} \rightarrow L_{\mathrm{cl}} = \ln(2mv_0^3/\Gamma Ze^2\omega), \tag{6.184}$$

obtained earlier in Equation (6.15).

In the soft-photon limit, the arguments of both the Born and classical logarithms (6.178) and (6.184) are large but they are different, being related by

$$L_{\mathrm{B}} = L_{\mathrm{cl}} + \gamma + \ln n. \tag{6.185}$$

Clearly, then, for finite $\hbar\omega/E_0$, it matters which one is employed; although this comment seems trivial, there has been some confusion in the literature on this point. Finally, it might be noted, as was done earlier, that the factor f_{S} is of no use in correcting the Born formula in the soft-photon limit since $f_{\mathrm{S}} \rightarrow 1$ in that limit. The correction factor f_{S} is an approximate one and is best applied away from the soft-photon limit.

6.5 RELATIVISTIC COULOMB BREMSSTRAHLUNG

At relativistic energies, the problem of photon production in charged-particle scattering is much more complicated. Even with the simplified methods of modern QED, a considerable amount of mathematical manipulations are required for the derivation of the cross section for arbitrary energy. The basic problem is that of the production of a bremsstrahlung photon in the scattering of an electron in the Coulomb field of a heavy nucleus (Z). At relativistic energies for the incident electron, it is no longer a good approximation, as in the non-relativistic problem, to neglect the momentum carried away by the outgoing photon; however, except when Z is large, the Born approximation is completely valid. The effects of spin and magnetic moment are important, and the bremsstrahlung cross section for scattering in a fixed Coulomb field is different for spin-0 and spin-$\frac{1}{2}$ particles. As in the problem of Compton scattering treated in the previous chapter, the case of spin-0 is much simpler than the more important one of spin-$\frac{1}{2}$. The cross section in the former case is similar (though different) to that for the latter, and it is useful to outline the derivation of both results, first treating the problem of spin 0. In fact, because of complexity of the mathematical details, the derivation of the spin-$\frac{1}{2}$ formula will not be carried through comletely after the fundamental steps have been set up.

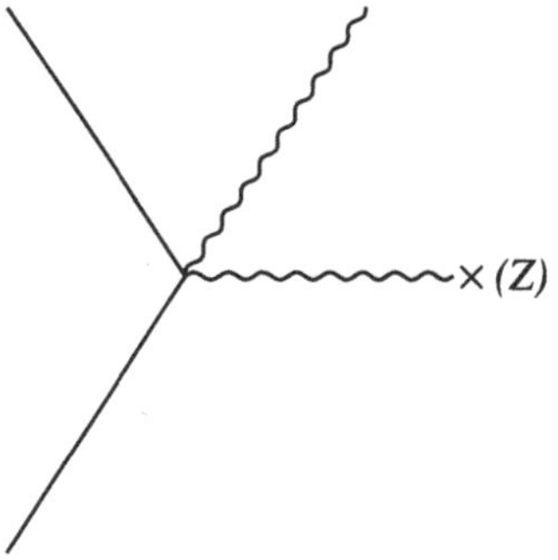

Figure 6.3 Additional diagram for bremsstrahlung in scattering by a spin-0 charge.

Bremsstrahlung can also be treated by the Weizsäcker-Williams method (see Section 8, Chapter 2) and this will be done later in this section. The W-W method provides some useful insight into the problem and to related phenomena. One closely related problem is that of *e–e* bremsstrahlung and this is treated at the end of the Section.

6.5.1 Spin-0 Problem

The Feynman diagrams for bremsstrahlung in the scattering of scalar charges in an external Coulomb potential are the same[25] as the graphs in Figure 6.1. The two corresponding amplitudes for the process are of the form

$$A_I + A_{II} = H' P_F(I) V + V P_F(II) H'. \tag{6.186}$$

Here H' is the invariant factor associated with the photon-emission vertex and V is that associated with the scattering-potential perturbation; P_F is the Feynman propagator. The perturbations H' and V are basically of the same nature, being proportional to the invariant $A_\mu(p_a + p_b)_\mu$, where $p_{a\mu}$ and $p_{b\mu}$ are, respectively, the incoming and outgoing four-momenta of the charges involved in the vertex and A_μ is the four-vector potential of the corresponding electromagnetic field [see Chapter 3, Equation (3.74)]. For the photon-emission vertex, in the momentum representation the A_μ would yield a factor ε_μ for the photon polarization four-vector and the matrix-element amplitude is of the form

$$H'_{ba} = \varepsilon_\mu (p_a + p_b)_\mu. \tag{6.187}$$

The vector potential associated with the Coulomb field of the central static charge Z has only a time component and the perturbation associated with the scattering is,

[25]Actually, there is another diagram (see Figure 6.3). This corresponds to a coupling $A_\mu A_{Z\mu}$, where A_μ refers to the outgoing photon and $A_{Z\mu}$ to the field of the fixed heavy nucleus Z. However, the latter has only the time component (scalar potential $i\Phi_Z$) and the former only spatial components (by choice of gauge $\varepsilon_0 = 0$). Thus, the diagram does not contribute.

in the momentum representation [see Chapter 4, Equations (4.115), (4.116)],

$$V_{ba} = (E_b + E_a)/\boldsymbol{q}^2; \tag{6.188}$$

here $\boldsymbol{q}$ is the momentum transfer to the scattering center.

In this section, since we are dealing with relativistic charges, let us take $\hbar = c = 1$. For scalar particles the Feynman propagator is [Chapter 3, Equation (3.86)]

$$P_F(p) = (p^2 + m^2)^{-1}, \tag{6.189}$$

where p_μ is the four-momentum in the intermediate state. Let $\boldsymbol{p}_0$ and $\boldsymbol{p}$ refer to the initial and final momenta of the charge and $\boldsymbol{k}$ that of the outgoing photon. The momentum transfer to the scattering center is then

$$\boldsymbol{q} = \boldsymbol{p}_0 - \boldsymbol{p} - \boldsymbol{k}. \tag{6.190}$$

A gauge can be chosen [see Chapter 3, Equation (3.158)] such that ε_μ has a time component $\varepsilon_0 = 0$ and, since the outgoing photon is transverse,

$$\varepsilon \cdot k = \varepsilon_\mu k_\mu = \boldsymbol{\varepsilon} \cdot \boldsymbol{k}. \tag{6.191}$$

There is four-momentum conservation at the vertices in Figure 6.1, so that, for each component of μ,

$$p_I = p + k, \quad p_{II} = p_0 - k. \tag{6.192}$$

The scattering center absorbs momentum but not energy, then, if E_0 and E refer to the initial and final energies of the charge,

$$E_I = E_0, \quad E_{II} = E, \quad E_0 = E + k. \tag{6.193}$$

The amplitudes A_I and A_{II} can now be evaluated:

$$\begin{aligned} A_I &= \frac{\boldsymbol{\varepsilon} \cdot (\boldsymbol{p} + \boldsymbol{p}_I)}{(p+k)^2 + m^2} \frac{E_I + E_0}{\boldsymbol{q}^2} = \frac{2E_0(\boldsymbol{\varepsilon} \cdot \boldsymbol{p})}{\boldsymbol{q}^2 (p \cdot k)}, \\ A_{II} &= \frac{(E + E_{II})}{\boldsymbol{q}^2} \frac{\boldsymbol{\varepsilon} \cdot (\boldsymbol{p}_{II} + \boldsymbol{p}_0)}{(p_0 - k)^2 + m^2} = -\frac{2E(\boldsymbol{\varepsilon} \cdot \boldsymbol{p}_0)}{\boldsymbol{q}^2 (p_0 \cdot k)}. \end{aligned} \tag{6.194}$$

The result[26] can be compared with the non-relativistic expression (6.49) and with the relativistic soft-photon amplitude (3.154) of Chapter 3. To compute the cross section, the non-covariant general relation (6.55) can be employed with the differential number of final states given by[27]

$$dN_f = (2\pi)^{-6}(\hbar^{-3}) p^2 dp\, d\Omega\, k^2 dk\, d\Omega_k, \tag{6.195}$$

and the δ-function equal to $\delta(E_0 - E - E_k)$. The squared perturbation Hamiltonian will be given by something of the form $[E_k = (\hbar c)k]$

$$|H_{f0}|^2 \to |\mathcal{M}|^2 / E_0 E E_k, \tag{6.196}$$

with the invariant amplitude $\mathcal{M}$ of the form

$$\mathcal{M} = C(A_I + A_{II}), \tag{6.197}$$

[26] We again employ the notation used before for relativistic formulations. The dot products in the denominators without boldface are the invariants $p \cdot k = p_\mu k_\mu$, etc.

[27] The factor $\hbar^{-3}$ would be included if c.g.s. units are employed.

where the multiplying constant C can be determined by a comparison with the non-relativistic formula already derived. If we can take the forms (6.194) with (c.g.s. units) the particle momenta equal to $p_\mu = \gamma mc(i, \boldsymbol{\beta})$ and the photon momentum equal to $\hbar k_\mu = \hbar k(i, \boldsymbol{n})$, with $\boldsymbol{n} = \boldsymbol{k}/k$ the unit vector in the direction of the photon motion,

$$\mathcal{M}/C = -\frac{2mc^2}{\boldsymbol{q}^2}\left(\gamma_0 \frac{\boldsymbol{\varepsilon}\cdot\boldsymbol{\beta}}{1-\boldsymbol{n}\cdot\boldsymbol{\beta}} - \gamma\frac{\boldsymbol{\varepsilon}\cdot\boldsymbol{\beta}_0}{1-\boldsymbol{n}\cdot\boldsymbol{\beta}_0}\right). \tag{6.198}$$

The non-relativistic limit (E, $E_0 \to mc^2$) can now be taken for comparison with the result (6.52) already derived. If the charge ze is scattered by a fixed central Coulomb potential charge (charge Ze), from the comparison (6.196) we readily obtain

$$C^2 = (2\pi)^3 z^4 Z^2 e^6. \tag{6.199}$$

The cross section is evaluated from

$$v_0 d\sigma = \frac{2\pi}{\hbar}\int \frac{|\mathcal{M}|^2}{E_0 E E_k}\delta(E_0 - E - E_k)dN_f, \tag{6.200}$$

and, since $E\,dE = (c^2)p\,dp$, the δ-function can be eliminated by integrating over dE (in dN_f) to give ($\varkappa = mc/\hbar = 1/\Lambda$)

$$d\sigma = \frac{\alpha^3}{\pi^2}\varkappa^2 z^4 Z^2 \frac{p}{p_0}\frac{dk}{k}\left[\gamma_0\frac{\boldsymbol{\varepsilon}\cdot\boldsymbol{\beta}}{1-\boldsymbol{n}\cdot\boldsymbol{\beta}} - \gamma\frac{\boldsymbol{\varepsilon}\cdot\boldsymbol{\beta}_0}{1-\boldsymbol{n}\cdot\boldsymbol{\beta}_0}\right]^2 \frac{d\Omega\, d\Omega_k}{\boldsymbol{q}^4}, \tag{6.201}$$

with the bracket expression in terms of the initial and final (dimensionless) β and γ. This is the most general result, valid for arbitrary polarization $\boldsymbol{\varepsilon}$ and differential in dk and the outgoing charge and photon solid angle elements. It has been expressed in a form that clearly indicates its "order" (α^3) and correct dimensionality ($|\boldsymbol{q}|$ has the dimensions of inverse length). A sum over polarizations can be performed using the identity

$$\sum_{\boldsymbol{\varepsilon}}(\boldsymbol{\varepsilon}\cdot\boldsymbol{a})(\boldsymbol{\varepsilon}\cdot\boldsymbol{b}) = \boldsymbol{a}\cdot\boldsymbol{b} - (\boldsymbol{n}\cdot\boldsymbol{a})(\boldsymbol{n}\cdot\boldsymbol{b}), \tag{6.202}$$

where $\boldsymbol{a}$ and $\boldsymbol{b}$ are arbitrary vectors. In terms of the angles[28] designated in Figure 6.4 the polarization sum of the squared brackets in Equation (6.201) is

$$\sum_{\boldsymbol{\varepsilon}}[\ldots]^2 \equiv W_0 = \gamma_0^2\frac{\beta^2\sin^2\theta}{(1-\beta\cos\theta)^2} + \gamma^2\frac{\beta_0^2\sin^2\theta_0}{(1-\beta\cos\theta_0)^2} - 2\gamma_0\gamma\beta_0\beta\frac{\sin\theta_0\sin\theta\cos\phi}{(1-\beta\cos\theta)(1-\beta_0\cos\theta_0)}. \tag{6.203}$$

In terms of the angles θ_0, θ, and ϕ, the solid angle element for the outgoing charge and photon is

$$d\Omega\, d\Omega_k = 2\pi\sin\theta_0\sin\theta\, d\theta_0\, d\theta\, d\phi. \tag{6.204}$$

[28]These are the convenient angles for the problem; θ is the angle between $\boldsymbol{p}$ and $\boldsymbol{k}$ while θ_0 is that between $\boldsymbol{p}_0$ and $\boldsymbol{k}$; ϕ is the angle between the $(\boldsymbol{p}_0, \boldsymbol{k})$ and $(\boldsymbol{p}, \boldsymbol{k})$ planes.

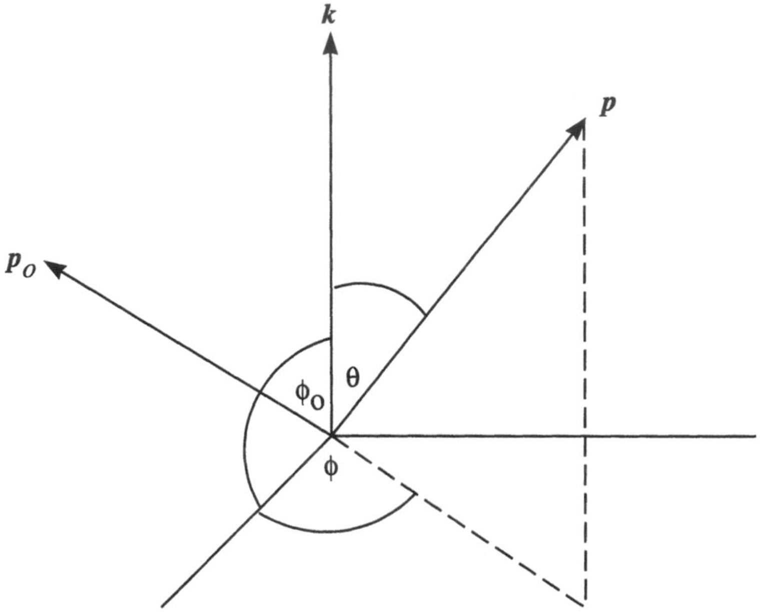

Figure 6.4 Angles describing the outgoing electron and photon.

The squared momentum transfer is

$$q^2 = p_0^2+p^2+k^2-2p_0p(\cos\theta_0\cos\theta+\sin\theta_0\sin\theta\cos\phi)-2p_0k\cos\theta_0+2pk\cos\theta. \quad (6.205)$$

Thus, the cross-section formula is rather complex and the integration over the solid angle element (6.204) to give $d\sigma$ differential in dk is a tedious problem. In fact, even in the extreme relativistic limit the integration involves a certain amount of analysis. Instead of performing these integrations, let us proceed directly to the more important spin-$\frac{1}{2}$ problem. The expression (6.201) is actually the fundamental result for spin–0 and was derived mainly to illustrate the methods of covariant QED and for comparison with the spin-$\frac{1}{2}$ result.

6.5.2 Spin-$\frac{1}{2}$: Bethe-Heitler Formula

The cross section for bremsstrahlung in the scattering of a relativistic spin-$\frac{1}{2}$ particle in a Coulomb potential was first calculated by Bethe and Heitler and independently by Sauter.[29] As in the spin-0 case considered above, there are two Feynman diagrams for the process, again as in Figure 6.1, and the total amplitude is obtained from an expression of the form (6.186). However, now the photon-coupling has the form $\not{\epsilon}$ and the Feynman propagator equals $(\not{p} - im)^{-1}$; further, the Coulomb-scattering factor has the form, say, $(Z/q^2)\gamma_0$, where γ_0 is now a Dirac matrix [see Chapter 4,

[29] H. A. Bethe and W. Heitler, *Proc. Roy. Soc.* **A146**, 83 (1934); F. Sauter, *Ann. Phys.* **20**, 404 (1934).

Equation (4.181)]. Let us write the Coulomb-scattering factor as

$$V = (Z/q^2)\not{a}, \tag{6.206}$$

where

$$a_\mu = (i, 0). \tag{6.207}$$

The amplitude for the process is then of the form

$$\begin{aligned} M &= \frac{Z}{q^2}\left[\not{\varepsilon}\frac{1}{\not{p}_I - im}\not{a} + \not{a}\frac{1}{\not{p}_{II} - im}\not{\varepsilon}\right], \\ &= \frac{Z}{q^2}\left[\not{\varepsilon}\frac{\not{p} + \not{k} + im}{(p+k)^2 + m^2}\not{a} + \not{a}\frac{\not{p}_0 - \not{k} + im}{(p_0 - k)^2 + m^2}\not{\varepsilon}\right], \\ &= \frac{Z}{q^2}\left[\frac{\not{\varepsilon}(\not{p} + \not{k} + im)\not{a}}{2p\cdot k} - \frac{\not{a}(\not{p}_0 - \not{k} + im)\not{\varepsilon}}{2p_0\cdot k}\right]. \end{aligned} \tag{6.208}$$

This expression can be simplified further. Since M is placed between the spinors $\overline{u}$ and u_0 to form

$$\mathcal{M} = \overline{u}Mu_0, \tag{6.209}$$

and since $\overline{u}\not{p} = im\overline{u}$ and $\not{p}_0 u_0 = imu_0$, in the first term in the brackets (6.208) $\not{p}$ can be moved to the left by employing the identity (see Chapter 3)

$$\not{\varepsilon}\not{p} = 2\varepsilon\cdot p - \not{p}\not{\varepsilon}. \tag{6.210}$$

Similarly, in the second term, $\not{p}_0$ can be moved to the right, and the amplitude (6.208) simplifies to

$$\frac{M}{Z/q^2} = \frac{(2\varepsilon\cdot p + \not{\varepsilon}\not{k})\not{a}}{2p\cdot k} - \frac{\not{a}(2\varepsilon\cdot p_0 - \not{k}\not{\varepsilon})}{2p_0\cdot k}. \tag{6.211}$$

The amplitude $M' = \gamma_0 M^\dagger \gamma_0$ is also needed [see Chapter 5, Equations (5.80)-(5.83)]:

$$\frac{M'}{Z/q^2} = -\frac{2\varepsilon\cdot p\not{a} + \not{a}\not{k}\not{\varepsilon}}{2p\cdot k} + \frac{2\varepsilon\cdot p_0\not{a} + \not{\varepsilon}\not{k}\not{a}}{2p_0\cdot k}. \tag{6.212}$$

Summing over initial and final spins, the invariant trace is formed (see Section 4.1):

$$X = \frac{1}{2}\mathrm{Tr}\left(M'\frac{\not{p} + im}{2im}M\frac{\not{p}_0 + im}{2im}\right). \tag{6.213}$$

This quantity corresponds to the factor $|\mathcal{M}|^2$ in the formula (6.200) for the cross section when the formula is applied in the spin-$\frac{1}{2}$ problem. Then, if we are not interested in the polarization of the outgoing photon a sum over $\boldsymbol{\varepsilon}$ components can be performed. As in the spin-0 problem, the multiplying constant for $d\sigma_{1/2}$ can be fixed through a comparison in the non-relativistic limit.

The result for the cross section—the Bethe-Heitler formula—can be expressed very concisely by comparing it with the spin-0 formula (6.201) [and (6.203)]. Both $d\sigma_0$ and $d\sigma_{1/2}$ can be written in the form

$$d\sigma = \frac{1}{\pi^2}\alpha^3\varkappa^2 z^4 Z^2 \frac{p}{p_0}\frac{dk}{k}\frac{d\Omega\, d\Omega_k}{\boldsymbol{q}_4} W, \tag{6.214}$$

where W is a dimensionless factor, equal to the expression (6.203) for spin 0. The factors W_0 and $W_{1/2}$ are basically functions of the momenta $\boldsymbol{p}_0$, $\boldsymbol{p}$, and $\boldsymbol{k}$ and are easily expressed in terms of the energies E_0, E, and k and certain dimensionless vectors. For example, the spin-0 formula can be written

$$m^2 W_0 = (E_0\boldsymbol{v} - E\boldsymbol{v}_0)^2, \tag{6.215}$$

in which

$$\boldsymbol{v} = (\boldsymbol{k}\times\boldsymbol{p})/(k\cdot p), \quad \boldsymbol{v}_0 = (\boldsymbol{k}\times\boldsymbol{p}_0)/(k\cdot p_0). \tag{6.216}$$

Note that $\boldsymbol{v}^2$, $\boldsymbol{v}_0^2$, and $\boldsymbol{v}\cdot\boldsymbol{v}_0$ are ratios of spatial-rotation scalars to Lorentz invariants. For the spin-$\frac{1}{2}$ problem, evaluation of the trace (6.213) and summation over polarizations yields

$$m^2 W_{1/2} = m^2 W_0 - \tfrac{1}{4}\boldsymbol{q}^2(\boldsymbol{v} - \boldsymbol{v}_0)^2 + \tfrac{1}{2}tk^2 \tag{6.217}$$

where

$$t = \frac{(\boldsymbol{k}\times\boldsymbol{q})^2}{(k\cdot p)(k\cdot p_0)} \tag{6.218}$$

is, again, a ratio of a spatial-rotation scalar to a Lorentz invariant.

The expression (6.217) is a compact way of writing the Bethe-Heitler formula; it is also convenient for comparison with the spin-0 case, especially in the low- and high-energy limits, and for an analysis of the contribution to the cross sections in these limits. In terms of the angles θ, θ_0, and ϕ (see Figure 6.4), the individual coefficients in the expressions for W_0 and $W_{1/2}$ are

$$\begin{aligned} r = \boldsymbol{v}^2 &= \frac{\beta^2\sin^2\theta}{(1-\beta\cos\theta)^2}, \\ r_0 = \boldsymbol{v}_0^2 &= \frac{\beta_0^2\sin^2\theta_0}{(1-\beta_0\cos\theta_0)^2}, \\ s = \boldsymbol{v}\cdot\boldsymbol{v}_0 &= \frac{\beta\beta_0\sin\theta\sin\theta_0\cos\phi}{(1-\beta\cos\theta)(1-\beta_0\cos\theta_0)}, \end{aligned} \tag{6.219}$$

$$t = \frac{(\gamma/\gamma_0)\beta^2\sin^2\theta + (\gamma_0/\gamma)\beta_0^2\sin^2\theta_0}{(1-\beta\cos\theta)(1-\beta_0\cos\theta_0)} - 2s. \tag{6.220}$$

The squared momentum transfer is given in Equation (6.205).

The cross section, differential only in dk, is obtained by performing the integration [see Equation (6.214)]

$$I = \iint d\Omega\, d\Omega_k\, W/\boldsymbol{q}^4, \tag{6.221}$$

and this is a tedious job. The result is a rather complicated formula that is given in the original Bethe-Heitler paper and in many textbooks. Let us note some features of this integration without actually carrying it out. First, in the non-relativistic limit $E, E_0 \to m \gg p, p_0, k, q$, and $W_{1/2} \to W_0$; the effects of spin are negligible at low energies. In this limit we have seen in Sections 6.2 and 6.3 how the cross section differential in dk can be obtained by casting the integral corresponding to the relativistic expression (6.221) into one differential only in dq. This cannot be done in general in the relativistic problem, although Bethe[30] has shown how it can be accomplished in the ultra-relativistic limit. The limit of very high energies is very important and we can at least consider some aspects of the problem in that case. When $E, E_0, k \gg m$, in the integration over $d\Omega\, d\Omega_k$ the main contribution comes from small momentum transfers, as a result of the factor $1/\boldsymbol{q}^4$ in the integrand (6.221). The minimum value of q is indeed very small and corresponds to the case where $\boldsymbol{p}_0$, $\boldsymbol{p}$, and $\boldsymbol{k}$ are in the same direction (θ and $\theta_0 \to 0$). Since $E_0 = E + k$ and $p = (E^2 - m^2)^{1/2} \approx E - m^2/2E$,

$$q_{\min} = p_0 - p - k \approx m^2 k/2E_0 E \ll m. \tag{6.222}$$

In the evaluation of the integrated cross section, the second term ($\propto \boldsymbol{q}^2$) on the right of Equation (6.217) can be neglected. The third term cannot be neglected, however; although the form (6.218) indicates a quadratic dependence on q, $\boldsymbol{k}$ is primarily perpendicular to $\boldsymbol{q}$, and the cross product (6.218) is not small. The contribution from this term is comparable to that from the first term, except in the soft-photon limit, for which it is negligible and[31]

$$d\sigma_{1/2} \to d\sigma_0 \qquad \text{(soft-photon limit)}. \tag{6.223}$$

For general k in the extreme relativistic limit the integration (6.221) yields, for spin-$\frac{1}{2}$,

$$d\sigma_{1/2} = 4\alpha^3 Z^2 \Lambda^2 \frac{dk}{k}\left(1 + \frac{E^2}{E_0^2} - \frac{2E}{3E_0}\right)\left(\ln \frac{2E_0 E}{mk} - \frac{1}{2}\right). \tag{6.224}$$

The energy-integrated cross section [see Equation (6.35)] is obtained by setting $E = E_0 - k$ and evaluating the integral

$$\Omega_{1/2} = \int_0^{E_0} k\, d\sigma_{1/2} = 4\alpha^3 Z^2 \Lambda^2 E_0 \left(\ln \frac{2E_0}{m} - \frac{1}{3}\right). \tag{6.225}$$

Another useful result is the correction to the non-relativistic bremsstrahlung cross section, to be applied when the incident electron is mildly relativistic (say, $E \sim$ 100 keV). By expanding the integrated Bethe-Heitler formula we have[32]

$$d\sigma = \alpha^3 Z^2 \Lambda^2 \frac{d\omega}{\omega}\left[\frac{16}{3}\left(\frac{mc}{p_0}\right)^2 \ln X + \frac{8}{3}\frac{2p_0^2 - p_\omega^2}{p_0^2} \ln X - \frac{4(4p_0^2 - 3p_\omega^2)}{3p_0(p_0^2 - p_\omega^2)^{1/2}}\right], \tag{6.226}$$

[30] H. A. Bethe, *Proc. Camb. Philos. Soc.* **30**, 524 (1934); see also K. S. Suh and H. A. Bethe, *Phys. Rev.* **115**, 672 (1959). We shall refer again to this work.

[31] In connection with this see Chapter 3, Section 5.2.

[32] Since this is basically a (corrected) non-relativistic formula, we insert the factors of c, the photon energy is $\hbar\omega$, and the variable $p_\omega = (2m\hbar\omega)^{1/2}$ is introduced.

where

$$X = \frac{p_0 + (p_0^2 - p_\omega^2)^{1/2}}{p_0 - (p_0^2 - p_\omega^2)^{1/2}} = X(p_\omega; p_0), \tag{6.227}$$

while the corrected energy-conservation relation is

$$\hbar\omega \equiv p_\omega^2/2m = (p_0^2 - p^2)/2m - (p_0^4 - p^4)/8m^3c^2 + \cdots . \tag{6.228}$$

Here, p_0 and p are the *relativistic* momenta; that is, the first term in brackets in Equation (6.226) is (numerically) slightly different from the Born formula (6.80). The second and third terms are small, and therein it is valid to employ non-relativistic kinematic relations. The factor $d\omega/\omega$ equals $2dp_\omega/p_\omega$ and the form (6.226) is expressed in terms of p_ω as variable with p_0 as a parameter. Integrating over p_ω, the energy-integrated cross section can be obtained:

$$\Omega = \int (p_\omega^2/2m)d\sigma; \tag{6.229}$$

the integration limits are 0 and

$$(p_\omega)_{\max} = p_0(1 - p_0^2/8m^2c^2), \tag{6.230}$$

the latter corresponding to p_0 [see Equation (6.228)]. In the integral (6.229) the relativistic correction (6.230) need be applied only for the first term in brackets in the cross section (6.226); however, since this term goes to zero as $p_\omega \to p_0$, the relativistic kinematic correction to $(p_\omega)_{\max}$ does not contribute. We then obtain

$$\Omega = (16/3)\alpha^3 Z^2 \Lambda^2 mc^2\left[1 + \tfrac{1}{3}(p_0/mc)^2\right], \tag{6.231}$$

thereby modifying the non-relativistic Born result (6.79). The correction in brackets is due to relativistic and magnetic-moment effects and is (when multiplied by the factor in front) comparable to the e–e value (6.158).

6.5.3 Relativistic Electron-Electron Bremsstrahlung

The calculation of the cross section for relativistic e–e bremsstrahlung presents a difficult problem. There are eight Feynman diagrams for the process; these would be the diagrams in Figure 4.6 for Møller scattering with external photon lines from any of the four legs representing the incoming and outgoing charges. Similarly, the eight diagrams for e^+–e^- bremsstrahlung would be obtained from the two in Figure 4.7 for Bhabha scattering with photon lines from the four legs. In addition to the task of evaluating the trace associated with these diagrams, there is the problem of integrating over angles for the outgoing particles to obtain an expression for $d\sigma$ differential only in the photon energy. The most recent and definitive calculation of $d\sigma_{ee}$ is perhaps that of Haug,[33] in which only the photon angular integration is

[33] E. Haug, *Z. Naturforsch.* **30a**, 1099 (1975). This paper gives references to most of the previous work on the relativistic e–e problem. An additional paper, not mentioned by Haug, is that of V. L. Lyuboshitz, *Zh. Eksp. Teor. Fiz.* **37**, 1727 (1959) [*Sov. Phys. –JETP* **10**, 1221 (1960)] and is devoted to polarization phenomena; see also, H. A. Olsen, *Applications of Quantum Electrodynamics*, Vol. **44**, pp. 83–201, Berlin: Springer, 1968.

done numerically, the other integrations being done analytically from the (lengthy) expression for the trace.

The relativistic calculation of e–e bremsstrahlung confirms, numerically, the non-relativistic formula (6.154) in that limit. In the intermediate-energy domain ($E_e \sim mc^2$) the cross section $d\sigma_{ee}$ is determined only numerically and not in terms of an analytic formula. However, in the extreme relativistic (ER) limit the e–e cross section does simplify, and can be expressed in terms of a formula. The formula in that limit is very specific and there is a physical reason for the equality that deserves some elaboration. The cross section has the following property: In a frame (K_r) in which one electron is initally at rest the cross section satisfies

$$d\sigma_{ee} \to d\sigma_{ep} \qquad (K_r\text{: ER limit}). \tag{6.232}$$

That is, in a frame with a target electron initially at rest, the cross section is the same as if the target were a proton (or, in fact, an anti-proton ot a positron). The identity (6.232) has its explanation in the momentum transfer distribution for the e–p problem, as discussed in part b of this section. In particular, the result that the main contribution comes from small q means that in the e–e problem in the frame K_r the role of the target electron (say, electron 1) is the same as that of a static Coulomb potential. Then if E_r is the energy of the incident electron and k_r is the photon energy, the cross section in the ER limit in the frame K_r is [see Equation (6.224)]

$$d\sigma_{ee} = 4\alpha^3 \Lambda^2 \frac{dk_r}{k_r}\left(\frac{4}{3} - \frac{4k_r}{3E_r} + \frac{k_r^2}{E_r^2}\right)\left(\ln \frac{2E_r(E_r - k_r)}{mk_r} - \frac{1}{2}\right). \tag{6.233}$$

The photon emission in K_r is primarily in the forward direction, that is, in the direction of the incident electron. However, there is a small component of emission in the backward direction that, while unimportant in K_r, contributes in a transformation to another frame. For example, in the c.m. frame (K'), due to the symmetry of the problem therein, the emission is primarily in the two opposite directions defined by $\boldsymbol{p}_1'$ and $\boldsymbol{p}_2' = -\boldsymbol{p}_1'$.

It is often necessary to have $d\sigma_{ee}$ expressed in lab-frame (K) quantities. Let us assume that in K both electrons are highly relativistic and that their inital momenta $\boldsymbol{p}_1$ and $\boldsymbol{p}_2$ are at an angle θ. In this frame the photon emission is primarily in two narrow cones around these momenta and the cross section $d\sigma_{ee}$ must include the contributions from both of these beams. The expression for $d\sigma_{ee}$ in terms of K-frame quantities can be obtained from a transformation of the K_r-frame formula (6.233) and from the addition of a term to account for the transformed emission in the backward direction in K_r. This term is established by requiring that the total $d\sigma_{ee}$ be a symmetric function of the Labels 1 and 2 designating the electrons before scattering. If the K_r-frame expression (6.233) refers to the case where electron 2 is incident on electron 1 (initially at rest in K_r), the transformation of this expression will yield the K_r-frame contribution to $d\sigma_{ee}$ corresponding to emission in the direction of $\boldsymbol{p}_2$. The two frames K and K_r have relative motion $\beta = v_1/(c) \to 1$ and Lorentz γ equal to E_1/m where E_1 is the lab frame energy (of electron 1). If $\boldsymbol{p}_1$ is, say, along the x-axis of K and $\boldsymbol{p}_2$ is at an angle θ, in K_r electron 2 is incident on electron 1 essentially in the direction of $-x_r$. The photon ($\boldsymbol{k}$) and $\boldsymbol{p}_2$ are at the angle θ in K

and the Lorentz transformations of these quantities are

$$\begin{aligned} k_r &= k(E_1/m)(1-\cos\theta), \\ E_r &= (E_2)_r = E_2(E_1/m)(1-\cos\theta). \end{aligned} \tag{6.234}$$

Then

$$dk_r/k_r = dk/k, \tag{6.235}$$

and

$$k_r/E_r = k/E_2. \tag{6.236}$$

Since, in going from K_r to K, the transformation velocity is along the $(-x_r)$-axis, $d\sigma_{ee}(K) = d\sigma_{ee}(K_r)$, being differential in only the photon energy (and not in angular distribution). Further, in the argument of the logarithm in Equation (6.233) the factor $(E_r - k_r)$ will transform as

$$E_r - k_r = (E_2 - k)(E_1/m)(1-\cos\theta). \tag{6.237}$$

Transforming the cross section (6.233), we then get, in K, the component of $d\sigma_{ee}$ associated with emission in the direction of $\boldsymbol{p}_2$:

$$\begin{aligned} d\sigma_{ee}(2) = 4\alpha^3\Lambda^2\frac{dk}{k}\left(\frac{4}{3} - \frac{4k}{3E_2} + \frac{k^2}{E_2^2}\right) \\ \times\left(\ln\frac{2E_2(E_2-k)E_1(1-\cos\theta)}{m^2k} - \frac{1}{2}\right). \end{aligned} \tag{6.238}$$

The contribution from emission in the direction of $\boldsymbol{p}_1$ is obtained by simply interchanging the indices 1 and 2 to give a total cross section:

$$d\sigma_{ee} = d\sigma_{ee}(2) + d\sigma_{ee}(2 \leftrightarrow 1). \tag{6.239}$$

In the special case of the c.m. frame, $\theta = \pi$, $E_1 = E_2 = E$, and the cross section is

$$d\sigma_{ee}(\text{c.m.}) = 8\alpha^3\Lambda^2\frac{dk}{k}\left(\frac{4}{3} - \frac{4k}{3E} + \frac{k^2}{E^2}\right)\left(\ln\frac{4E^2(E-k)}{m^2k} - \frac{1}{2}\right). \tag{6.240}$$

The K-frame photon-energy integrated cross section can also be obtained easily. The contribution from emission in the direction of $\boldsymbol{p}_2$ will be

$$\Omega_{ee}(2) = \int_0^{E_2} k\,d\sigma_{ee}(2) = 4\alpha^3\Lambda^2E_2\left(\ln\frac{2E_1E_2(1-\cos\theta)}{m^2} - \frac{1}{3}\right). \tag{6.241}$$

[see Equation (6.225)]. The total, including emission in the cone around $\boldsymbol{p}_1$, will be

$$\begin{aligned} \Omega_{ee} &= \Omega_{ee}(2) + \Omega_{ee}(2 \leftrightarrow 1) \\ &= 4\alpha^3\Lambda^2(E_1+E_2)\left(\ln\frac{2E_1E_2(1-\cos\theta)}{m^2} - \frac{1}{3}\right). \end{aligned} \tag{6.242}$$

The above expressions, in particular Equations (6.239)– (6.242) hold when both electrons are highly relativistic, and it should be noted that the results are based on the identity (6.232). The general lab-frame expressions were obtained through the use of kinematic relations. In these equations the invariant property of a certain factor might also be noted. In a frame where both electrons are highly relativistic,

$$E_1 E_2(1 - \cos\theta) \equiv \nu = E_1 E_2 - \boldsymbol{p}_1 \cdot \boldsymbol{p}_2 = -(p_1 \cdot p_2) = \text{inv.} \tag{6.243}$$

Again, this factor is an invariant only among frames where both particles are highly relativistic.

Finally, the reader may be reminded that all of the above results for e–e bremsstrahlung in the ER limit would also hold for e^+– e^+ and $e^+ - e^-$ bremsstrahlung. That is, the same cross-section formulas would hold, independent of the signs of the combination of charges. The identities hold, once again, because of the negligible contributions from large momentum transfers. For the same reason, exchange effects do not come into the cross section formulas. At energies $\sim mc^2$ this would not be the case, however, and the e^+–e^- cross section[34] would differ from $d\sigma_{ee}$. At non-relativistic energies (see Chapter 6, Section 2.4) the e^+–e^- cross section is, of course, much larger than $d\sigma_{ee}$, since the lowest-order amplitude is dipole for the former and quadrupole for the latter.

6.5.4 Weizsäcker-Williams Method

The W-W method was developed in Section 2.8 and, in fact, brief mention was made therein of its application to the bremsstrahlung problem. The idea for the method in this application is that the process in viewed as Compton scattering of the virtual photons of the static (lab frame K) Coulomb field of the nucleus Z by the incident electron. In the frame (K') of the electron, the time-varying electromagnetic fields originating from the charge Ze are equivalent to a flux of photons. The fields or photons, through the interaction associated with the Compton scattering, then become radiation-field outgoing bremsstrahlung photons. Viewed in this manner, the Compton and bremsstrahlung processes are seen to be closely related, and in his original paper von Weizsäcker showed how the highly relativistic limit of the Bethe-Heitler formula can be obtained from the Klein-Nishina formula for Compton scattering.

Let us, in fact, first consider the non-relativistic case and derive the bremsstrahlung cross section from the Thomson-limit formula for Compton scattering. In the NR limit the outgoing photon has the same energy in the K and K' frames and the energy change in the scattering is negligible. The scattering removes the photons from the virtual field to the radiation field and the bremmsstrahlung cross section is computed from the basic relation [see Chapter 2, Equation (2.173)]

$$d\sigma = \int dN\, d\sigma_{\text{sc}}. \tag{6.244}$$

If the light particle (electron) of charge ze is incident on the heavy scattering center (charge Ze) at impact parameter b and with velocity βc, in the frame (K') of the

[34]Relativistic e^+–e^- bremsstrahlung was computed by S. M. Swanson, *Phys. Rev.* **154**, 1601 (1967); the results in this paper are not as useful as the e–e calculations of Haug, however, since all of the angular integrations are not performed.

former the differential number of virtual photons incident on z is

$$dN = \frac{2\alpha}{\pi}\left(\frac{Z}{\beta}\right)^2 \frac{d\omega'}{\omega'}\frac{db'}{b'}. \tag{6.245}$$

In the NR limit, for $d\sigma_{\text{sc}}$ we can take the total Thomson cross section

$$d\sigma_{\text{sc}} \to \sigma_{\text{T}} = (8\pi/3)\alpha^2 \Lambda^2 z^4, \tag{6.246}$$

while ω' may be set equal to the K-frame frequency ω. The bremsstrahlung cross section is then, very simply,

$$d\sigma = \frac{16}{3}\alpha^3 \Lambda^2 \frac{z^4 Z^2}{\beta^2}\frac{d\omega}{\omega} \ln \frac{b'_{\text{max}}}{b'_{\text{min}}}, \tag{6.247}$$

obtained by integrating over impact parameters.

Considerations of b'_{max} and b'_{min} are similar to those already made in our previous approaches to the bremsstrahlung problem. For example, in the classical ($\beta \ll \alpha$) and soft-photon limit (see Chapter 6, Section 1.1), $b'_{\text{max}} \sim v/\omega$ while $zZe^2/b'_{\text{min}} \sim mv^2$. In the classical limit and *away from* soft frequencies, the details of the (upper) extent of the virtual photon spectrum are important and the W-W approach is not so simple and useful. However, in the Born limit, as we have seen, instead of impact parameters, the momentum transfer is the more convenient integration variable, and

$$b'_{\text{max}}/b'_{\text{min}} \to q_{\text{max}}/q_{\text{min}} = (1+\zeta)/(1-\zeta), \tag{6.248}$$

with ζ given by Equation (6.71). Also, the result holds for general ω and not just for soft photons; in the Born limit the details of the upper part of the virtual photon spectrum are not important. Kinematics ($\hbar\omega < E_0$) is applied to eliminate the unphysical higher frequencies. The classical maximum frequency is $\omega_{\text{cl}} \sim v/b_{\text{min}} \sim mv^3/zZe^2$, while quantum mechanics requires $\omega_q < \omega_{\text{max}} \sim mv^2/\hbar$. Then

$$\omega_{\text{cl}}/\omega_q \sim \hbar v/zZe^2 \sim (E_0/z^2 Z^2 \text{Ry})^{1/2}, \tag{6.249}$$

which is very large in the Born limit.

The application of the W-W method to bremsstrahlung is more interesting at relativistic energies and, in particular, in the ER limit. In this case it is convenient to employ the following dimensionless units for the photon energy:

$$\begin{aligned} &(K) \quad \epsilon : \text{units of } E_0 = \gamma_0 mc^2, \\ &(K') \quad \epsilon' : \text{units of } mc^2. \end{aligned} \tag{6.250}$$

Then in the electron (initial) rest frame K' the energy of the scattered photon is

$$\epsilon' = \frac{\epsilon'_0}{1+\epsilon'_0(1-\cos\theta')}, \tag{6.251}$$

where θ' is the scattering angle. In the frame K and with the units (6.250), the outgoing bremsstrahlung photon energy is the Lorentz transformation of the energy (6.251):

$$\epsilon = \epsilon'(1-\cos\theta'). \tag{6.252}$$

The relations (6.251) and (6.252) can be combined to relate the three energies through

$$\epsilon_0' = \frac{\epsilon}{(1-\cos\theta')(1-\epsilon)} = \frac{\epsilon'}{1-\epsilon}. \tag{6.253}$$

The K'-frame solid angle element for the outgoing photon can be expressed as a differential in ϵ with the help of these relations:

$$d\Omega' = 2\pi d(1-\cos\theta') = \frac{2\pi}{\epsilon_0'}\frac{d\epsilon}{(1-\epsilon)^2}. \tag{6.254}$$

Similarly,

$$\cos\theta_1 = 1 - \frac{1}{\epsilon_0'}\frac{\epsilon}{1-\epsilon}, \tag{6.255}$$

and the Compton cross section can be written as a function of ϵ with ϵ_0' as a parameter. For example, the spin-$\frac{1}{2}$ Klein-Nishina cross section [Chapter 5, Equation (5.89)] is then

$$d\sigma_{\mathrm{KN}} = \pi\alpha^2\Lambda^2\frac{1}{\epsilon_0'}\left[1-\epsilon+\frac{1}{1-\epsilon}-\frac{2}{\epsilon_0'}\frac{\epsilon}{1-\epsilon}+\frac{1}{\epsilon_0'^2}\left(\frac{\epsilon}{1-\epsilon}\right)^2\right]d\epsilon. \tag{6.256}$$

The differential incident photon flux in K' is [see Equation (6.245)]

$$dN = \frac{2\alpha Z^2}{\pi}\frac{db'}{b'}\frac{d\epsilon_0'}{\epsilon_0'}, \tag{6.257}$$

and the bremsstrahlung cross section is obtained by taking the product of the expressions (6.256) and (6.257) and integrating over db' and $d\epsilon_0'$. Given ϵ, the minimum ϵ_0' is ($\theta' = \pi$)

$$(\epsilon_0')_{\min} = \frac{\epsilon}{2(1-\epsilon)} \ll (\epsilon_0')_{\max} \sim \gamma_0. \tag{6.258}$$

The main contribution comes from the lower limit of ϵ_0' and we have

$$d\sigma_{1/2} = 4\alpha^3\Lambda^2 Z^2\left[\frac{4}{3}(1-\epsilon)+\epsilon^2\right]\frac{d\epsilon}{\epsilon}\ln\frac{b'_{\max}}{b'_{\min}}, \tag{6.259}$$

where the argument of the logarithmic factor is large. We can take $b'_{\min} \sim \hbar/mc$, while $b'_{\max}$ may be found from the relation (2.168) of Chapter 2; that is,

$$b'_{\max} \sim \frac{\gamma c}{\omega'_{\min}} \sim \frac{\gamma\hbar}{mc\epsilon'_{\min}} \sim \frac{2\gamma(\hbar/mc)(1-\epsilon)}{\epsilon}. \tag{6.260}$$

The cross section is then

$$d\sigma_{1/2} = 4\alpha^3\Lambda^2 Z^2\left[\frac{4}{3}(1-\epsilon)+\epsilon^2\right]\frac{d\epsilon}{\epsilon}\ln\frac{2\gamma(1-\epsilon)}{\epsilon}, \tag{6.261}$$

which is essentially the same as the result (6.224).

For comparison, we can also easily obtain the cross section for bremsstrahlung in the scattering of a highly relativistic spin-0 charge ze by a heavy nucleus. Here we must employ the corresponding Compton formula given by Equation (5.67).

Expressed in the form (6.256), that is, differential in ϵ with ϵ_0' as a parameter, this cross section is

$$d\sigma_C' = \pi\alpha^2 \Lambda^2 z^4 \frac{d\epsilon}{\epsilon_0'}\left[2 - \frac{2}{\epsilon_0'}\frac{\epsilon}{1-\epsilon} + \frac{1}{\epsilon_0'^2}\left(\frac{\epsilon}{1-\epsilon}\right)^2\right]. \tag{6.262}$$

The calculation of the bremsstrahlung cross section is carried out in exactly the same manner as in the spin-$\frac{1}{2}$ case and we find

$$d\sigma_0 = \frac{16}{3}\alpha^3 \Lambda^2 z^4 Z^2 (1-\epsilon)\frac{d\epsilon}{\epsilon}\ln\frac{2\gamma(1-\epsilon)}{\epsilon}. \tag{6.263}$$

We see that for $z = 1$ the result is the same as the spin-$\frac{1}{2}$ Bethe-Heitler expression (6.261), except that the ϵ^2 term in brackets is missing. We have discussed this characteristic of the bremsstrahlung formulas earlier [see Equation (6.223)]; effects of spin are unimportant in the soft-photon limit.

One of the most significant features of the evaluation of the bremsstrahlung cross section by the W-W method is the result that the main contribution from the spectrum of virtual photons comes from its low-energy end. This, in itself, explains some characteristics of the formulas, such as the fundamental identity (6.232) involving the *e–e* cross section. Earlier, we have explained this identity in terms of the small contribution from large momentum transfers. This is equivalent to the property of the W-W calculation that the soft end of the ϵ_0' spectrum gives the principal contribution. In the frame where both electrons are highly relativistic, *e–e* bremsstrahlung can be viewed as resulting from each electron scattering the other's virtual photons, the two contributions being additive and uncoupled.

6.6 ELECTRON-ATOM BREMSSTRAHLUNG

By "atom" we mean a general atomic system with nucleus Z and N_e bound electrons; that is, the system may be an ion ($N_e \neq Z$) or, if there is more than one nucleus, a molecule or molecular ion. The general topic is complicated, simplifying only in certain energy domains, so that we cannot give useful and accurate formulations that encompass all energies. Basically, there are two difficult energies: $E_0 \sim \text{Ry}$ and $E_0 \sim mc^2$, for which the theory is either complicated or incomplete.

6.6.1 Low Energies

Consider first the case of bremsstrahlung in the scattering of an electron of energy $E_0 \ll \text{Ry}$ off a neutral atom. The photon energies are then

$$\hbar\omega < E_0 \ll \text{Ry} \sim e^2/a_0, \tag{6.264}$$

where a_0 is the Bohr radius; a_0 is also the characteristic range of the scattering potential. The inequality (6.264) also implies that

$$\lambda\!\!\!^{-}/a_0 = c/\omega a_0 \gg \hbar c/e^2 = 1/\alpha \approx 137; \tag{6.265}$$

that is, the photon wavelength is large compared with the range of the scattering potential. Also, if $\lambdabar_0$ is the de Broglie wavelength of the incident electron,

$$\lambdabar_0/a_0 \sim (\mathrm{Ry}/E_0)^{1/2} \gg 1, \tag{6.266}$$

so that the electron does not distinguish the details of the scattering center (atom).

The limit $E_0 \ll \mathrm{Ry}$ means that only s-waves contribute to the scattering so that the atom scatters like a hard sphere[35]; that is, the *radiationless* scattering cross section is of the form

$$d\sigma_{\mathrm{sc}}/d\Omega = \sigma_{\mathrm{el}}/4\pi, \tag{6.267}$$

where σ_{el} is the total elastic cross section. Acting like a hard sphere, the time interval of the atom's scattering perturbation is $\tau_{\mathrm{coll}} \sim a_0/v_0$. Therefore, the inequality $\hbar\omega < E_0 \ll \mathrm{Ry}$ implies $\omega\tau_{\mathrm{coll}} < (E_0/\mathrm{Ry})^{1/2} \ll 1$, which is equivalent to the classical soft-photon limit. In other words, for this short-range scattering problem the soft-photon formulas apply for arbitrary $\hbar\omega$ as in the Born limit. However, if we evaluate the bremsstrahlung cross section by means of an integration over the electron scattering angle θ, we must be more careful than in the Born limit [where the integration variable is the momentum transfer q—see Equation (6.61)]. It is now necessary to insert explicitly a factor accounting for the reduction in the number of final states available to the outgoing electron. The factor arises, for fixed ω, from the number $dN_f/dE_f \propto v_f$, where v_f is the final electron velocity. The reduction factor is then

$$\phi_f = v_f/v_0, \tag{6.268}$$

which can also be regarded as the ratio of the outgoing to incoming electron fluxes. In other words, if we wanted the photon emission probability *per state* for the outgoing electron, we would employ[36]

$$dw = \frac{2\alpha}{3\pi}\left(\frac{\Delta \boldsymbol{v}}{c}\right)^2 \frac{d\omega}{\omega}. \tag{6.269}$$

In terms of the electron initial and final velocities and the scattering angle,

$$(\Delta \boldsymbol{v})^2 = v_0^2 + v_f^2 - 2v_0 v_f \cos\theta, \tag{6.270}$$

where v_f and v_0 are related by

$$\frac{1}{2}m(v_0^2 - v_f^2) = \hbar\omega. \tag{6.271}$$

The bremsstrahlung cross section is computed from

$$d\sigma_{\mathrm{b}} = \int d\sigma_{\mathrm{el}}\phi_f dw, \tag{6.272}$$

---

[35]For finite E_0 (say, $E_0 \sim 1$ eV) the anisotropy in $d\sigma_{\mathrm{sc}}/d\Omega$ is suprisingly large, due primarily to exchange effects and the polarization of the atom by the incident electron. It should be emphasized that the treatment of electron-atom bremsstrahlung in the low-energy limit as outlined here is, in a certain sense, an oversimplification. For scattering by some simple atomic systems (H, He, H_2), more elaborate treatments have indeed been given. References may be found in the paper R. J. Gould, *Astrophys. J.* **302**, 205 (1986) on which this simplified treatment is based; the paper also gives a more extensive and critical evaluation of the present simplification.

[36]This is the photon-emission probability summed over polarizations and integrated over emission angles [see Chapter 2, Equation (2.143)]. If we were interested in polarizations and the angular distribution of emission, we should use the more general expression [Equation (6.54)].

and on integrating over $d\Omega = 2\pi \sin\theta d\theta$, we obtain[37]

$$d\sigma_b = \frac{4\alpha}{3\pi}\sigma_{\rm el}\frac{2E_0 - \hbar\omega}{mc^2}\left(1 - \frac{\hbar\omega}{E_0}\right)^{1/2}\frac{d\omega}{\omega}, \tag{6.273}$$

and an integrated cross section

$$\Omega = \int \hbar\omega\, d\sigma_{\rm b} = \frac{64}{45}\frac{\alpha}{\pi}\frac{\sigma_{\rm el}}{mc^2}E_0^2. \tag{6.274}$$

The photon spectrum extends from $\hbar\omega = 0$ to E_0 and it might be noted that $d\sigma_{\rm b}/d\omega \to 0$ at $\omega_{\rm max} = E_0/\hbar$, a result of the velocity-space or phase-space reduction factor going to zero. In the soft-photon limit $d\sigma_b$ exhibits the usual infrared divergence characteristic.

Now let us consider the problem of bremsstrahlung in the scattering by an ion, that is, a nucleus Z plus N_e ($< Z$) bound electrons. However, again we confine the energies to the domain

$$E_0 \ll (Z - N_e)^2 \text{Ry}, \tag{6.275}$$

that is, to the classical domain for Coulomb scattering. The factor $Z - N_e$ represents the screened nuclear charge, corresponding to complete screening by the bound electrons. In the determination of the classical Coulomb bremsstrahlung cross section, the characteristic radial distance involved in the integration over impact parameters is given by Equation (6.24)

$$r_{\rm cl} = Ze^2/mv^2. \tag{6.276}$$

However, with complete screening, the Z in $r_{\rm cl}$ should really be replaced by the effective value

$$Z_e = Z - N_e. \tag{6.277}$$

The condition (6.275) can then be written, since Ry $\sim e^2/a_0$, where a_0 is the Bohr radius,

$$r_{\rm cl} \gg a_0/Z_e. \tag{6.278}$$

But a_0/Z_e is roughly the size of the bound-electron charge distribution in the ion; the inequality (6.278) then establishes the self-consistency of the assertion that in the limit (6.275) there is essentially complete screening by the bound electrons. In other words, the cross section for electron-ion bremsstrahlung would be given by the expression (6.15) in the soft-photon limit and the result (6.33) for most of the photon spectrum with, in each formula, Z replaced by Z_e.

6.6.2 Born Limit—Non-Relativistic

The limit

$$E_0 \gg Z^2\text{Ry} \tag{6.279}$$

corresponds to the domain of the Born approximation and, as we shall see. to the occurrence of *weak screening* by the bound electrons that might be surrounding the

[37] The soft-photon limit of the result (6.272) had been derived earlier by the author by similar methods: *Australian J. Phys.* **22**, 189 (1969).

nucleus. In the limit (6.279) the problem is inherently quantum mechanical and conclusions regarding screening are dictated by the magnitude of the ratio of the de Broglie wavelength λ_0 of the incident electron to the size a_0/Z of the bound-electron cloud. Since $\lambda_0 \sim \hbar/(mE_0)^{1/2}$, we see that the ratio is

$$\frac{\lambda_0}{a_0/Z} \sim \left(\frac{Z^2\hbar^2/ma_0^2}{E_0}\right)^{1/2} \sim \left(\frac{Z^2\mathrm{Ry}}{E_0}\right)^{1/2} \ll 1. \tag{6.280}$$

That is, the incident electron basically "sees" the unscreened nucleus and the unscreened bound electrons individually. In other words, there is no screening of the nuclear charge by the bound electrons, and the bremsstrahlung cross section is the same as if the bound electrons were not there.

The above results should be qualified slightly because, as the incident energy becomes close to relativistic values, the effects of the bound electrons begin to come in again. This occurs through the phenomenon of electron-electron bremsstrahlung (see Section 6.3), but the relative magnitude of the contribution will be $\sim N_e(E_0/mc^2)$, comparable to the relativistic correction to Coulomb bremmstrahlung. The effect is then small at non-relativistic energies.

6.6.3 Intermediate Energies—Non-Relativistic

When $E_0 \sim Z^2\mathrm{Ry}$ the bremsstrahlung problem is extremely complicated and has not been treated. Here we try to outline a prescription for an appropriate evaluation of the cross section in this energy domain which is intermediate between that of the classical and Born approximations. The result will provide a kind of bridge between the two domains that has, at least, some physical basis that would suggest a reasonable accuracy (say 20%) at the halfway point ($Z^2\mathrm{Ry}$) of the energy range and a greater accuracy at lower and higher energies.

One procedure for evaluating the cross section would be to employ the pure-Coulomb formulas (6.33) and (6.34) with a Gaunt factor as described in Chapter 6, Section 4.1. That is, the Coulomb formulas are used but with an effective central charge corresponding to a screened value

$$Z_e' = Z - s, \tag{6.281}$$

reduced from that of the nucleus by an amount s due to electron screening. The problem is then reduced to an evaluation of this screening parameter which must have the following limits:

$$s \longrightarrow \begin{cases} N_e & E_0 \ll Z^2\mathrm{Ry}, \\ 0 & E_0 \gg Z^2\mathrm{Ry}. \end{cases} \tag{6.282}$$

The Gaunt factor would also be evaluated for the effective value Z_e' to yield a cross section proportional to $Z_e'^2 g(E_0, \omega; Z_e')$.

For the purpose of finding a prescription for the screening parameter s, consider the radiationless scattering of an electron by an atomic system; the scattering is also treated in the Born approximation. The scattering potential is

$$V(\boldsymbol{r}) = Ze^2/r - e^2 \sum_j |\boldsymbol{r} - \boldsymbol{r}_j|^{-1}, \tag{6.283}$$

where $\boldsymbol{r}$ refers to the free electron and $\boldsymbol{r}_j$ to the bound electrons. If the initial and final atomic states are $|0\rangle$ and $|n\rangle$, respectively, the scattering cross section will be proportional to the square of the matrix element

$$V_{n0} = \langle n| \int d^3 \boldsymbol{r} e^{i\boldsymbol{q}\cdot\boldsymbol{r}} V(\boldsymbol{r})|0\rangle. \tag{6.284}$$

The integral can be evaluated,[38] yielding

$$V_{n0} = 4\pi e^2 q^{-2} \langle n|\mathcal{Z}|0\rangle, \tag{6.285}$$

where

$$\mathcal{Z} = Z - \sum_j e^{i\boldsymbol{q}\cdot\boldsymbol{r}_j}. \tag{6.286}$$

If it is specified that the atomic system is unchanged[39] in the scattering ($|n\rangle = |0\rangle$), the relevant matrix element is V_{00}, proportional to

$$\langle 0|\mathcal{Z}|0\rangle = Z - F, \tag{6.287}$$

where

$$F = \langle 0| \sum_j e^{i\boldsymbol{q}\cdot\boldsymbol{r}_j} |0\rangle = F(q) \tag{6.288}$$

is the *atomic form factor*. Since the main contribution to the bremsstrahlung cross section comes from small momentum transfers, a reasonable prescription for the screening parameter would be

$$s \approx F(q_{\min}) = \sum_j F_j(q_{\min}). \tag{6.289}$$

For the minimum momentum transfer we can set $q_{\min} = k_0 - k_f$, where

$$k_f^2 = k_0^2 - (2m/\hbar^2)(\hbar\omega + \overline{E}_{\text{ex}}), \tag{6.290}$$

with $\overline{E}_{\text{ex}}$ equal to some mean bound-state excitation energy[40] of the atomic system.

The contribution to F from the individual nl bound electrons is determined by the one-electron wave functions. For a hydrogenic $1s$ state we have

$$F_{1s}(q) = [1 + (a_1 q/2)^2]^{-2}, \tag{6.291}$$

where $a_1 = a_0/Z_1$ is the corresponding Bohr radius of the $1s$ state with inner-shell effective Z equal to Z_1. The Z-values for atomic electrons may be taken very simply from the Slater rules.[41] For the higher shells let us take squared wave

---

[38] In the integral over the e–e term $|\boldsymbol{r} - \boldsymbol{r}|^{-1}$ make the variable change $\boldsymbol{\rho} = \boldsymbol{r} - \boldsymbol{r}_j$ with $d^3\boldsymbol{r} = d^3\boldsymbol{\rho}$.

[39] It is more appropriate to sum over n and the result is a somewhat different expression (see Section 6.6c).

[40] This excitation energy could be set equal to the value that appears in stopping-power calculations (cf. H. A. Bethe and R. Jackiw, *Intermediate Quantum Mechanics*, 2nd. ed., New York: W. A. Benjamin, Inc., 1968).

[41] J. C. Slater, *Phys. Rev.* **36**, 57 (1930); *Quantum Theory of Atomic Structure*, 2 Vols., New York: McGraw-Hill, 1960; *Quantum Theory of Matter* (New York: McGraw-Hill, 1968). More recently, better screening constants have been introduced [see Y-D. Jung and R. J. Gould, *Phys. Rev.* **A44**, 11 (1991)].

functions approximated by δ-functions at the shell radius:

$$u^2(\mathbf{r}) \approx (4\pi r^2)^{-1}\delta(r - r_{nl}), \tag{6.292}$$

where

$$r_{nl} = n^2 a_0 / Z_{nl}. \tag{6.293}$$

The 1-electron form factors are then

$$F_{nl}(q) \approx \sin q r_{nl} / q r_{nl} \quad (n \geq 2). \tag{6.294}$$

Both forms (6.291) and (6.294) exhibit the required asymptotic behavior (6.282) when employed to compute the screening parameter.

Obviously the above treatment is very rough, but it may be taken as a starting point for future more accurate calculations and for present approximate calculations as suggested. Some shortcomings are already obvious, such as the lack of an exchange contribution, perhaps a 5–10% effect. Another difficulty is that the effect of the cancellation of the photon fields from coupling to the bound and free electrons has not been included properly. Only coupling to the latter has been included, since the bound electrons are taken to provide an external scattering potential [see Equation (6.283)]. We have seen in Section 6.3 how the cancellation occurs in e–e bremsstrahlung. Actually, the procedure employing the potential (6.283) *is* valid in electron-atom bremsstrahlung at ultrarelativistic electron energies. In this limit the problem of screening is much simpler than at energies $\sim Z^2$Ry and we now turn to this case.

6.6.4 Relativistic Energies—Formulation

For a reason that will become clear, it is convenient to approach the problem of relativistic electron-atom bremsstrahlung from the extreme relativistic end. In this limit, as we have seen in Section 5.3 of Chapter 6, in scattering off a free electron $d\sigma_{ee} \to d\sigma_{ep}$; the target electron provides practically a static Coulomb potential and exchange effects are negligible. The fundamental reason for this equivalence can be ascribed to the very small value of the minimum momentum transfer (6.222) and to the fact that the main contribution to the bremsstrahlung cross section comes from small q-values. We use the symbol δ for this parameter (6.222), in relativistic units ($\hbar = c = 1$) equal to

$$\delta = q_{\min} = m^2 k / 2E_0 E = m^2 k / 2E_0(E_0 - k), \tag{6.295}$$

in terms of the photon energy k and the incident-electron energy E_0. For $k \sim E_0$, $\delta \ll m$, and some of the formulation begun earlier is now very relevant for the electron-atom problem.

With the bound electron providing an additional component to the scattering potential, the Coulomb-potential matrix element would then be replaced by the expression (6.285). This factor is, in turn, determined by the matrix element of the "effective nuclear charge" (6.286) between the initial (0) and final (n) atomic states. If we are not interested in the final atomic state, a summation over n should be taken;

the atomic factor in the differential bremsstrahlung cross section is then[42]

$$\zeta_{\mathrm{sc}}(\boldsymbol{q}) = \sum_n |\langle n|\mathcal{Z}|0\rangle|^2. \tag{6.296}$$

The eigenfunctions satisfy a *closure relation*

$$\sum_n |n\rangle\langle n| = 1, \tag{6.297}$$

by virtue of the completeness of their set, where the above sums include the continuum states (target system left ionized). Because of the identity (6.297), the scattering factor (6.296) is

$$\zeta_{\mathrm{sc}}(\boldsymbol{q}) = \sum_n \langle 0|\mathcal{Z}^*|n\rangle\langle n|\mathcal{Z}|0\rangle = \langle 0|\mathcal{Z}^*\mathcal{Z}|0\rangle; \tag{6.298}$$

that is, it is determined by the initial-state expectation value of $\mathcal{Z}^*\mathcal{Z}$. This product is[43]

$$\mathcal{Z}^*\mathcal{Z} = Z^2 - 2Z\sum_j \mathrm{Re}\, e^{i\boldsymbol{q}\cdot\boldsymbol{r}_j} + \sum_j\sum_k \mathrm{Re}\, e^{i\boldsymbol{q}\cdot(\boldsymbol{r}_j-\boldsymbol{r}_k)}, \tag{6.299}$$

and it is convenient to separate terms in this expression. For example, if there are N_e atomic electrons, in the double sum in the last term in $\mathcal{Z}^*\mathcal{Z}$ there are N_e terms for which $j = k$, each yielding unity, and $N_e(N_e - 1)$ terms with $j \neq k$ having oscillatory exponential form. In the expectation value (6.298) it is convenient to rearrange terms and introduce the functions

$$G(\boldsymbol{q}) = \langle 0|N_e - \sum_j \mathrm{Re}\, e^{i\boldsymbol{q}\cdot\boldsymbol{r}_j}|0\rangle, \tag{6.300}$$

$$H(\boldsymbol{q}) = \langle 0|N_e(N_e-1) - \sum_{j\neq k} \mathrm{Re}\, e^{i\boldsymbol{q}\cdot(\boldsymbol{r}_j-\boldsymbol{r}_k)}|0\rangle. \tag{6.301}$$

Then if we add and subtract terms $2ZN_e$ and $N_e(N_e - 1)$ and add the term N_e from $j = k$ in the double sum, the scattering factor (6.298) can be written

$$\zeta_{\mathrm{sc}}(\boldsymbol{q}) = (Z - N_e)^2 + 2ZG(\boldsymbol{q}) - H(\boldsymbol{q}). \tag{6.302}$$

The functions G and H have been introduced in order to make use of the properties

$$G(\boldsymbol{q})/N_e,\ H(\boldsymbol{q})/N_e(N_e-1) \longrightarrow \begin{cases} 0 & q \to 0, \\ 1 & q \to \infty. \end{cases} \tag{6.303}$$

There is a characteristic q-value that defines the region intermediate between the limiting forms (6.303); this value (q_s) is associated with effects of screening by the atomic electrons, and is essentially the inverse of the characteristic spatial size of the bound-electron wave functions:

$$q_s \sim r_j^{-1} \sim Z_j/a_0. \tag{6.304}$$

[42] The effects of the slight differences in the excitation energies of the states n on the kinematics associated with the remaining factors [in the factor W in the expression (6.221)] in the cross section are neglected. Since these excitation energies ($\sim Z^2$Ry) are much smaller than E_0, this is a very good approximation.

[43] The "Re" in the double sum in the last term could be left out; j and k both run from 1 to N_e (number of bound electrons) and can be interchanged in the double sum. This sum can then be replaced by half the i, j and $i \leftrightarrow j$ sums.

Here Z_j is the effective nuclear charge seen by the jth electron, reduced from Z by the mutual screening of the other electrons. The value of q_s representing the transition domain between the limits (6.303) is then determined by the atomic wave functions, and would be given roughly by the expression (6.304) with Z_j equal to some average value, say, $\overline{Z}_j \sim Z - \frac{1}{2}N_e$, for the bound electrons. The dimensionless momentum transfer in units of mc is, at the value q_s,

$$\delta_s = \hbar q_s/mc \sim \overline{Z}_j \Lambda/a_0 = \alpha \overline{Z}_j \approx \overline{Z}_j/137, \tag{6.305}$$

which is small compared with unity except for high-Z atoms.

An important comparison is that of δ_s with the minimum momentum transfer (6.295) for the case $k = \frac{1}{2}E_0$ (middle of photon spectrum); setting $\delta_s = \delta(k = \frac{1}{2}E_0)$, yields a characterisitc energy $E_0 = E_s$ associated with screening:

$$E_s = mc^2/2\alpha \overline{Z}_j. \tag{6.306}$$

The energy E_s has the following meaning: for $E_0 \ll E_s$, $\delta \gg \delta_s$, and the large-q limit (6.303) applies, for which

$$\zeta_{\mathrm{sc}} \xrightarrow[E_0 \ll E_s]{} (Z - N_e)^2 + 2ZN_e - N_e(N_e - 1) = Z^2 + N_e. \tag{6.307}$$

That is, the domain $E_0 \ll E_s$ is that of weak screening for which the effects of atomic binding are unimportant and for which the bremsstrahlung cross section is a sum of (unscreened) contributions from the nucleus Z and the N_e electrons. In other words, the nucleus and the electrons act as if they were separated and independent scatterers. It should be added that it is still required that $E_0 \gg mc^2$, since the identity (6.232) has been assumed. However, as the energy E_0 is lowered to $\sim mc^2$ it is still a valid conclusion that screening effects can be neglected and the electron-nucleus and electron-electron contributions to the total bremsstrahlung cross section are additive (and independent); the *e–e* contribution must then be evaluated more carefully at these energies (see Section 5.3 of Chapter 6).

The domain $E_0 \gg E_s$ is that of strong screening and the evaluation of the cross section in this limit requires a development necessary to treat the (intermediate-screening) domain $E_0 \sim E_s$, providing a transition from weak to strong screening. The mathematical approximations required were obtained long ago by Bethe (see Footnote 30), and we shall quote his results.[44] The procedure is designed for the evaluation of the total bremsstrahlung cross section, integrated over angles for the outgoing electron and photon, but differential in the photon energy. That is, the integral (6.221) is transformed and integrated to the point where it is cast into a form whereby the only remaining integration is over dq. This is particularly appropriate for electron-atom bremsstrahlung beacause the matrix element of the total scattering potential is a function of only the momentum transfer q. The result of this work[30,45] is that the bremsstrahlung cross section can be written, for scattering by a general atomic system,

$$d\sigma_b = \alpha^3 \Lambda^2 \frac{dk}{k}\left[\left(1 + \frac{E^2}{E_0^2}\right)\phi_1 - \frac{2E}{3E_0}\phi_2\right]. \tag{6.308}$$

---

[44] See also R. J. Gould, *Phys. Rev.* **185**, 72(1969); the treatment here follows closely that in this paper.

[45] See also R. J. Gould, *Phys. Rev.* **185**, 72 (1969).

Here $E = E_0 - k$, and the ϕ-functions are evaluated from the scattering factor[46] ζ_{sc}:

$$\phi_a = 4 \int f_a(q; E_0, k)\zeta_{sc}(q)\, dq. \tag{6.309}$$

The functions f_a are different for $a = 1$ and 2, and have a simple form only in the domain of small q.

The q-integration in the expressions (6.306) extends from the minimum value δ [Equation (6.295)] to a maximum q_{max}, but the precise value of the latter need not be specified, other than that

$$q_{max}/\Lambda \gg 1. \tag{6.310}$$

Because of the q-dependence of the integrands (6.309), the upper limit contributes a negligible amount in the extreme relativistic limit ($E_0 \gg mc^2$). It is only in this limit where screening effects can be treated in a simple manner, but, as we have seen [see Equations (6.303), (6.304)], at the lower energies screening is unimportant. The complexity in the integrations (6.309) involves only the terms involving G and H in ζ_{sc}, and it is convenient to break up the integrals by introducing an intermediate value q_0:

$$\int_{q_{min}}^{q_{max}} = \int_{q_{min}}^{q_0} + \int_{q_0}^{q_{max}}. \tag{6.311}$$

The value of q_0 is chosen so that it is within an overlapping region defined by the domains A and B such that

$$q/\Lambda \begin{cases} \ll 1 & (A) \\ \gg \delta, \delta_s & (B). \end{cases} \tag{6.312}$$

Since q_0 is within both A and B, in the evaluation of the integrals (6.309) it will not appear in the final result. The procedure is convenient because

$$\begin{aligned} &\text{in } A: \quad f_a \to f_a' \quad \text{(simplified expressions)} \\ &\text{in } B: \quad G \to N_e;\ H \to N_e(N_e - 1). \end{aligned} \tag{6.313}$$

That is, in domain A the functions f_a simplify because of the limit of small q, while in B the functions G and H simplify although the expressions for f_a are complicated (f_a''). Then we can write

$$\begin{aligned} \int_{q_{min}}^{q_{max}} f_a \begin{Bmatrix} G \\ H \end{Bmatrix} dq &= \int_{q_{min}}^{q_0} f_a' \begin{Bmatrix} G \\ H \end{Bmatrix} dq + \int_{q_0}^{q_{max}} f_a'' \begin{Bmatrix} N_e \\ N_e(N_e - 1) \end{Bmatrix} dq \\ &= \int_{q_{min}}^{q_{max}} f_a' \begin{Bmatrix} G \\ H \end{Bmatrix} dq + \int_{q_0}^{q_{max}} (f_a'' - f_a') \begin{Bmatrix} N_e \\ N_e(N_e - 1) \end{Bmatrix} dq. \end{aligned} \tag{6.314}$$

It is convenient to express q in units of Λ to evaluate the (dimensionless) functions ϕ_1 and ϕ_2. Then the first integral on the right can be broken up to

$$\int_{\delta}^{q_{max}} = \int_{\delta}^{1} + \int_{1}^{q_{max}}, \tag{6.315}$$

[46]The scattering factor ζ_{sc} is always a function of the magnitude $q = |\boldsymbol{q}|$ when the atomic system is spherically symmetric or averaged over azimuthal states m_l.

and over the 1 to $q_{\max}$ part $G \to N_e$ and $H \to N_e(N_e - 1)$. The two simplified functions are[30,44]

$$f_1' = q^{-3}(q - \delta)^2, \tag{6.316}$$

$$f_2' = q^{-4}[q^3 - 6\delta^2 q \ln(q/\delta) + 3\delta^2 q - 4\delta^3]. \tag{6.317}$$

In the limit (6.310)

$$\int_1^{q_{\max}} f_1'(q)\, dq \to \int_1^{q_{\max}} f_2'(q)\, dq \to \ln q_{\max}. \tag{6.318}$$

This term is canceled by one from the last integral on the right-hand side of Equation (6.314); it is found that

$$\int_{q_0}^{q_{\max}} (f_a'' - f_a')dq = \alpha_a - \ln q_{\max}, \tag{6.319}$$

with

$$\alpha_1 = 1; \quad \alpha_2 = \tfrac{5}{6}. \tag{6.320}$$

(Note that q_0 does not appear.) These developments allow an explicit expression for ϕ_1 and ϕ_2 involving only the *simplified* kinematic functions f_1' and f_2' and the purely atomic functions G and H.

First we should note that when the q-integration is applied to the unscreened constant term $(Z - N_e)^2$ in ζ_{sc} [Equation (6.302)], the result is, for $a = 1$ and 2,

$$\int_\delta^1 f_a'(q)dq + \alpha_a = \ln(1/\delta) - \tfrac{1}{2}, \tag{6.321}$$

yielding the unscreened functions

$$\phi_u = 4\left[\ln(2E_0E/k) - \tfrac{1}{2}\right]. \tag{6.322}$$

The total ϕ_1 and ϕ_2 to be substituted into the cross-section formula (6.308) are then given by

$$\phi_a = (Z - N_e)^2\phi_u + 8Z\left[\alpha_a N_e + \int_\delta^1 G(q) f_a'(q;\delta)\, dq\right]$$
$$- 4\left[\alpha_a N_e(N_e - 1) + \int_\delta^1 H(q) f_a'(q;\delta)\, dq\right]. \tag{6.323}$$

This is a general formula valid in the limits of weak, intermediate, and strong screening. For example, we see that in the weak-screening limit $\delta \gg \delta_s$, where $G \to N_e$ and $H \to N_e(N_e - 1)$, the expression (6.321) yields

$$\phi_a \xrightarrow[\delta \gg \delta_s]{} (Z^2 + N_e)\phi_u, \tag{6.324}$$

corresponding to the result (6.307) already obtained.

6.6.5 Relativistic Energies—Results and Discussion

Perhaps one of the first things that should be emphasized is the form of the momentum transfer distribution in its contribution to the bremsstrahlung cross section. Despite the q^{-4} factor in the integral (6.221), we find, in the end, that the small-q form of the integrand is as dq/q, as is seen from the expressions (6.316) and (6.317). This form is, of course, the same as in the non-relativistic formula (6.74) and applies in the relativistic case for weak or strong screening. For the screening parts of ϕ_a [that is, *not* the first term in the expression (6.323)], there are dominant terms proportional to $\ln(1/\delta_{\text{eff}})$, where

$$\delta_{\text{eff}} \approx \max(\delta, \delta_s). \tag{6.325}$$

Above $q \sim 1$ the momentum transfer distribution steepens such that the upper end toward $q_{\max}$ contributes negligibly to the bremsstrahlung cross section.

Concerning the general expression for the cross section at arbitrary energy, we see from the expression (6.323) for ϕ_a that, unless $N_e = Z$, the first term yields a contribution that is the same as that for a bare nucleus, except that the nuclear charge is the effective value $Z - N_e$. There are, in addition, the contributions involving the terms in brackets in ϕ_a, and these are associated with screening and atomic binding, but the form of these terms is very different from the pure Coulomb term $(Z - N_e)^2\phi_u$. In the limit of strong screening ($\delta \ll \delta_s$), because of the dependence (6.303) for G and H, that is, that

$$G/N_e,\; H/N_e(N_e - 1) \approx \begin{cases} 0 & q < \delta_s, \\ 1 & q > \delta_s, \end{cases} \tag{6.326}$$

the forms (6.316) and (6.317) for f_a' indicate the essential results

$$\left.\begin{aligned} \int G(q) f_a'\, dq &\approx N_e \ln(1/\delta_s), \\ \int H(q) f_a'\, dq &\approx N_e(N_e - 1)\ln(1/\delta_s) \end{aligned}\right\} \quad (\delta \ll \delta_s). \tag{6.327}$$

These expressions, valid in the high-energy limit ($E_0 \gg E_s$), give terms in ϕ_a that are independent of energy, as opposed to the weak-screening or unscreened terms that have a logarithmic energy dependence.

A more accurate evaluation of the screening terms in the cross section requires an evaluation of the integrals in Equation (6.323). For few-electron systems these integrals can be computed easily, and it is convenient to employ analytical forms for atomic wave functions in the calculation of the expectation values G and H. In the case of a one-electron system in the ground state, the evaluation of ϕ_1 and ϕ_2 is particularly simple. Then $N_e = 1$ and $H = 0$; ϕ_a reduces to

$$\phi_a(N_e = 1) = (Z - 1)^2\phi_u + 8Z\left(\alpha_a + \int_\delta^1 G f_a'\, dq\right), \tag{6.328}$$

with

$$G = 1 - F(q). \tag{6.329}$$

For the system in the ground $1s$ state, the atomic form factor $F(q)$ is given by the result (6.291) and the integral in the expression (6.328) can be evaluated. It is useful to express the results in terms of the dimensionless parameter

$$\Delta = \tfrac{1}{2}a_0\delta = (1/4\alpha)(kmc^2/E_0E), \tag{6.330}$$

equal to, at the midpoint of the photon spectrum,

$$\Delta(k = \tfrac{1}{2}E_0) = 34.26(mc^2/E_0) = \overline{\Delta}. \tag{6.331}$$

At the energy (6.306) of intermediate screening the chracteristic value (6.331) equals $\frac{1}{2}Z$; that is, the domain of intermediate screening corresponds to $\Delta \sim \Delta_{\text{int}} \sim \frac{1}{2}Z$. For $Z = 1$, $\Delta \ll 1$ means strong screening, while $\Delta \gg 1$ is the domain of weak screening. For several values of Z the integrals

$$I_a = \int_\delta^1 Gf_a'\,dq \tag{6.332}$$

are given in graphical form in the author's paper[44] as a function of Δ. Some limiting forms are of interest, such as that for strong screening:

$$I_1 \to I_2 \to I(0) = \ln(1/2\alpha Z) + \tfrac{1}{2} \qquad (\Delta \ll Z). \tag{6.333}$$

For large Δ the unscreened result (6.324) is approached and it is interesting how fast this happens; it is found[44] that, even for $\Delta \approx Z$, the I_a are close to the unscreened values

$$I_a \to I_{au} = \tfrac{1}{4}\phi_u - \alpha_a. \tag{6.334}$$

The differences are

$$\begin{aligned} I_{1u} - I_1 &\to Z^4/60\Delta^4 \qquad (\Delta/Z \gtrsim 1),\\ I_{2u} - I_2 &\to Z^4/84\Delta^4 \qquad (\Delta/Z \gtrsim 1). \end{aligned} \tag{6.335}$$

Two-electron systems are a little more complicated, but simplify somewhat when the ground-state atomic wave function can be approximated by the form

$$\psi(1,2) = \psi_1(1)\psi_1(2), \tag{6.336}$$

where ψ_1 is a one-electron function. The scattering factor (6.302) is then given by

$$\zeta_{\text{sc}} = (Z-2)^2 + 4Z(1-F) - 2(1-F^2), \tag{6.337}$$

where F is now the one-electron form factor (evaluated with the function ψ_1). That is, we can write

$$\zeta_{\text{sc}} = c_0 + \sum_p c_p(1 - F^p), \tag{6.338}$$

with $p = 1, 2$. The factors $1 - F$ and $1 - F^2$ both have the property to yield the forms (6.303), with the result that

$$\phi_a = c_0\phi_u + 4\sum_p c_p\big(\alpha_a + I_a^{(p)}\big), \tag{6.339}$$

with

$$I_a^{(p)} = \int_\delta^1 f_a'(1 - F^p)\, dq. \tag{6.340}$$

The simplest two-electron wave function for the ground state is what might be called the "one-parameter Hylleraas function" in which the only variational parameter is the effective Z in a hydrogenic function for ψ_1. This is the well-known "$Z - \frac{5}{16}$" function, hydrogenic with an effective value

$$Z_{\text{eff}} = Z - \tfrac{5}{16}, \tag{6.341}$$

for which the one-electron form factor is again the expression (6.291) with Z replaced by Z_{eff}. The advantage of this type of approximation is that it gives results for arbitrary Z and the approximation is, in fact, better for larger Z. The integrals $I_a^{(1)}$ and $I_a^{(2)}$ have been evaluated in this manner and results expressed graphically[44] for several Z-values.

Summarizing the results for electron-atom bremsstrahlung, we can say, in general, that with the cross section given by equation (6.308) the ϕ_a have the form

$$\phi_a = (Z - N_e)^2 \phi_u + \phi_a', \tag{6.342}$$

having an unscreened and a screened part (ϕ_a'). At low energies ($E_0 \ll E_s$)

$$\phi_a' \to N_e(2Z + 1 - N_e)\phi_u \qquad \text{(low energy)}, \tag{6.343}$$

to give the limit (6.321). At high energies ($E_0 \gg E_s$)

$$\phi_a' \to \phi_a'(0) = \text{const.} \qquad \text{(high energy)}. \tag{6.344}$$

For example, for one-electron systems the functions become [see Equations (6.328), (6.332), (6.333)]

$$\begin{aligned}\phi_a' = 8Z(\alpha_a + I_a) \xrightarrow[\text{(high energy)}]{} & 8Z\big(\alpha_a + I(0)\big) = \phi_a'(0) \\ & = 8Z\big[\ln(1/2\alpha Z) + \alpha_a + \tfrac{1}{2}\big].\end{aligned} \tag{6.345}$$

In the case of atomic hydrogen $\phi_1'(0) = 45.82$ and $\phi_2'(0) = 44.48$; for He$^+$ $\phi_1'(0) = 80.54$ and $\phi_2'(0) = 77.87$.

The energy-integrated cross section ($\Omega = \int k\, d\sigma_b$) is obtained by integrating over the photon spectrum. At low energies the result can be written[47]

$$\Omega \to \Omega_{ep}(Z^2 + N_e) \qquad \text{(low energy)}, \tag{6.346}$$

where Ω_{ep} is the value for electron-proton bremsstrahlung [Equation (6.245) with $Z = 1$]. For general energies, when screening effects are important, corresponding to the separation (6.342), the cross section has the form

$$d\sigma_b = (Z - N_e)^2 d\sigma_{ep} + d\sigma_b', \tag{6.347}$$

[47] We again neglect the excitation energies of atomic levels (see Footnote 42).

where $d\sigma_b'$ would be given by the expression (6.308) with the ϕ_a' part in Equation (6.342). The energy-integrated cross section would then be

$$\Omega = \Omega_{ep}(Z - N_e)^2 + \Omega'. \tag{6.348}$$

In the high-energy or strong-screening limit

$$\Omega' \to \alpha^3 \Lambda^2 E_0\left[\tfrac{4}{3}\phi_1'(0) - \tfrac{1}{3}\phi_2'(0)\right] \qquad \text{(high energy).} \tag{6.349}$$

At intermediate energies it is a good approximation to neglect the weak energy dependence in the ϕ_a' factors and write

$$\Omega' \approx \alpha^3 \Lambda^2 E_0 \overline{\phi}\,', \tag{6.350}$$

where [see Equation (6.349)]

$$\overline{\phi}\,' = \tfrac{4}{3}\overline{\phi_1}\,' - \tfrac{1}{3}\overline{\phi_2}\,', \tag{6.351}$$

with the $\overline{\phi_a}\,'$ evaluated at $\Delta = \overline{\Delta}$ [Equation (6.331)].

Bibliographical Notes

The basic reference for a quantum-mechanical treatment of Coulomb bremsstrahlung at general non-relativistic energies is

1. A. Sommerfeld, *Atombau und Spektrallinien*, 2. Aufl., Bd. 2 Braunschweig: F. Vieweg und Sohn, 1939.

Usually the Born limit at non-relativistic energies is obtained from a limit of the more general relativistic expression. The derivation in Section 6.2, instead starts from purely mathematical QED, treating both the scattering potential and electromagnetic interaction (single-photon vertex) as perturbations. For a different approach see

2. H. A. Bethe and E. E. Salpeter, *Quantum Mechanics of One- and Two-Electron Atoms*, Sect. 77, Berlin: Springer-Verlag, 1957.

The method described therein treates only the electromagnetic interaction as a perturbation, employing first-order perturbation theory with, however, the wave functions for the incoming and outgoing electron corrected (first Born approximation) for the effects of the scattering potential.

Relativistic Coulomb bremsstrahlung is covered in the standard texts on quantum electrodynamics such as the works of Jauch and Rohrlich, Bjorken and Drell, Feynman, Dyson, and Heitler (References 3, 5, 7, 8, and 9 at the end of Chapter 3). Feynman, in particular, considers the spin-0 problem, which is much simpler than the for spin-$\frac{1}{2}$ (see Section 5.1 of Chapter 6).

Screening effects in relativistic bremsstrahlung are discussed as in Section 6.6 by

3. G. R. Blumenthal and R. J. Gould, *Rev. Mod. Phys.* **42**, 237 (1970).

Index

absorption and stimulated emission 70–72
spontaneous emission 71
stimulated scattering 73
angular distributions 17
angle transformations 18
anti-particle states 92
asymptotic amplitude for Coulomb scattering 151
atomic size 3

barn (definition) 4
binary collision rate 18–20
fundamental relative velocity generalization 20
Bohr radius 3
bremsstrahlung, classical non-relativistic case 211–217
energy-integrated cross section 217
general case, Gaunt factor 216–217
soft-photon limit 211–213
WKB series 240
bremsstrahlung, non-relativistic Born limit 217–233
Born correction, Sommerfeld–Elwert factor 223–225
electron–positron bremsstrahlung 226–228
electron–electron bremsstrahlung 228–236
direct and exchange amplitudes 228–233
photon–emission probability, with and without exchange 232–234
intermediate energies (classical-to-Born) 236–240
soft-photon limit 239–240
bremsstrahlung, electron–atom 254–267
Born limit: non-relativistic 256
intermediate energies: non-relativistic 257–259
low energies 254–256
electon–ion bremsstrahlung 256
phase-space reduction factor 255
short-range scattering 254
relativistic energies, formulation 259–263
results and discussion 264–267
bremsstrahlung, relativistic Coulomb 240
relativistic electron–electron bremsstrahlung 248–251
spin-0 problem 241–244
spin-$\frac{1}{2}$: Bethe–Heitler formula 244–248
extreme relativistic limit 249
non-relativistic limit 247–248
Weizsäcker–Williams method 251–253

channels for reactions 131–133
examples of channels 133
characteristic lengths, times, energies, etc. 1
Classical Electrodynamics 37
classical electron radius 3
Compton effect 76
Compton scattering, classical limit 177
Thomson cross section 178–181
validity of the classical limit 181
Compton scattering, non-relativistic QED treatment 182–186
photon scattering by a magnetic moment 186–188
Compton scattering, relativistic spin-0 case 188–191
Compton scattering, relativistic spin-$\frac{1}{2}$ case 191–196
invariant forms 194
Klein-Nishina formula 193
limiting forms and comparisons 195–196
Compton wavelength 3, 178
electron Compton wavelength 3
pion Compton wavelength 4
Coulomb scattering, classical 135–142
"classical Born approximation" 136
exact treatment 138–139
distance of closest approach 139
two-body problem and relative motion 139–140
impact parameter 135
reduced mass 140
validity of classical limit 141
Coulomb scattering, non-relativistic Born approximation 142
non-relativistic Born approximation 142

normalization volume for plane-wave states 143
perturbation matrix element 143
screened Coulomb potential 144
validity of the Born approximation 154
Coulomb scattering, non-relativistic exact treatment 138
two-body problem 148–149
Coulomb scattering of identical particles, non-relativistic 150–154
arbitrary spin 154
spin-0 Born formula 152
spin-0 exact formula 152
spin-$\frac{1}{2}$ formula 154
Coulomb scattering, relativistic 156
scattering of relativistic spin-$\frac{1}{2}$ particles 166–171
Møller and Bhabha scattering 171–174
scattering by a fixed center 170–171
scattering potential 144
Fourier transform of 144
momentum transer in scattering 144
spin-0 case 156–158
two-body problem 158
distinguishable particles 158–161
scattering of charged anti-particles 163–165
two identical charges 162–163
covariance 8
covariant Lagrangian in QED 97
covariant QED 77, 95
charge renormalization 130
mass renormalization 77, 127–130
relationship to classical electrodynamics 79
crossing symmetry 132–133
"channels" for processes 130–131
examples of channels for various processes 133

density transformation 20
Dirac equation 103
Dirac matrices 104, 169
normalization of Dirac wave functions 167
double Compton scattering 199
extreme relativistic limit 207–209
non-relativistic case, arbitrary energy 202–207
non-relativistic case, soft-photon limit 199–191

Einstein-de Broglie relation 10
Euler's constant 213, 239

Fermi Golden Rule 21, 87
Feynman propagator 89
intermediate states 88–89
Feynman–Schwinger–Dyson formulation of QED 77
fields of a moving charge 49
accelerated charge 53–54
charge in uniform motion 49–52
intermediate states 88–89
fine structure constant 2
role in QED 79
Fourier spectra 46–49
four-vectors 8
4-vector force 14
4-vector potential 13
momentum 4-vector 9
propagation 4-vector 10
surface 4-vector 11
tensors 7
trace of a tensor 9
velocity 4-vector 8

gauge invariance 38, 81
gauge transformation 38
Lorentz gauge condition 38

Hamiltonian 76
Hermitian character 87
intermediate states 88
photon-coupling Hamiltonian 80, 84, 90
single-photon and two-photon vertices 90–92
vertices and diagrams 88
higher-order couplings 78
divergence problems 78–79
Lamb shift 77
radiative corrections 77
hypergeometric function 146

intermediate-state photons 82
invariant flux 195
invariant transition rate 102–103

kinematic effects 14
threshold energies in collision phenomena 15
kinematic invariants in QED 130–131
Kronecker delta function 6

Lagrangian 13
covariant Lagrangian 13
Levi-Civita (Ricci) symbol 15
Liénard-Wiechert potentials 50–52
localization distance 3
Lorentz–Dirac equation 60
Lorentz transformation 5, 6

Maxwell equations 37
covariant Maxwell equations 15

Maxwell Lagrangian 100
metric choice 6
Mott formula 170–171
multipole expansion 42–46
  multipole radiated power 43–46

non-relativistic QED 80
nucleon Rydberg energy 4
pair annihilation and production 197–199
  Coulomb focusing 198–199
perturbation diagrams 89
perturbation matrix element 93–96
  in covariant theory (Feynman diagrams) 95
  in non-covariant and covariant theory 96
    conservation of momentum and energy at a vertex 96–97
phase-space factors 21–35
  general theorems 26
  one-particle distributions 32
    Maxwellian distribution 33
  simple examples 23–25
phase-space volume 21
  for massless particles 22
  invariant phase space 34
  limit of large N 22
  phase space integrals 26–32
  symmetry factor 23
  with and without momentum conservation 23
photon concept 10, 76
photon-coupling vertices 90–92
  couplings 84
  Klein-Gordon Lagrangian 98
  matrix elements 100
  propagator 89, 102, 106
  source–field equation 103
  with particles without spin 100
photon states 83
  photon amplitude 83–84
  polarization and wave vector 83
  Poynting vector 83
positronium 198
potentials and gauges 39–41
  Coulomb gauge 39
  gauge transformations 38, 81
  Lorentz gauge 39
principle of minimal electromagnetic coupling 81
Principle of Relativity 6
probability for a process 88
  total amplitude 95

quantum electrodynamics (QED) 75
  history of QED 76

radiation reaction 57–59
radiation from a relativistic charge 54–57
radiative corrections and renormalization 127–130
  self-energy diagrams 128
reduced mass 16, 140
relativistic covariance 5
relativistic invariants 5
  4-dimensional invariant volume 10
  invariant momentum space elements 11
  invariant number density 12
  invariant phase space 11, 34–35
  invariant time interval 8
  occupation number 12
retarded potentials 39–41
Riemann ζ-function 239
Rydberg energy 3

scattering amplitude 150
  direct and exchange amplitude 151–152
  interference amplitude 152
Sommerfeld–Elwert factor 223, 238
spin-$\frac{1}{2}$ QED 103–106
spin wave functions, two electron 153
soft-photon emission 61–65, 83
  multipole formulation 61
soft-photon emission in QED 109–122
  emission from spin transitions 113–116
  non-relativistic theory 109–113
  photon-emission probability 112
  relativistic particles without spin 116–118
  relativistic spin-$\frac{1}{2}$ particles 119–122
spacetime transformation 5
spin and magnetic moment 10
spin sums 166–169
  projection operators 168
  trace theorems 169–170
strong interaction cross section 4
superposition principle 150–152, 154

time-dependent perturbation theory 84–87
time-dependent Schrödinger equation 76

units 1, 4

WKB series (in soft-photon non-relativistic bremsstrahlung) 239
Weizsäcker-Williams method 65–70
  in bremsstrahlung 251–253
  equivalent photon fluxes 68–70

Milton Keynes UK
Ingram Content Group UK Ltd.
UKHW020640290824
447545UK00007B/209